21st Century Homestead
Sustainable Environmental Design

Contents

Chapter 1

Sustainable design

See also: Sustainable engineering and Ecological design

Sustainable design (also called environmental design, environmentally sustainable design, environmentally conscious design, etc.) is the philosophy of designing physical objects, the built environment, and services to comply with the principles of social, economic, and ecological sustainability.[1]

1.1 Theory

The intention of sustainable design is to "eliminate negative environmental impact completely through skillful, sensitive design".[1] Manifestations of sustainable design require renewable resources, impact the environment minimally, and connect people with the natural environment.

Beyond the "elimination of negative environmental impact", sustainable design must create projects that are meaningful innovations that can shift behaviour. A dynamic balance between economy and society, intended to generate long-term relationships between user and object/service and finally to be respectful and mindful of the environmental and social differences.[2]

1.1.1 Conceptual problems

Diminishing returns

The principle that all directions of progress run out, ending with diminishing returns, is evident in the typical 'S' curve of the technology life cycle and in the useful life of any system as discussed in industrial ecology and life cycle assessment. Diminishing returns are the result of reaching natural limits. Common business management practice is to read diminishing returns in any direction of effort as an indication of diminishing opportunity, the potential for accelerating decline and a signal to seek new opportunities

elsewhere. (see also: law of diminishing returns, marginal utility and Jevons paradox.)

Unsustainable Investment

A problem arises when the limits of a resource are hard to see, so increasing investment in response to diminishing returns may seem profitable as in the Tragedy of the Commons, but may lead to a collapse. This problem of increasing investment in diminishing resources has also been studied in relation to the causes of civilization collapse by Joseph Tainter among others.[3] This natural error in investment policy contributed to the collapse of both the Roman and Mayan, among others. Relieving over-stressed resources requires reducing pressure on them, not continually increasing it whether more efficiently or not[4]

Waste prevention

Negative Effects of Waste

About 80 million tonnes of waste in total are generated in the U.K. alone, for example, each year.[5] And with reference to only household waste, between 1991/92 and 2007/08, each person in England generated an average of 1.35 pounds of waste per day.[6]

Experience has now shown that there is no completely safe method of waste disposal. All forms of disposal have negative impacts on the environment, public health, and local economies. Landfills have contaminated drinking water. Garbage burned in incinerators has poisoned air, soil, and water. The majority of water treatment systems change the local ecology. Attempts to control or manage wastes after they are produced fail to eliminate environmental impacts.

The toxic components of household products pose serious health risks and aggravate the trash problem. In the U.S., about eight pounds in every ton of household garbage contains toxic materials, such as heavy metals like nickel, lead, cadmium, and mercury from batteries, and organic com-

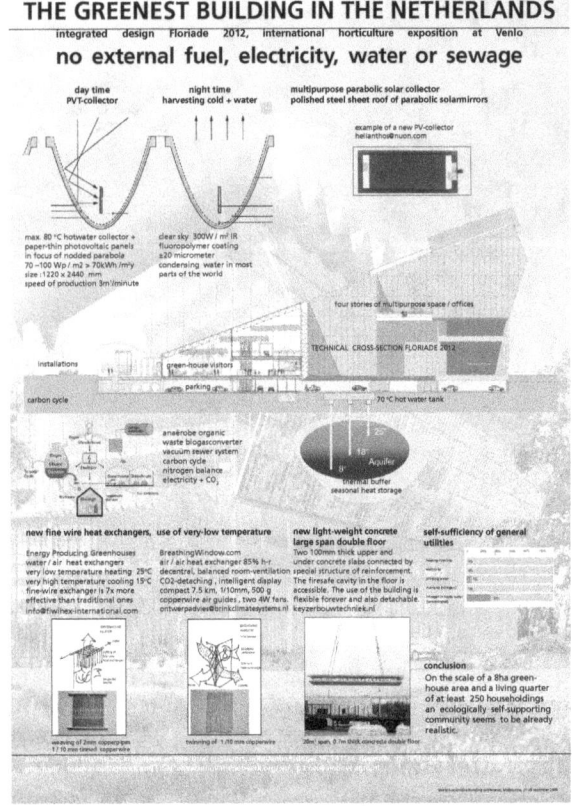

Plans for Floriade 2012 in Venlo, the Netherlands: "The Greenest Building in the Netherlands - no external fuel, electricity, water or sewage."

pounds found in pesticides and consumer products, such as air freshener sprays, nail polish, cleaners, and other products.[7] When burned or buried, toxic materials also pose a serious threat to public health and the environment.

The only way to avoid environmental harm from waste is to prevent its generation. Pollution prevention means changing the way activities are conducted and eliminating the source of the problem. It does not mean doing without, but doing differently. For example, preventing waste pollution from litter caused by disposable beverage containers does not mean doing without beverages; it just means using refillable bottles.

Waste prevention strategies In planning for facilities, a comprehensive design strategy is needed for preventing generation of solid waste. A good garbage prevention strategy would require that everything brought into a facility be recycled for reuse or recycled back into the environment through biodegradation. This would mean a greater reliance on natural materials or products that are compatible with the environment.

Any resource-related development is going to have two

basic sources of solid waste — materials purchased and used by the facility and those brought into the facility by visitors. The following waste prevention strategies apply to both, although different approaches will be needed for implementation:[8]

- use products that minimize waste and are nontoxic

- compost or anaerobically digest biodegradable wastes

- reuse materials onsite or collect suitable materials for offsite recycling

1.1.2 Sustainable design principles

The California Academy of Sciences, San Francisco, California, is a sustainable building designed by Renzo Piano. It opened on September 27, 2008

.

One Central Park, Sydney

While the practical application varies among disciplines, some common principles are as follows:

- Low-impact materials: choose non-toxic, sustainably produced or recycled materials which require little energy to process

- Energy efficiency: use manufacturing processes and produce products which require less energy

- Emotionally durable design: reducing consumption and waste of resources by increasing the durability of relationships between people and products, through design

- Design for reuse and recycling: "Products, processes, and systems should be designed for performance in a commercial 'afterlife'."[9]

- Design impact measures for total carbon footprint and life-cycle assessment for any resource used are increasingly required and available.^[10] Many are complex, but some give quick and accurate whole-earth estimates of impacts. One measure estimates any spending as consuming an average economic share of global energy use of 8,000 BTU (8,400 kJ) per dollar and producing CO_2 at the average rate of 0.57 kg of CO_2 per dollar (1995 dollars US) from DOE figures.[11]

- Sustainable design standards and project design guides are also increasingly available and are vigorously being developed by a wide array of private organizations and individuals. There is also a large body of new methods emerging from the rapid development of what has become known as 'sustainability science' promoted by a wide variety of educational and governmental institutions.

- Biomimicry: "redesigning industrial systems on biological lines ... enabling the constant reuse of materials in continuous closed cycles..."[12]

- Service substitution: shifting the mode of consumption from personal ownership of products to provision of services which provide similar functions, e.g., from a private automobile to a carsharing service. Such a system promotes minimal resource use per unit of consumption (e.g., per trip driven).[13]

- Renewability: materials should come from nearby (local or bioregional), sustainably managed renewable sources that can be composted when their usefulness has been exhausted.

- Robust eco-design: robust design principles are applied to the design of a pollution sources.[14]

- The physics principle [15] that accounts for the urge to have sustainability, and for the evolutionary design in general, is the constructal_law.

Bill of Rights for the Planet

A model of the new design principles necessary for sustainability is exemplified by the "Bill of Rights for the Planet" or "Hannover Principles" - developed by William McDonough Architects for EXPO 2000 that was held in Hannover, Germany.

The Bill of Rights:

1. Insist on the right of humanity and nature to co-exist in a healthy, supportive, diverse, and sustainable condition.

2. Recognize Interdependence. The elements of human design interact with and depend on the natural world, with broad and diverse implications at every scale. Expand design considerations to recognizing even distant effects.

3. Respect relationships between spirit and matter. Consider all aspects of human settlement including community, dwelling, industry, and trade in terms of existing and evolving connections between spiritual and material consciousness.

4. Accept responsibility for the consequences of design decisions upon human well-being, the viability of natural systems, and their right to co-exist.

5. Create safe objects of long-term value. Do not burden future generations with requirements for maintenance or vigilant administration of potential danger due to the careless creations of products, processes, or standards.

6. Eliminate the concept of waste. Evaluate and optimize the full life-cycle of products and processes, to approach the state of natural systems in which there is no waste.

7. Rely on natural energy flows. Human designs should, like the living world, derive their creative forces from perpetual solar income. Incorporating this energy efficiently and safely for responsible use.

8. Understand the limitations of design. No human creation lasts forever and design does not solve all problems. Those who create and plan should practice humility in the face of nature. Treat nature as a model and mentor, not an inconvenience to be evaded or controlled.

9. Seek constant improvement by the sharing of knowledge. Encourage direct and open communication between colleagues, patrons, manufacturers and users to link long term sustainable considerations with ethical

responsibility, and re-establish the integral relationship between natural processes and human activity.

These principles were adopted by the World Congress of the International Union of Architects (UIA) in June 1993 at the American Institute of Architects' (AIA) Expo 93 in Chicago. Further, the AIA and UIA signed a "Declaration of Interdependence for a Sustainable Future." In summary, the declaration states that today's society is degrading its environment and that the AIA, UIA, and their members are committed to:

- Placing environmental and social sustainability at the core of practices and professional responsibilities

- Developing and continually improving practices, procedures, products, services, and standards for sustainable design

- Educating the building industry, clients, and the general public about the importance of sustainable design

- Working to change policies, regulations, and standards in government and business so that sustainable design will become the fully supported standard practice

- Bringing the existing built environment up to sustainable design standards.

In addition, the Interprofessional Council on Environmental Design (ICED), a coalition of architectural, landscape architectural, and engineering organizations, developed a vision statement in an attempt to foster a team approach to sustainable design. ICED states: The ethics, education and practices of our professions will be directed to shape a sustainable future. . . . To achieve this vision we will join . . . as a multidisciplinary partnership."

These activities are an indication that the concept of sustainable design is being supported on a global and interprofessional scale and that the ultimate goal is to become more environmentally responsive. The world needs facilities that are more energy efficient and that promote conservation and recycling of natural and economic resources.[16]

1.2   Applications

Applications of this philosophy range from the microcosm — small objects for everyday use, through to the macrocosm — buildings, cities, and the Earth's physical surface. It is a philosophy that can be applied in the fields of architecture, landscape architecture, urban design, urban planning, engineering, graphic design, industrial design, interior design, fashion design and human-computer interaction.

Sustainable design is mostly a general reaction to global environmental crises, the rapid growth of economic activity and human population, depletion of natural resources, damage to ecosystems, and loss of biodiversity.[17] In 2013, eco architecture writer Bridgette Meinhold surveyed emergency and long-term sustainable housing projects that were developed in response to these crises in her book, "Urgent Architecture: 40 Sustainable Housing Solutions for a Changing World."[18][19] Featured projects focus on green building, sustainable design, eco-friendly materials, affordability, material reuse, and humanitarian relief. Construction methods and materials include repurposed shipping containers, straw bale construction, sandbag homes, and floating homes.[20]

The limits of sustainable design are reducing. Whole earth impacts are beginning to be considered because growth in goods and services is consistently outpacing gains in efficiency. As a result, the net effect of sustainable design to date has been to simply improve the efficiency of rapidly increasing impacts. The present approach, which focuses on the efficiency of delivering individual goods and services, does not solve this problem. The basic dilemmas include: the increasing complexity of efficiency improvements; the difficulty of implementing new technologies in societies built around old ones; that physical impacts of delivering goods and services are not localized, but are distributed throughout the economies; and that the scale of resource use is growing and not stabilizing.

1.3   Examples

1.3.1   Beauty and sustainable design

Because standards of sustainable design appear to emphasize ethics over aesthetics, some designers and critics have complained that it lacks inspiration. Pritzker Architecture Prize winner Frank Gehry has called green building "bogus,"[21] and National Design Awards winner Peter Eisenman has dismissed it as "having nothing to do with architecture."[22] In 2009, *The American Prospect* asked whether "well-designed green architecture" is an "oxymoron."[23]

Others claim that such criticism of sustainable design is misguided. A leading advocate for this alternative view is architect Lance Hosey, whose book *The Shape of Green: Aesthetics, Ecology, and Design* (2012) was the first dedicated to the relationships between sustainability and beauty. Hosey argues not just that sustainable design needs to be aesthetically appealing in order to be successful, but also that following the principles of sustainability to their logical conclusion requires reimagining the shape of everything designed, creating things of even greater beauty. Reviewers have suggested that the ideas in *The Shape of Green* could "revolutionize what it means to be sustainable."[24] Small and large buildings are beginning to successfully incorporate principles of sustainability into award-winning designs. Examples include One Central Park and Architectus Science Faculty Building UTS.

1.3.2 Emotionally durable design

Stain Teacups: Bethan Laura Wood, 2009

According to Jonathan Chapman of the University of Brighton, UK, emotionally durable design reduces the consumption and waste of natural resources by increasing the resilience of relationships established between consumers and products."[25] Essentially, product replacement is delayed by strong emotioal ties.[26] In his book, *Emotionally Durable Design: Objects, Experiences & Empathy*, Chapman describes how "the process of consumption is, and has always been, motivated by complex emotional drivers, and is about far more than just the mindless purchasing of newer and shinier things; it is a journey towards the ideal or desired self, that through cyclical loops of desire and disappointment, becomes a seemingly endless process of serial destruction".[27] Therefore, a product requires an attribute, or number of attributes, which extend beyond utilitarianism. [28]

According to Chapman, 'emotional durability' can be achieved through consideration of the following five elements:

- **Narrative**: How users share a unique personal history with the product.

- **Consciousness**: How the product is perceived as autonomous and in possession of its own free will.

- **Attachment**: Can a user be made to feel a strong emotional connection to a product?

- **Fiction**: The product inspires interactions and connections beyond just the physical relationship.

- **Surface**: How the product ages and develops character through time and use.

As a strategic approach, "emotionally durable design provides a useful language to describe the contemporary relevance of designing responsible, well made, tactile products which the user can get to know and assign value to in the long-term."[29] According to Hazel Clark and David Brody of Parsons The New School for Design in New York, "emotionally durable design is a call for professionals and students alike to prioritise the relationships between design and its users, as a way of developing more sustainable attitudes to, and in, design things."[30]

1.3.3 Eco fashion and home accessories

Creative designers and artists are perhaps the most inventive when it comes to upcycling or creating new products from old waste. A growing number of designers upcycle waste materials such as car window glass and recycled ceramics, textile offcuts from upholstery companies, and even decommissioned fire hose to make belts and bags. Whilst accessories may seem trivial when pitted against green scientific breakthroughs; the ability of fashion and retail to influence and inspire consumer behaviour should not be underestimated. Eco design may also use bi-products of industry, reducing the amount of waste being dumped in landfill, or may harness new sustainable materials or production techniques e.g. fabric made from recycled PET plastic bottles or bamboo textiles.

1.3.4 Sustainable architecture

Main article: Sustainable architecture

Sustainable architecture is the design of sustainable buildings. Sustainable architecture attempts to reduce the collective environmental impacts during the production of building components, during the construction process, as well

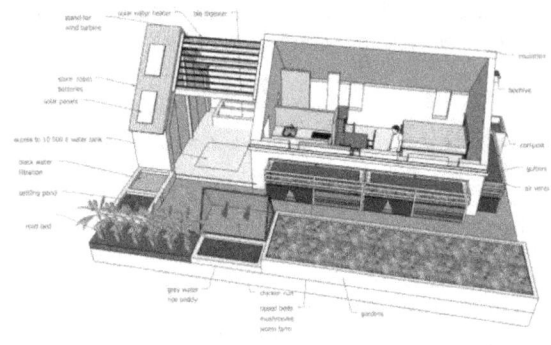

Sustainable building design

as during the lifecycle of the building (heating, electricity use, carpet cleaning etc.) This design practice emphasizes efficiency of heating and cooling systems; alternative energy sources such as solar hot water, appropriate building siting, reused or recycled building materials; on-site power generation - solar technology, ground source heat pumps, wind power; rainwater harvesting for gardening, washing and aquifer recharge; and on-site waste management such as green roofs that filter and control stormwater runoff. This requires close cooperation of the design team, the architects, the engineers, and the client at all project stages, from site selection, scheme formation, material selection and procurement, to project implementation.[31]

Sustainable architects design with sustainable living in mind.[32] Sustainable vs green design is the challenge that designs not only reflect healthy processes and uses but are powered by renewable energies and site specific resources. A test for sustainable design is — can the design function for its intended use without fossil fuel — unplugged. This challenge suggests architects and planners design solutions that can function without pollution rather than just reducing pollution. As technology progresses in architecture and design theories and as examples are built and tested, architects will soon be able to create not only passive, null-emission buildings, but rather be able to integrate the entire power system into the building design. In 2004 the 59 home housing community, the Solar Settlement, and a 60,000 sq ft (5,600 m^2) integrated retail, commercial and residential building, the Sun Ship, were completed by architect Rolf Disch in Freiburg, Germany. The Solar Settlement is the first housing community worldwide in which every home, all 59, produce a positive energy balance.[33]

An essential element of Sustainable Building Design is indoor environmental quality including air quality, illumination, thermal conditions, and acoustics. The integrated design of the indoor environment is essential and must be part of the integrated design of the entire structure. ASHRAE Guideline 10-2011 addresses the interactions among in-

door environmental factors and goes beyond traditional standards.[34]

Concurrently, the recent movements of New Urbanism and New Classical Architecture promote a sustainable approach towards construction, that appreciates and develops smart growth, architectural tradition and classical design.[35][36] This in contrast to modernist and globally uniform architecture, as well as leaning against solitary housing estates and suburban sprawl.[37] Both trends started in the 1980s. The Driehaus Architecture Prize is an award that recognizes efforts in New Urbanism and New Classical Architecture, and is endowed with a prize money twice as high as that of the modernist Pritzker Prize.[38]

1.3.5   Sustainable planning

Cohousing community illustrating greenspace preservation, tightly clustered housing, and parking on periphery, Ann Arbor, Michigan, 2003.

Urban planners that are interested in achieving sustainable development or sustainable cities use various design principles and techniques when designing cities and their infrastructure. These include Smart Growth theory, Transit-oriented development, sustainable urban infrastructure and New Urbanism. Smart Growth is an urban planning and transportation theory that concentrates growth in infill sites within the existing infrastructure of a city or town to avoid urban sprawl; and advocates compact, transit-oriented development, walkable, bicycle-friendly land use, including mixed-use development with a range of housing choices. Transit-oriented development attempts to maximise access to public transport and thereby reduce the need for private vehicles. Public transport is considered a form of Sustainable urban infrastructure, which is a design approach which promotes protected areas, energy-efficient buildings, wildlife corridors and distributed, rather than centralized, power generation and waste water treatment. New Urbanism is more of a social and aesthetic urban design movement than a green one, but it does emphasize diversity of land use and population, as well as walkable communities which inherently reduce the need for automotive travel.

Both urban and rural planning can benefit from including sustainability as a central criterion when laying out roads, streets, buildings and other components of the built environment. Conventional planning practice often ignores or discounts the natural configuration of the land during the planning stages, potentially causing ecological damage such as the stagnation of streams, mudslides, soil erosion, flooding and pollution. Applying methods such as scientific modelling to planned building projects can draw attention to problems before construction begins, helping to minimise damage to the natural environment.

Cohousing is an approach to planning based on the idea of intentional communities. Such projects often prioritize common space over private space resulting in grouped structures that preserve more of the surrounding environment.

Watershed assessment of carrying capacity; estuary, riparian zone restoration and groundwater recharge for hydrologic cycle viability; and other opportunities and issues about Water and the environment show that the foundation of smart growth lies in the protection and preservation of water resources. The total amount of precipitation landing on the surface of a community becomes the supply for the inhabitants. This supply amount then dictates the carrying capacity - the potential population - as supported by the "water crop."

1.3.6 Sustainable landscape and garden design

Main articles: Sustainable landscape architecture, Sustainable landscaping and Sustainable gardening

Sustainable landscape architecture is a category of sustainable design and energy-efficient landscaping concerned with the planning and design of outdoor space. Plants and materials may be bought from local growers to reduce energy used in transportation. Design techniques include planting trees to shade buildings from the sun or protect them from wind, using local materials, and on-site composting and chipping not only to reduce green waste hauling but to increase organic matter and therefore carbon in the soil.

Some designers and gardeners such as Beth Chatto also use drought-resistant plants in arid areas (xeriscaping) and elsewhere so that water is not taken from local landscapes and habitats for irrigation. Water from building roofs may be collected in rain gardens so that the groundwater is recharged, instead of rainfall becoming surface runoff and increasing the risk of flooding.

Areas of the garden and landscape can also be allowed to grow wild to encourage bio-diversity. Native animals may also be encouraged in many other ways: by plants which provide food such as nectar and pollen for insects, or roosting or nesting habitats such as trees, or habitats such as ponds for amphibians and aquatic insects. Pesticides, espe-

cially persistent pesticides, must be avoided to avoid killing wildlife.

Soil fertility can be managed sustainably by the use of many layers of vegetation from trees to ground-cover plants and mulches to increase organic matter and therefore earthworms and mycorrhiza; nitrogen-fixing plants instead of synthetic nitrogen fertilizers; and sustainably harvested seaweed extract to replace micronutrients.

Sustainable landscapes and gardens can be productive as well as ornamental, growing food, firewood and craft materials from beautiful places.

Sustainable landscape approaches and labels include organic farming and growing, permaculture, agroforestry, forest gardens, agroecology, vegan organic gardening, ecological gardening and climate-friendly gardening.

1.3.7 Sustainable graphic design

Main article: Sustainable graphic design

Sustainable graphic design considers the environmental impacts of graphic design products (such as packaging, printed materials, publications, etc.) throughout a life cycle that includes: raw material; transformation; manufacturing; transportation; use; and disposal. Techniques for sustainable graphic design include: reducing the amount of materials required for production; using paper and materials made with recycled, post-consumer waste; printing with low-VOC inks; and using production and distribution methods that require the least amount of transport.

1.3.8 Sustainable agriculture

Main: Organic farming

Sustainable agriculture adheres to three main goals:

- Environmental Health,
- Economic Profitability,
- Social and Economic Equity.

A variety of philosophies, policies and practices have contributed to these goals. People in many different capacities, from farmers to consumers, have shared this vision and contributed to it. Despite the diversity of people and perspectives, the following themes commonly weave through definitions of sustainable agriculture.

There are strenuous discussions — among others by the agricultural sector and authorities — if existing pesticide protocols and methods of soil conservation adequately protect topsoil and wildlife. Doubt has risen if these are sustainable, and if agrarian reforms would permit an efficient agriculture with fewer pesticides, therefore reducing the damage to the ecosystem.

For more information on the subject of sustainable agriculture: "UC Davis: Sustainable Agriculture Research and Education Program".[39]

1.3.9   Domestic machinery and furniture

Main article: Home appliances

Automobiles, home appliances and furnitures can be designed for repair and disassembly (for recycling), and constructed from recyclable materials such as steel, aluminum and glass, and renewable materials, such as Zelfo, wood and plastics from natural feedstocks. Careful selection of materials and manufacturing processes can often create products comparable in price and performance to non-sustainable products. Even mild design efforts can greatly increase the sustainable content of manufactured items.

Improvements to heating, cooling, ventilation and water heating

- Absorption refrigerator
- Annualized geothermal solar
- Earth cooling tubes
- Geothermal heat pump
- Heat recovery ventilation
- Hot water heat recycling
- Passive cooling
- Renewable heat
- Seasonal thermal energy storage (STES)
- Solar air conditioning
- Solar hot water

1.3.10   Disposable products

Detergents, newspapers and other disposable items can be designed to decompose, in the presence of air, water and common soil organisms. The current challenge in this area is to design such items in attractive colors, at costs as low as competing items. Since most such items end up in landfills,

protected from air and water, the utility of such disposable products is debated.

1.3.11   Energy sector

Main articles: Solar energy and Alternative energy
See also: Solar thermal collector, Solar panel and Passive solar building design

Sustainable technology in the energy sector is based on utilizing renewable sources of energy such as solar, wind, hydro, bioenergy, geothermal, and hydrogen. Wind energy is the world's fastest growing energy source; it has been in use for centuries in Europe and more recently in the United States and other nations. Wind energy is captured through the use of wind turbines that generate and transfer electricity for utilities, homeowners and remote villages. Solar power can be harnessed through photovoltaics, concentrating solar, or solar hot water and is also a rapidly growing energy source.[40]

The availability, potential, and feasibility of primary renewable energy resources must be analyzed early in the planning process as part of a comprehensive energy plan. The plan must justify energy demand and supply and assess the actual costs and benefits to the local, regional, and global environments. Responsible energy use is fundamental to sustainable development and a sustainable future. Energy management must balance justifiable energy demand with appropriate energy supply. The process couples energy awareness, energy conservation, and energy efficiency with the use of primary renewable energy resources.[41]

1.3.12   Water sector

Main articles: Reclaimed water, Rainwater harvesting and Stormwater harvesting
 Sustainable water technologies have become an important industry segment with several companies now providing important and scalable solutions to supply water in a sustainable manner.

Beyond the use of certain technologies, Sustainable Design in Water Management also consists very importantly in correct implementation of concepts. Among one of these principal concepts is the fact normally in developed countries 100% of water destined for consumption, that is not necessarily for drinking purposes, is of potable water quality. This concept of differentiating qualities of water for different purposes has been called "fit-for-purpose".[42] This more rational use of water achieves several economies, that are not only related to water itself, but also the consumption of energy, as to achieve water of drinking quality can

A 35,003 litre rainwater harvesting tank in Kerala

be extremely energy intensive for several reasons.

1.4 Terminology

In some countries the term *sustainable design* is known as ecodesign, green design or environmental design. Victor Papanek, embraced social design and social quality and ecological quality, but did not explicitly combine these areas of design concern in one term. *Sustainable design* and *design for sustainability* are more common terms, including the triple bottom line (people, planet and profit).

In the EU, the concept of sustainable design is referred to as ecodesign. Little discussions have however taken place over the importance of this concept in the run-up to the circular economy package, that the European Commission will be tabling by the end of 2015. To this effect, an Ecothis.EU campaign was launched to raise awareness about the economic and environmental consequences of not including eco-design as part of the circular economy package.[43]

1.5 Sustainable technologies

Sustainable technologies use less energy, fewer limited resources, do not deplete natural resources, do not directly or indirectly pollute the environment, and can be reused or recycled at the end of their useful life.[44] There is significant overlap with appropriate technology, which emphasizes the suitability of technology to the context, in particular considering the needs of people in developing countries. However, the most appropriate technology may not be the most sustainable one; and a sustainable technology may have high cost or maintenance requirements that make it unsuitable as an "appropriate technology," as that term is commonly

used.

1.6 See also

- Active daylighting
- Active solar
- Agroforestry
- BREEAM
- Bright green environmentalism
- Building Information Modeling
- Building services engineering
- Circles of Sustainability
- Climate-friendly gardening
- Cool roof
- Cradle to Cradle
- Daylighting
- Eco-innovation
- Ecodistrict
- Ecological Restoration
- Ecosa Institute
- Ecosystem services
- Energy plus house
- Environmentally friendly
- Green chemistry
- Green library
- Green transport
- History of passive solar building design
- Landscape ecology
- Leadership in Energy and Environmental Design
- List of energy storage projects
- List of low-energy building techniques
- List of sustainable agriculture topics
- Metadesign
- Passive solar

- Passive solar design
- Principles of Intelligent Urbanism
- Renewable resource
- Source reduction
- Superinsulation
- Sustainable art
- Terreform ONE
- Transition Design
- Vertical garden
- Water conservation
- Watershed
- Zero energy building

1.6.1   Advocates and practitioners

- Bill Mollison
- Buckminster Fuller
- Daniel A. Vallero
- David Holmgren
- Hellmuth, Obata and Kassabaum
- J. Baldwin
- Jonathan Chapman
- Ken Fern, founder of Plants for a Future
- Ken Yeang
- Lance Hosey
- Martin Crawford
- Michael Braungart
- Michael Hendrix
- Mike Reynolds
- Mitchell Joachim
- Paolo Soleri
- Paul Hawken
- Robert Hart
- Sim Van der Ryn

- Tom Bender
- Vandana Shiva
- Victor Papanek
- William McDonough

1.6.2   Events, conferences, workshops, classes, and organizations

- Agroforestry Research Trust
- AIA Committee on the Environment
- American Society of Landscape Architects
- Center for the Built Environment
- Compostmodern
- EC3 Global
- Ecoweek
- o2 Global Network
- Open Source Ecology
- Permaculture Association
- The Zeitgeist Movement
- United States Green Building Council
- Worldchanging

1.7   References

[1] McLennan, J. F. (2004), The Philosophy of Sustainable Design

[2] "Sustainable Design Research".

[3] JA Tainter 1988 The Collapse of Complex Societies Cambridge Univ. Press ISBN 978-0521386739

[4] Buzz Holling 1973 Resilience and Stability of Ecological Systems

[5] Waste and recycling, DEFRA

[6] Household waste, Office for National Statistics.

[7] US EPA, "Expocast"

[8] Various. "Guiding Principles of Sustainable Design." Chapter 9: Waste Prevention.

[9] Anastas, P. L. and Zimmerman, J. B. (2003). "Through the 12 principles of green engineering". Environmental Science and Technology. March 1. 95-101A.

[10] D. Vallero and C. Brasier (2008), Sustainable Design: The Science of Sustainability and Green Engineering. John Wiley and Sons, Inc., Hoboken, NJ, ISBN 0470130628.

[11] US DOE 20 yr Global Product & Energy Study.

[12] Paul Hawken, Amory B. Lovins, and L. Hunter Lovins (1999). Natural Capitalism: Creating the Next Industrial Revolution. Little, Brown. ISBN 978-0-316-35316-8

[13] Ryan, Chris (2006). "Dematerializing Consumption through Service Substitution is a Design Challenge". Journal of Industrial Ecology. 4(1). doi:10.1162/108819800569230

[14] Ben-Gal I., Katz R. and Bukchin J., "Robust Eco-Design: A New Application for Quality Engineering", IIE Transactions, Vol. 40 (10), p. 907 - 918.

[15] Bejan, Adrian (October 2015). "Sustainability: The Water and Energy Problem, and the Natural Design Solution". *European Review* **23** (4): 481-488. doi:10.1017/S1062798715000216.

[16] Various. "Guiding Principles of Sustainable Design". THE PRINCIPLES OF SUSTAINABILITY.

[17] Fan Shu-Yang, Bill Freedman, and Raymond Cote (2004). "Principles and practice of ecological design". Environmental Reviews. 12: 97–112.

[18] Meinhold, Bridgette. *Urgent Architecture: 40 Sustainable Housing Solutions for a Changing World*. W. W. Norton & Company, Inc. Retrieved 26 May 2014.

[19] Vidal, John. "Humanitarian intent: Urgent Architecture from ecohomes to shelters – in pictures". theguardian.com. Retrieved 26 May 2014.

[20] "URGENT ARCHITECTURE: Inhabitat Interviews Author Bridgette Meinhold About Her New Book". YouTube.com. Retrieved 26 May 2014.

[21] Michael Arndt, "Architect Gehry on LEED Buildings: Humbug", Bloomberg Businessweek, April 07, 2010]

[22] Intercontinental Curatorial Project, Interview with Peter Eisenman, June 18, 2009

[23] Kriston Capps, "Green Building Blues," *The American Prospect,* February 12, 2009

[24] Claire Easley. "Not Pretty? Then It's Not Green". *Builder.*

[25] Chapman, J., 'Design for [Emotional] Durability', *Design Issues*, vol xxv, Issue 4, Autumn, pp29-35, 2009 doi:10.1162/desi.2009.25.4.29

[26] Page, Tom (2014). "Product attachment and replacement: implications for sustainable design" (PDF). *Int. J. Sustainable Design*. Retrieved 9/1/15. Check date values in: |access-date= (help)

[27] Chapman, J., *Emotionally Durable Design: Objects, Experiences and Empathy*, Earthscan, London, 2005

[28] Page, Tom (2014). "Product attachment and replacement: implications for sustainable design" (PDF). *Int. J. Sustainable Design*. Retrieved 09/01/15. Check date values in: |access-date= (help)

[29] Lacey, E. (2009). Contemporary ceramic design for meaningful interaction and emotional durability: A case study. *International Journal of Design*, 3(2), 87-92

[30] Clark, H. & Brody, D., Design Studies: A Reader, Berg, New York, US, 2009, p531 ISBN 9781847882363

[31] Ji Yan and Plainiotis Stellios (2006): Design for Sustainability. Beijing: China Architecture and Building Press. ISBN 7-112-08390-7

[32] Holm, Ivar (2006). *Ideas and Beliefs in Architecture and Industrial design: How attitudes, orientations, and underlying assumptions shape the built environment*. Oslo School of Architecture and Design. ISBN 82-547-0174-1.

[33] "Rolf Disch - SolarArchitektur". *more-elements.com.*

[34] ASHRAE Guideline 10-2011: "Interactions Affecting the Achievement of Acceptable Indoor Environments"

[35] "Charter of the New Urbanism". *cnu.org.*

[36] "Beauty, Humanism, Continuity between Past and Future". Traditional Architecture Group. Retrieved 23 March 2014.

[37] Issue Brief: Smart-Growth: Building Livable Communities. American Institute of Architects. Retrieved on 2014-03-23.

[38] "Driehaus Prize". Together, the $200,000 Driehaus Prize and the $50,000 Reed Award represent the most significant recognition for classicism in the contemporary built environment.. Notre Dame School of Architecture. Retrieved 23 March 2014.

[39] Feenstra, G (December 1997). What Is Sustainable Architecture?. Retrieved June 27, 2009, UC SAREP Web site

[40] "Renewable Energy Policy Project & CREST Center for Renewable Energy and Sustainable Technology"

[41] Various. "Guiding Principles of Sustainable Design". Chapter 7: Energy Management.

[42] "Water recycling & alternative water sources". *health.vic.gov.au.* Archived from the original on 7 January 2010.

[43] "The Ecothis.eu campaign website". ecothis.eu. Retrieved August 3, 2015.

[44] "Sustainable Roadmap - Open Innovation". *connect.innovateuk.org.* 2012. Retrieved December 3, 2012.

1.8   External links

- Sustainability in the Desert: A review of Sustainable Design in the Middle East. Via Carboun

- Chris Hendrickson, Noellette Conway-Schempf, Lester Lave and Francis McMichael. "Introduction to Green Design."

- Material Review: One to Watch

Chapter 2

Environmental design and planning

Environmental design and planning is the moniker used by several Ph.D. programs that take a multidisciplinary approach to the built environment. Typically environmental design and planning programs address architectural history or design (interior or exterior), city or regional planning, landscape architecture history or design, environmental planning, construction science, cultural geography, or historic preservation. Social science methods are frequently employed; aspects of sociology or psychology can be part of a research program.

The concept of "environmental" in these programs is quite broad and can encompass aspects of the natural, built, work, or social environments.

2.1 Areas of research

- Architecture
- Construction science
- Ecology
- Environmental impact design
- Environmental planning
- Environmental psychology
- Environmental sociology
- Historic preservation
- Landscape architecture
- Sociology of architecture
- Sustainability
- Urban planning

2.2 Academic programs

The following universities offer a Ph.D. in environmental design and planning:

- Clemson University, College of Architecture, Arts and Humanities (Now called "Planning, Design, and the Built Environment")
- Arizona State University, College of Design
- Kansas State University
- University of Calgary (technically the Ph.D. is in "environmental design," but encompasses the same scope as the other programs)

Virginia Tech until recently offered the degree program, but has since replaced it with programs in "architecture and design research" and "planning, governance, and globalization".

2.3 Related programs

- University of Missouri, Columbia: Ph.D. in Human Environmental Sciences (PDF file) with emphasis in Architectural Studies.
- Texas A & M University offers a Ph.D. in architecture that emphasizes environmental design.

Chapter 3

Adaptive management

Adaptive management (AM), also known as **adaptive resource management** (ARM), is a structured, iterative process of robust decision making in the face of uncertainty, with an aim to reducing uncertainty over time via system monitoring. In this way, decision making simultaneously meets one or more resource management objectives and, either passively or actively, accrues information needed to improve future management. Adaptive management is a tool which should be used not only to change a system, but also to learn about the system (Holling 1978). Because adaptive management is based on a learning process, it improves long-run management outcomes. The challenge in using the adaptive management approach lies in finding the correct balance between gaining knowledge to improve management in the future and achieving the best short-term outcome based on current knowledge (Allan & Stankey 2009).

3.1 Objectives

There are a number of scientific and social processes which are vital components of adaptive management, including:

1. Management is linked to appropriate temporal and spatial scales

2. Management retains a focus on statistical power and controls

3. Use of computer models to build synthesis and an embodied ecological consensus

4. Use of embodied ecological consensus to evaluate strategic alternatives

5. Communication of alternatives to political arena for negotiation of a selection

The achievement of these objectives requires an open management process which seeks to include past, present and future stakeholders. Adaptive management needs to at least maintain political openness, but usually aims to create it. Adaptive management must therefore be a scientific and social process. It must focus on the development of new institutions and institutional strategies in balance with scientific hypothesis and experimental frameworks (resilliance.org).

Adaptive management can proceed as either **passive adaptive management** or **active adaptive management**, depending on how learning takes place. Passive adaptive management values learning only insofar as it improves decision outcomes (i.e. passively), as measured by the specified utility function. In contrast, active adaptive management explicitly incorporates learning as part of the objective function, and hence, decisions which improve learning are valued over those which do not (Holling 1978; Walters 1986). In both cases, as new knowledge is gained, the models are updated and optimal management strategies are derived accordingly. Thus, while learning occurs in both cases, it is treated differently. Often, deriving actively adaptive policies is technically very difficult, which prevents it being more commonly applied.[1]

3.2 Features

Key features of both passive and active adaptive management are:

- Iterative decision-making (evaluating results and adjusting actions on the basis of what has been learned)

- Feedback between monitoring and decisions (learning)

- Explicit characterization of system uncertainty through multi-model inference

- Bayesian inference

- Embracing risk and uncertainty as a way of building understanding

However, a number of process failures related to information feedback can prevent effective adaptive management decision making (Elzinga et al. 1998):

1. data collection is never completely implemented;

2. data are collected but not analyzed;

3. data are analyzed but results are inconclusive;

4. data are analyzed and are interesting, but are not presented to decision makers;

5. data are analyzed and presented, but are not used for decision-making because of internal or external factors

3.3   History

The use of adaptive management techniques can be traced back to peoples from ancient civilisations. For example, the Yap people of Micronesia have been using adaptive management techniques to sustain high population densities in the face of resource scarcity for thousands of years (Falanruw 1984). In using these techniques, the Yap people have altered their environment creating, for example, coastal mangrove depressions and seagrass meadows to support fishing and termite resistant wood (Stankey and Shinder 1997).

The origin of the adaptive management concept can be traced back to ideas of scientific management pioneered by Frederick Taylor in the early 1900s (Haber 1964). While the term 'adaptive management' evolved in natural resource management workshops through decision makers, managers and scientists focussing on building simulation models to uncover key assumptions and uncertainties (Bormann *et al.* 1999).

Two ecologists at The University of British Columbia, C.S. Holling (1978) and C.J Walters (1986) further developed the adaptive management approach as they distinguished between passive and active adaptive management practice. Kai Lee, notable Princeton physicist, expanded upon the approach in the late 1970s and early 1980s while pursuing a post-doctorate degree at UC Berkeley. The approach was further developed at the International Institute for Applied Systems Analysis (IIASA) in Vienna, Austria, while C.S. Holling was director of the Institute. In 1992, Hilbourne described three learning models for federal land managers, around which adaptive management approaches could be developed, these are reactive, passive and active.

Adaptive management has probably been most frequently applied in Yap, Australia and North America, initially applied in fishery management, but received more broad application in the 1990s and 2000s. One of the most successful applications of adaptive management has been in the area of waterfowl harvest management in North America, most notably for the mallard (Johnson et al., 1993; Nichols et al., 2007).

Adaptive management in a conservation project and program context can trace its roots back to at least the early 1990s, with the establishment of the Biodiversity Support Program (BSP) in 1989. BSP was a USAID-funded consortium of WWF, The Nature Conservancy (TNC), and World Resources Institute (WRI). Its Analysis and Adaptive Management Program sought to understand the conditions under which certain conservation strategies were most effective and to identify lessons learned across conservation projects. When BSP ended in 2001, TNC and Foundations of Success (FOS, a non-profit which grew out of BSP) continued to actively work in promoting adaptive management for conservation projects and programs. The approaches used included Conservation by Design (TNC) and Measures of Success (FOS).

In 2004, the Conservation Measures Partnership (CMP) – which includes several former BSP members – developed a common set of standards and guidelines for applying adaptive management to conservation projects and programs.

3.4   Use in environmental practices

Applying adaptive management in a conservation project or program involves the integration of project/program design, management, and monitoring to systematically test assumptions in order to adapt and learn. The three components of adaptive management in environmental practice are:

- **Testing Assumptions** is about systematically trying different actions to achieve a desired outcome. It is not, however, a random trial-and-error process. Rather, it involves using knowledge about the specific site to pick the best known strategy, laying out the assumptions behind how that strategy will work, and then collecting monitoring data to determine if the assumptions hold true.

- **Adaptation** involves changing assumptions and interventions to respond to new or different information obtained through monitoring and project experience.

- **Learning** is about explicitly documenting a team's planning and implementation processes and its successes and failures for internal learning as well as learning across the conservation community. This learning enables conservation practitioners to design and manage projects better and avoid some of the perils others have encountered (Stankey et al. 2005).Learning about a managed system is only useful in cases where management decisions are repeated (Rout et al. 2009).

3.5   Application to environmental projects and programs

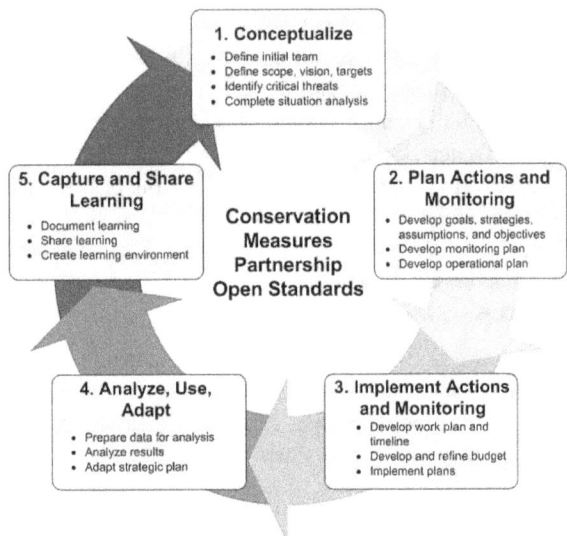

Figure 1: CMP Adaptive Management Cycle

Open Standards for the Practice of Conservation lay out 5 main steps to an adaptive management project cycle (see Figure 1). The *Open Standards* represent a compilation and adaptation of best practices and guidelines across several fields and across several organizations within the conservation community. Since the release of the initial *Open Standards* (updated in 2007 and 2013), thousands of project teams from conservation organizations (e.g., TNC, Rare, and WWF), local conservation groups, and donors alike have begun applying these *Open Standards* to their work. In addition, several CMP members have developed training materials and courses to help apply the Standards.

Some recent write-ups of adaptive management in conservation include: wildlife protection (SWAP, 2008), forests ecosystem protection (CMER, 2010), coastal protection and restoration (LACPR, 2009), natural resource management (water, land and soil), species conservation especially, fish conservation from overfishing (FOS, 2007) and climate change (DFG, 2010). In addition, some other examples follow:

- In 2006-2007, FOS worked with The National Fish and Wildlife Foundation (NFWF) to develop an evaluation system help NFWF gauge impact across the various coral reef habitat and species conservation projects;

- In 2007, FOS worked with the Ocean Conservancy (OC) to evaluate the effectiveness of this Scorecard in helping to end overfishing in domestic fisheries.

- Between 1999-2004, FOS worked for WWF's Asian Rhino and Elephant Action Strategy (AREAS) Program to ensure that Asian elephants and rhinos thrive in secure habitats within their historical range and in harmony with people.

- The Department of Fish and Game (DFG) is developing and implementing adaptation strategies to help protect, restore and manage fish and wildlife, with the understanding that some level of climate change will occur and that it will have profound effects on ecosystems in the United States.

- The Adaptive Management program was created by CMR to provide science-based recommendations and technical information to assist the Forest Practices Board. In April 2010, the Forest Practices Adaptive Management Annual Science Conference was held in Washington.

- In 2009, The Louisiana Coastal Protection and Restoration (LACPR) Technical Report has been developed by the United States Army Corps of Engineers (USACE) according to adaptive management process.

- Since 2009, the Kenya Wildlife Service has been managing its marine protected areas using adaptive management in an ongoing process of learning through the Science for Active Management (SAM) Program.

3.6   Tools and guidance for conservation practitioners

The following resources offer guidance on designing and planning conservation projects (Steps 1 and 2 of the *Open Standards*), as well as more general guidance on the adaptive management process.

Step 1 Conceptualize (Describing the Project's Context)

- IUCN-CMP Unified Classifications of Direct Threats and Conservation Actions: A standardized taxonomy of direct threats to biodiversity and conservation actions that helps conservation teams speak a common language across projects to facilitate learning.

- Using Conceptual Models to Document a Situation Analysis: A guide that explains how to build a conceptual model to clearly portray what drives threats to biodiversity within a project site.

- Instruction Manual for completing Step 1 of the WWF Standards – Define: An online course manual to walk users through the conceptualization phase of the adaptive management project cycle.

Step 2 Plan Actions and Monitoring

- Using Results Chains to Improve Strategy Effectiveness: A guide that explains how to build results chains – a tool for clarifying a project team's assumptions about how their actions will contribute to reducing threats and conserving biodiversity.

- Instruction Manual for completing Step 2 of the WWF Standards – Design: An online course manual used by WWF's Online Campus to walk users through the actions and monitoring plan phase of the adaptive management project cycle.

3.7 Use in other practices as a tool for sustainability

Adaptive management as a systematic process for improving environmental management policies and practices is the traditional application however, the adaptive management framework can also be applied to other sectors seeking sustainability solutions such as business and community development. Adaptive management as a strategy emphasizes the need to change with the environment and to learn from doing. Adaptive management applied to ecosystems makes overt sense when considering ever changing environmental conditions. The flexibility and constant learning of an adaptive management approach is also a logical application for organizations seeking sustainability methodologies. Businesses pursuing sustainability strategies would employ an adaptive management framework to ensure that the organization is prepared for the unexpected and geared for change. By applying an adaptive management approach the business begins to function as an integrated system adjusting and learning from a multi-faceted network of influences not just environmental but also, economic and social (Dunphy, Griffths, & Benn, 2007). The goal of any sustainable organization guided by adaptive management principals must be to engage in active learning to direct change towards sustainability (Verine, 2008). This "learning to manage by managing to learn" (Bormann BT, 1993) will be at the core of a sustainable business strategy.

Sustainable community development requires recognition of the relationship between environment, economics and social instruments within the community. An adaptive management approach to creating sustainable community policy and practice also emphasizes the connection and confluence of those elements. Looking into the cultural mechanisms which contribute to a community value system often highlights the parallel to adaptive management practices, "with [an] emphasis on feedback learning, and its treatment of uncertainty and unpredictability"(Berkes, Colding, & Folke, 2000). Often this is the result of indigenous knowledge and historical decisions of societies deeply rooted in ecological practices (Berkes, Colding, & Folke, 2000). By applying an adaptive management approach to community development the resulting systems can develop built in sustainable practice as explained by the Environmental Advisory Council (2002), "active adaptive management views policy as a set of experiments designed to reveal processes that build or sustain resilience. It requires, and facilitates, a social context with flexible and open institutions and multi-level governance systems that allow for learning and increase adaptive capacity without foreclosing future development options" (p. 1121).

In an ever changing world, adaptive management appeals to many practices seeking sustainable solutions by offering a framework for decision making that proposes to support a sustainable future which, "conserves and nurtures the diversity — of species, of human opportunity, of learning institutions and of economic options"(The Environmental Advisory Council, 2002, p. 1121).

3.8 General resources

Information and guidance on the entire adaptive management process is available from CMP members' websites and other online sources:

- The Conservation Measures Partnership's *Open Standards for the Practice of Conservation* provide general guidance and principles for good adaptive management in conservation.

- Miradi Adaptive Management Software for Conservation Projects is user friendly software developed through a joint venture between CMP and Benetech. The software walks conservation teams through each step of the Open Standards.

- Foundations of Success (FOS) Resources and Training web pages list reference materials on adaptive management and monitoring and evaluation, as well as information about online or in-person courses in adaptive management.

- The Nature Conservancy's Conservation Action Planning (CAP) Resources page includes detailed guidance and tools for implementing the CAP adaptive management process. See also TNC's CAP Standards.

- The Wildlife Conservation Society's Living Landscapes page contains extensive guidance materials on WCS's approach to adaptive management.

- WWF's web page on the *WWF Standards of Conservation Project and Programme Management* contains detailed guidance, resources, and tools for the steps in WWF's adaptive management process.

- Measures of Success: Designing, Managing, and Monitoring Conservation and Development Projects, written in 1998 by Richard Margoluis and Nick Salafsky, was one of the first detailed manuals about applying adaptive management to conservation projects. Also available in Spanish.

- Foundations of Success (FOS) web pages list Asian Rhino and Elephant Program Evaluation in 2004.

- Foundations of Success (FOS) web pages list National Fish & Wildlife Foundation's Coral Fund in 2007.

- Foundations of Success (FOS) web pages list Ocean Conservancy's Overfishing Scorecard in 2007.

- The Department of Fish and Game(DFG) web pages list Adapting to Climate Change programme.

- U.S. Army Corps of Engineers web pages list Louisiana Coastal Protection and Restoration Final Technical Report in 2009.

- Washington State Department of Natural Resource(CMR) web pages list Forest Practices Adaptive Management Program in 2010.

3.9 See also

- Conservation biology
- Decision theory
- Ecology
- Fisheries
- Forestry
- Optimization
- Operations research
- Uncertainty

3.10 References

[1] Carey, G, Crammond, B, Malbon, E et al 2015, 'Adaptive policies for reducing inequalities in the social determinants of health.', International Journal of Health Policy and Management. http://www.ijhpm.com/article_3097_0.html

- Bormann, B.T.; Wagner, F.H.; Wood, G., Algeria, j.; Cunningham, P.G.; Brooks, M.H.; Friesema, P.; Berg, J.; Henshaw,J (1999). *Ecological Stewardship: A common reference for ecosystem management.* Amsterdam: Elsevier.

- Williams, Byron K.; Robert C. Szaro; Carl D. Shapiro (2007). *Adaptive Management: The U.S. Department of the Interior Technical Guide.* US Department of the Interior. ISBN 1-4113-1760-2.

- Holling, C. S. (ed.) (1978). *Adaptive Environmental Assessment and Management.* Chichester: Wiley. ISBN 0-471-99632-7.

- Falanrue, M. *People pressure and management of limited resources on Yap* (in McNeely, J.A.; Miller, and K.R (eds)). Washington DC: The Smithsonian Institution Press.

- Walters, C.J. (1986). *Adaptive Management of Renewable Resources.* New York, NY: Mc Graw Hill. ISBN 0-02-947970-3.

- Lee, Kai N. (1993). *Compass and Gyroscope: Integrating Science and Politics for the Environment.* Washington, D.C.: Island Press. ISBN 1-55963-197-X.

- Walters, Carl (1986). *Adaptive Management of Renewable Resources.* New York: Macmillan. ISBN 0-02-947970-3.

- Argyris, Chris; Donald A. Schön (1978). *Organizational Learning: A Theory of Action Perspective.* Reading, Massachusetts: Addison-Wesley. ISBN 0-201-00174-8.

- Gunderson, Lance H.; C. S. Holling; Stephen S. Light (eds.) (1995). *Barriers and Bridges to the Renewal of Ecosystems and Institutions.* New York: Columbia University Press. ISBN 0-231-10102-3.

- Margoluis, Richard; Nick Salafsky (1998). *Measures of Success: Designing, Managing, and Monitoring Conservation and Development Projects.* Washington, D.C: Island Press. ISBN 978-1-55963-612-4.

- Schön, Donald A. (1984). *The Reflective Practitioner: How Professionals Think In Action.* New York: Basic Books. ISBN 978-0-465-06878-4.

- Senge, Peter M. (2006). *The Fifth Discipline: The Art and Practice of the Learning Organization*. New York: Currency Doubleday. ISBN 978-0-385-26095-4.

- Catherine Allan and George H. Stankey (2009). *Adaptive Environmental Management: A Practitioner's Guide*. The Netherlands: Dordrecht. ISBN 978-90-481-2710-8.

- Johnson, F.A.; Williams, B.K.; Nichols, J.D.; Hines, J.El; Kendall, W.L.; Smith, G.W.; and Caithamer, D.F. (1993). "Developing an adaptive management strategy for harvesting waterfowl in North America". *Trans N Am Wildl Nat Resour Conf* (58): 565–583.

- Hilbourne, R (1992). "Can fisheries agencies learn from experience?". *Fisheries* **4** (17): 6–14.

- Nichols, J.D.; Runge, M.C.; Johnson, F.A.; and Williams, B.K. (2007). "Adaptive harvest management of North American waterfowl populations: a brief history and future prospects". *Journal of Ornithology* **148** (148): 343. doi:10.1007/s10336-007-0256-8.

- Nichols, J.D.; and Johnson, F.A.; and Williams, B.K. (1995). "Managing North American waterfowl in the face of uncertainty". *Annu. Rev. Ecol. Syst.* **26**: 177–199. doi:10.1146/annurev.es.26.110195.001141.

- Johnson, F.A.; and Williams, B.K. (1999). "Protocol and practice in the adaptive management of waterfowl harvests.". *Conservation Ecology* **3** (8).

- Margoluis, R.; Stem, C.; Salafsky, N.; Brown, M. (2009). "Using conceptual models as a planning and evaluation tool in conservation". *Evaluation and Program Planning* **32** (2): 138–147. doi:10.1016/j.evalprogplan.2008.09.007. PMID 19054560.

- Marmorek, D.R.; D.C.E. Robinson; C. Murray; L. Grieg (2006). *Enabling Adaptive Forest Management* (PDF). National Commission on Science for Sustainable Forestry. p. 94 pp.

- Murray, Carol; David Marmorek (2003). "Adaptive management and ecological restoration" (PDF). In in Peter Friederici (ed.),. *Ecological Restoration of Southwestern Ponderosa Pine Forests*. Washington, D.C.: Island Press. pp. 417–428. ISBN 1-55963 652-1.

- Peterman, Randall M.; Calvin N. Peters (1998). "Decision analysis: taking uncertainties into account in forest resource management". In in Vera Sit and Brenda Taylor (eds).,. *Statistical Methods for Adaptive Management Studies*. Victoria, B.C.: B.C. Ministry of Forests. pp. 105–127. ISBN 0-7726-3512-9.

- Salafsky, N.; Margoluis, R.; Redford, K.; Robinson, J. (2002). "Improving the practice of conservation: A conceptual framework and agenda for conservation science". *Conservation Biology* **16** (6): 1469–1479. doi:10.1046/j.1523-1739.2002.01232.x.

- Salafsky, N.; Salzer, D., Stattersfield, A.J.; Hilton-Taylor, C.; Neugarten, R.; Butchart, S.H.M.; Collen, B.; Cox, N.; Master, L.L.; O'Connor, S.; Wilkie, D. (2009). "A standard lexicon for biodiversity conservation: Unified classifications of threats and actions". *Conservation Biology* **22** (4): 897–911. doi:10.1111/j.1523-1739.2008.00937.x. PMID 18544093.

- Salzer, D.; Salafsky, N. (2006). "Allocating resources between taking action, assessing status, and measuring effectiveness of conservation actions". *Natural Areas Journal* **26** (3): 310–316. doi:10.3375/0885-8608(2006)26[310:ARBTAA]2.0.CO;2.

- Stankey, George H; Roger N. Clark; Bernard T. Bormann (2005). *Adaptive management of natural resources: theory, concepts, and management institutions*. Gen. Tech. Rep. PNW-GTR-654. Portland, OR: U.S. Department of Agriculture, Forest Service, Pacific Northwest Research Station. p. 73 p.

- Stem, C..; Margoluis, R.; Salafsky, N.; Brown, M. (2005). "Monitoring and evaluation in conservation: A review of trends and approaches". *Conservation Biology* **19** (2): 295–309. doi:10.1111/j.1523-1739.2005.00594.x.

- Virine, Lev; Michael Trumper (2008). *Project Decisions: The Art and Science*. Vienna, VA: Management Concepts. ISBN 978-1-56726-217-9.

- Elzinga, C.L., D. W. Salzer, J. W. Willoughby (1998). *Measuring and Monitoring Plant Populations* (PDF). Denver, CO: Bureau of Land Management. BLM Technical Reference 1730-1.

- Alana L. Moore and Michael A. (2009). "On Valuing Information in Adaptive-Management, Models". *Conservation Biology* **24** (4): 984–993. doi:10.1111/j.1523-1739.2009.01443.x. PMID 20136870.

- George H. Stankey, Roger N. Clark and Bernard T. Bormann (2005). *Adaptive Management of Natural Resources: Theory, Concepts, and Management Institutions* (PDF). Washington: United States Department of Agriculture (USDA).

- Gregory R, Ohlson D and Arvai J. (2006). "Deconstructing adaptive management:

criteria for application to environmental management". *Ecological Applications* **16** (6): 2411–2425. doi:10.1890/1051-0761(2006)016[2411:DAMCFA]2.0.CO;2. PMID 17205914.

- Australian Government Connected Water |url= http://www.connectedwater.gov.au/framework/ adaptive_management.html|

- Berkes, F., Colding, J., & Folke, C. (2000). "Rediscovery of Traditional Ecological Knowledge as Adaptive Management". *Ecological Applications* **10** (5): 1251–1262. doi:10.1890/1051-0761(2000)010[1251:ROTEKA]2.0.CO;2.

- Chaffee, E. E. (1985). "Three Models of Strategy". *The Academy of Management Review* **10** (1): 89–98. doi:10.5465/amr.1985.4277354.

- Dunphy, D., Griffths, A., & Benn, S (2007). *Organizational Change for Corporate Sustainability*. London: Routledge.

- The Environmental Advisory Council (2002). *Resilience and Sustainable Development: Building Adaptive Capacity in a World of Transformation*. stockholm: EDITA NORSTEDTS TRYCKERI AB.

- Verine, L. (2008). "Adaptive Project Management". *PM World Today* **10** (5): 1–9.

- Rout, T.M., Hauser, C.E., Possingham, H.P. (2009). "Optimal adaptive management for the translocation of a threatened species". *Ecological Applications* **19** (2): 515–516. doi:10.1890/07-1989.1. PMID 19323207.

- Shea, K., Possingham, H.P., Murdoch, WW., Roush, R. (2002). "Active Adaptive Management in Insect Pest and Weed Control: Intervention with a Plan for Learning". *Ecological Applications* **12** (3): 927–936. doi:10.1890/1051-0761(2002)012[0927:AAMIIP]2.0.CO;2.

Chapter 4

Appropedia

Appropedia is a wiki website for collaborative solutions in sustainability, poverty reduction and international development, with a particular focus on appropriate technology.[1] Appropedia has been used by a number of nonprofit organizations and individuals working in sustainable development, including Demotech[2] and the Full Belly Project.[3]

Appropedia runs on the software MediaWiki. Currently it houses over 6,000 articles and has received more than 24,000 uploaded files (11 August 2015).[4]

4.1 History

Appropedia wordle constructed from the 5,000 most popular content page names.

After years of other online and offline sustainable collaboration projects, Lonny Grafman started Appropedia in April 2006 with a focus on appropriate technology, defined very broadly. He was assisted by Aaron Antrim and Gabriel Krause in the decision to use the MediaWiki engine for Appropedia. Curt Beckmann then joined Appropedia later in 2006.

Contact with WikiGreen in December 2006[5] led to a quick decision to join forces. It was decided to use the name Appropedia. This merger led to a large increase in content, including permission from various publications, and some content from CD3WD (CDs for the 3rd World).[6]

Since then Appropedia has imported content from other sites:

- International Rivers Network gave permission for several pages worth of content.

- Practical Action gave permission for over 100 articles

- Demotech, an organization in the Netherlands, has built many pages.

- The wiki for the organization Village Earth was merged in March 2007.[7] This was the original wiki on appropriate technology.

- The How To Live Wiki, run by Vinay Gupta, merged much of its material into Appropedia in March 2007, and Vinay joined the Appropedia team.

- The Sgoals wiki, focused on sustainable business practices, joined in mid-2007.

- The Students for Global Sustainability Wiki merged in January 2008.

- CCAT, the Campus Center for Appropriate Technology, partnered with Appropedia to move their project pages to Appropedia in April 2008.

- In 2012, use of the Semantic MediaWiki extension was added.

4.2 Use as source

Papers and books that have used information from Appropedia as source material include:

1. Open-Source Lab: How to Build Your Own Hardware and Reduce Research Costs by Joshua M. Pearce was published in 2014 by Elsevier (ISBN 9780124104624).

2. Kreye, Melissa M. "Metal accumulation in gill epithelium and liver tissue in steelhead (Oncorhynchus mykiss) reared in reclaimed wastewater", Thesis (M.S.) – Humboldt State University, Natural Resources: Wastewater Utilization Program, 2008.

3. Urmila Balasubramaniyam, Llionel S. Zisengwe, Niccoló Meriggi, Eric Buysman, Biogas production in climates with long cold winters, Wageningen, May 2008.

4. James A. West and Margaret L. West, "Using Wikis for Online Collaboration: The Power of the Read-Write Web", Jossey-Bass (December 15, 2008).

4.3   Use as pedagogical tool

Appropedia has also been used by proponents of open source appropriate technology such as Prof.   Joshua Pearce in service learning,[8] including in language education[9][10][11] and engineering education.[12][13]

4.4   See also

- Ekopedia

- Open-source appropriate technology

- Wiser.org

- Hackteria

4.5   References

[1] Joshua M. Pearce, Lonny Grafman, Thomas Colledge, and Ryan Legg, "Leveraging Information Technology, Social Entrepreneurship and Global Collaboration for Just Sustainable Development" Proceedings of the 12th Annual National Collegiate Inventors and Innovators Alliance Conference, pp. 201–210, 2008. Full text:

[2] "Demotech, Design for self reliance". Demotech.org. Retrieved 2013-03-27.

[3] "Agricultural Development & Technology – Full Belly Project". Fullbellyproject.org. Retrieved 2012-01-21.

[4] "Appropedia". Appropedia.org.   2015-08-11.   Retrieved 2015-08-11.

[5] "User talk:Chriswaterguy – Appropedia: The sustainability wiki". Appropedia. Retrieved 2012-01-21.

[6] "www.cd3wd.com – alexweir1949@gmail.com – cd3wd – High Quality Technical Development Info for the Third World – and the SEEV fraud-proof voting system for the Third World – last updated 2011/03". cd3wd. Retrieved 2012-01-21.

[7] Village Earth. 2007. Village Earth joins the Appropedia wiki community. Appropriate Technology Project Blog.

[8] J. M. Pearce, L. Grafman, T. Colledge, and R. Legg, "Leveraging Information Technology, Social Entrepreneurship and Global Collaboration for Just Sustainable Development" Proceedings of the 12th Annual National Collegiate Inventors and Innovators Alliance Conference, pp. 201–210, 2008.

[9] E. ter Horst and J. M. Pearce, "Foreign Languages and the Environment: A Collaborative Instructional Project", The Language Educator, pp. 52–56, October, 2008.

[10] J. M. Pearce and E. ter Horst "Appropedia and Sustainable Development for Improved Service Learning", Proceedings of Association for the Advancement of Sustainability in Higher Education 2008.

[11] Joshua M. Pearce and Eleanor ter Horst, "Overcoming Language Challenges of Open Source Appropriate Technology for Sustainable Development in Africa", Journal of Sustainable Development in Africa, 11(3) pp. 230–245, 2010.

[12] Joshua M. Pearce, "Appropedia as a Tool for Service Learning in Sustainable Development", Journal of Education for Sustainable Development, 3(1), pp. 45–53, 2009. Q-Space pre-print

[13] S. Murphy and N. Saleh, "Information literacy in CEAB's accreditation criteria: the hidden attribute", In Proceedings of The Sixth International Conference on Innovation and Practices in Engineering Design and Engineering Education, 2009. Hamilton, ON July 27–29, 2009.

4.6   External links

- Official website

Chapter 5

Bioregionalism

Bioregionalism is a political, cultural, and ecological system or set of views based on naturally defined areas called bioregions, similar to ecoregions. Bioregions are defined through physical and environmental features, including watershed boundaries and soil and terrain characteristics. Bioregionalism stresses that the determination of a bioregion is also a cultural phenomenon, and emphasizes local populations, knowledge, and solutions.[1]

Bioregionalism is a concept that goes beyond national boundaries—an example is the concept of Cascadia, a region that is sometimes considered to consist of most of Oregon and Washington, the Alaska Panhandle, the far north of California and the West Coast of Canada, sometimes also including some or all of Idaho and western Montana.[2] Another example of a bioregion, which does not cross national boundaries, but does overlap state lines, is the Ozarks, a bioregion also referred to as the Ozarks Plateau, which consists of southern Missouri, northwest Arkansas, the northeast corner of Oklahoma, southeast corner of Kansas.[3]

5.1 Overview

The term was coined by Allen Van Newkirk, founder of the Institute for Bioregional Research, in 1975,[4] given currency by Peter Berg and Raymond Dasmann in the early 1970s,[5] and has been advocated by writers such as Kirkpatrick Sale.[6]

The bioregionalist perspective opposes a homogeneous economy and consumer culture with its lack of stewardship towards the environment. This perspective seeks to:

- Ensure that political boundaries match ecological boundaries.[7]

- Highlight the unique ecology of the bioregion.

- Encourage consumption of local foods where possible.

- Encourage the use of local materials where possible.

- Encourage the cultivation of native plants of the region.

- Encourage sustainability in harmony with the bioregion.[8]

5.2 Relationship to environmentalism

Bioregionalism, while akin to environmentalism in certain aspects, such as a desire to live in harmony with nature, differs in certain ways from classical, 20th century environmentalism.[9]

According to Peter Berg, bioregionalism is proactive, and is based on forming a harmony between human culture and the natural environment, rather than being protest-based like the original environmental movement. Also, while classical environmentalists saw human industry as the enemy of nature and nature as a victim needing to be saved; bioregionalists see humanity and its culture as a part of nature, focusing on building a positive, sustainable relationship with the environment, rather than a focus on preserving and segregating the wilderness from the world of humanity.[10]

5.3 In politics

North American Bioregional Assemblies have been biannual gatherings of bioregionalists throughout North America since 1984 and have given rise to national level Green Parties. In addition, bioregionalism spawned the sustainability movement. The tenets of bioregionalism are often used by green movements, which oppose political organizations whose boundaries conform to existing electoral districts. This problem is perceived to result in elected representatives voting in accordance with their constituents, some of whom may live outside a defined bioregion, and may run counter to the well-being of the bioregion.

At the local level, several bioregions have congresses that meet regularly. For instance, the Ozark Plateau bioregion hosts a yearly Ozark Area Community Congress, better known as OACC, which has been meeting every year since 1980,[11] most often on the first weekend in October. The Kansas Area Watershed, "KAW" was founded in 1982 and has been meeting regularly since that time.[12] KAW holds a yearly meeting, usually in the spring.

The government of the Canadian province of Alberta has recently made major to its land-use policies including a separate "land-use framework" document for each major river basin within the province. This is supported by local initiatives such as the Beaver Hills Initiative, which seeks to create a large biosphere reserve encompassing Elk Island National Park and the surrounding area.[13]

5.4   See also

- Deep Ecology

- Eco-Communalism

- Ecological footprint

- Bioregional decolonization

- Cascadia (independence movement)

- Grassroots democracy

- Green anarchism

- List of ecoregions

- Permaculture

- Social ecology

5.5   References

[1] "Bioregionalism: The Need For a Firmer Theoretical Foundation", Don Alexander, Trumpeter v13.3, 1996.

[2] "Cascadia: The New Frontier". Cascadia Prospectus. 2010-02-12. Retrieved 2012-11-08.

[3] "About OACC Ozark Area Community Congress". *OACC Ozark Area Community Congress*. Retrieved 30 December 2011.

[4] McGinnis, Michael Vincent (1999). *Bioregionalism*. London and New York: Routledge. p. 22. ISBN 041515444-8.

[5] Berg, Peter and Raymond Dasmann, "Reinhabiting California," *The Ecologist* 7, no. 10 (1977)

[6] Anderson, Walter Truett. There's no going back to nature, *Mother Jones* (September/October 1996)

[7] Davidson, S. (2007) "The Troubled Marriage of Deep Ecology and Bioregionalism," *Environmental Values*, vol. 16(3): 313-332

[8] Bastedo, Jamie. *Shield Country: The Life and Times of the Oldest Piece of the Planet*, Red Deer Press, 1994. ISBN 0-88995-191-8

[9] "Peter Berg of Planet Drum". Sustainable-city.org. 1998-02-12. Retrieved 2012-11-08.

[10] "Peter Berg of Planet Drum". Sustainable-city.org. 1998-02-12. Retrieved 2012-11-08.

[11] "About OACC Ozark Area Community Congress". *OACC Ozark Area Community Congress*. Retrieved 1 February 2013.

[12] "Kansas Area Watershed Council History". Retrieved 1 February 2013.

[13] beaerhills.ab.ca

5.6   Further reading

- Mike Carr, *Bioregionalism and Civil Society: Democratic Challenges to Corporate Globalism*, UBC Press, 2004. ISBN 978-0774809443.

- Peter Berg, editor. *Reinhabiting A Separate Country: A Bioregional Anthology of Northern California*. San Francisco: Planet Drum, 1978. ISBN 0-937102-00-8.

- Peter Berg, *Envisioning Sustainability*, Subculture Books, 2009. ISBN 978-0-9799194-8-0.

- Michael McGinnis, editor. *Bioregionalism*, Routledge, 1998. ISBN 0-415-15445-6.

- Kirkpatrick Sale, *Dwellers in the Land: The Bioregional Vision*. Random House, 1985. ISBN 0-8203-2205-9 (University of Georgia Press, 2000).

- Gary Snyder. *A Place in Space: Ethics, Aesthetics, and Watersheds*. Counterpoint, 1995. ISBN 1-887178-27-9

- Robert Thayer. *LifePlace: Bioregional Thought and Practice*, University of California Press, 2003. ISBN 0-520-23628-9

- Emanuele Guerrieri Ciaceri. *Bioregionalismo. La visione locale di un mondo globale*. Argo Edizioni, Italia 2006. ISBN 978-88-88659-19-0

- Doug Aberley, editor. *Boundaries of Home: Mapping for Local Empowerment*. New Society Publishers, 1998. ISBN 978-0-86571-272-0

5.7 External links

- Free Cascadia.org, the website belonging to Alexander Baretich, designer of the Cascadian flag, and advocate of Bioregionalism.

- Encyclopedia of Earth: Ecoregion

- North American Bioregional Congress

- Ozark Area Community Congress

- Planet Drum Foundation website.

- Putah-Cache Bioregion Project - interdisciplinary research and educational project at UC Davis.

Chapter 6

Bioretention

A bioretention cell, also called a rain garden, in the United States. It is designed to treat polluted stormwater runoff from an adjacent parking lot. Plants are in winter dormancy.

Bioretention is the process in which contaminants and sedimentation are removed from stormwater runoff. Stormwater is collected into the treatment area which consists of a grass buffer strip, sand bed, ponding area, organic layer or mulch layer, planting soil, and plants. Runoff passes first over or through a sand bed, which slows the runoff's velocity, distributes it evenly along the length of the ponding area, which consists of a surface organic layer and/or groundcover and the underlying planting soil. The ponding area is graded, its center depressed. Water is ponded to a depth of 15 cm (5.9 in) and gradually infiltrates the bioretention area or is evapotranspired. The bioretention area is graded to divert excess runoff away from itself. Stored water in the bioretention area planting soil exfiltrates over a period of days into the underlying soils.[1]

6.1   Filtration

Each of the components of the bioretention area is designed to perform a specific function. The grass buffer strip reduces incoming runoff velocity and filters particulates from the runoff. The sand bed also reduces the velocity, filters particulates, and spreads flow over the length of the bioretention area. Aeration and drainage of the planting soil are provided by the 0.5 m (20 in) deep sand bed. The ponding area provides a temporary storage location for runoff prior to its evaporation or infiltration. Some particulates not filtered out by the grass filter strip or the sand bed settle within the ponding area.[1]

The organic or mulch layer also filters pollutants and provides an environment conducive to the growth of microorganisms, which degrade petroleum-based products and other organic material. This layer acts in a similar way to the leaf litter in a forest and prevents the erosion and drying of underlying soils. Planted groundcover reduces the potential for erosion as well, slightly more effectively than mulch. The maximum sheet flow velocity prior to erosive conditions is 0.3 meters per second (1 foot per second) for planted groundcover and 0.9 meters per second (3 feet per second) for mulch.[2]

The clay in the planting soil provides adsorption sites for hydrocarbons, heavy metals, nutrients and other pollutants. Stormwater storage is also provided by the voids in the planting soil. The stored water and nutrients in the water and soil are then available to the plants for uptake. The layout of the bioretention area is determined after site constraints such as location of utilities, underlying soils, existing vegetation, and drainage are considered. Sites with loamy sand soils are especially appropriate for bioretention because the excavated soil can be backfilled and used as the planting soil, thus eliminating the cost of importing planting soil. An unstable surrounding soil stratum and soils with a clay content greater than 25 percent may preclude the use of bioretention, as would a site with slopes greater than 20 percent or a site with mature trees that would be removed during construction of the best management practices.[3]

6.2 Heavy metal remediation

Contaminant trace metals such as zinc, lead, and copper are found in stormwater runoff from impervious surfaces (e.g. roadways and sidewalks). Treatment systems such as rain gardens and stormwater planters utilize a bioretention layer to remove heavy metals in stormwater runoff. Dissolved forms of heavy metals may bind to sediment particles in the roadway that are then captured by the bioretention system. Additionally, heavy metals may adsorb to soil particles in the bioretention media as the runoff filters through.[4] In laboratory experiments, bioretention cells removed 94%, 88%, 95%, and >95% of zinc, copper, lead, and cadmium, respectively from water with metal concentrations typical of stormwater runoff. While this is a great benefit for water quality improvement, bioretention systems have a finite capacity for heavy metal removal. This will ultimately control the lifetime of bioretention systems, especially in areas with high heavy metal loads.[5]

Metal removal by bioretention cells in cold climates was similar or slightly lower than that in warmer environments. Plants are less active in colder seasons, suggesting that most of the heavy metals remain in the bioretention media rather than being taken up by plant roots.[6] Therefore, removal and replacement of the bioretention layer will become necessary in areas with heavy metal pollutants in stormwater runoff to extend the life of the treatment system.

6.3 See also

- Bioswale

- Groundwater recharge

- Rain garden

- Urban runoff

- Phytoremediation

6.4 References

[1] United States Environmental Protection Agency (EPA). Washington, DC (1999). "Storm Water Technology Fact Sheet: Bioretention." Document No. EPA-832-F-99-012.

[2] Clar, M. L., Barfield, B. J., & O'Connor, T. P. (2004). "Stormwater Best Management Practice Design Guide, Volume 2: Vegetative Biofilters." Cincinnati, OH: United States Environmental Protection Agency. Document no. EPA/600/R-04/121A.

[3] Prince George's County Department of Environmental Resources, Largo, MD (2007). "Bioretention Manual." Chapter 1.[Dead Link]

[4] Li, H., & Davis, A. P. (2008)."Heavy metal capture and accumulation in bioretention media." Environmental science & technology, 42(14), 5247-5253.

[5] Sun, X., & Davis, A. P. (2007)."Heavy metal fates in laboratory bioretention systems." Chemosphere, 66(9), 1601-1609.

[6] Muthanna, T. M., Viklander, M., Gjesdahl, N., & Thorolfsson, S. T. (2007)."Heavy metal removal in cold climate bioretention." Water, air, and soil pollution, 183(1-4), 391-402.

- Davis, Allen P. (2007). "Field Performance of Bioretention: Water Quality". *Environmental Engineering Science* **24** (8): 1048–1064. doi:10.1089/ees.2006.0190.

- Liu, Jia; Sample, David J.; Bell, Cameron; Guan, Yuntao (2014). "Review and Research Needs of Bioretention Used for the Treatment of Urban Stormwater". *Water* **6** (4): 1069–1099. doi:10.3390/w6041069.

- Traver, Robert G.; Davis, Allen P.; Hunt, William F. (October 2007). "Bioretention and Bioinfiltration BMPs: Three researchers' experience". *Stormwater* (Santa Barbara, CA: Forester Media). ISSN 1531-0574.

6.5 External links

- Combating Climate Change with Landscape Architecture; Resource Guide - American Society of Landscape Architects

Chapter 7

Blue-Green Cities

Blue-Green Cities aim to recreate a naturally oriented water cycle while contributing to the amenity of the city by bringing water management and green infrastructure together.[1] This is achieved by combining and protecting the hydrological and ecological values of the urban landscape while providing resilient and adaptive measures to address future changes in climate, landuse, water management, and socio-economic activity in the city. Designing and utilising the urban environment to manage water resources, water demand (including rainwater harvesting), and the interplay between flood and drought are key drivers. Integrating water management with urban green space provision plus the added value associated with the connection and interaction between blue and green assets[2] are key concepts of a Blue-Green City. Blue-Green Cities generate a multitude of environmental, ecological, socio-cultural and economic benefits through integrated planning and management [3] and may be key to future resilience and sustainability of urban environments and processes.

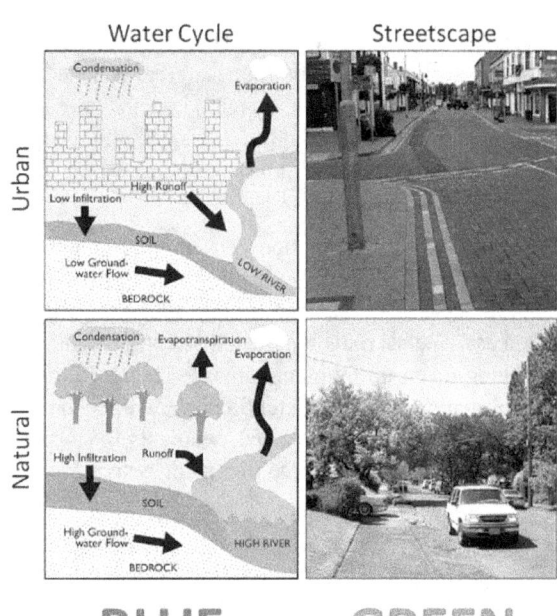

Comparison of hydrologic (water cycle) and environmental (streetscape) attributes in conventional (upper) and Blue-Green Cities.

7.1 Background

Blue-Green Cities aim to reintroduce the natural water cycle into urban environments and provide effective measures to manage fluvial (river), coastal, and pluvial (urban runoff or surface water) flooding [1] while championing the concept of multi-functional greenspace and landuse to generate multiple benefits for the environment, society, and the economy.

Visible water in cities has massively declined in the last century [4] and many areas are facing future water scarcity in response to changes in climate, landuse and population. The concept of Blue-Green Cities involves working with green and blue infrastructure components to secure a sustainable future and generate multiple benefits for the environmental, ecological, social and cultural spheres. This requires a coordinated approach to water resource and green space management from institutional organisations, industry, academia and local communities and neighbourhoods.

The natural water cycle is characterised by high evaporation, a high rate of infiltration, and low surface runoff.[1] This typically occurs in rural areas with abundant permeable surfaces (soils, green space), trees and vegetation, and natural meandering water courses. In contrast, in most urban environments there is more surface runoff, less infiltration and less evaporation. Green and blue spaces are often disconnected. The lack of infiltration in urban environments may reduce the amount of groundwater, which can have significant implications in some cities that experience drought. In urban environments water is quickly transported over the impermeable concrete, spending little time on the surface before being redirected underground into a network of pipes and sewers. However, these conventional systems ('grey' infrastructure) may not be sustainable, particularly in light of potential future climate change. They may be highly expensive and lack many of the multiple benefits associated with Blue-Green infrastructure.

Land planning and engineering design approaches in Blue-Green Cities aim to be cost effective, resilient, adaptable, and help mitigate against future climate change, while minimising environmental degradation and improving aesthetic and recreational appeal. Key functions in Blue-Green Cities include protecting natural systems and restoring natural drainage channels, mimicking pre-development hydrology, reducing imperviousness, and increasing infiltration, surface storage and the use of water retentive plants.[5] A key factor is interlinking the blue and green assets to create Blue-Green corridors through the urban environment.[2]

Blue-Green Cities favour the holistic approach and aim for interdisciplinary cooperation in water management, urban design, and landscape planning. Community understanding, interaction and involvement in the evolution of Blue-Green design are actively promoted. Blue-Green Cities typically incorporate sustainable urban drainage systems (SUDS), a term used in the United Kingdom, known as water-sensitive urban design (WSUD) in Australia, and low-impact development or best management practice (BMP) in the United States. Green infrastructure is also a term that is used to define many of the infrastructure components for flood risk management in Blue-Green Cities.

Water management components in Blue-Green Cities are part of a wider complex "system of systems" providing vital services for urban communities. The urban water system interacts with other essential infrastructure such as information and telecommunications, energy, transport, health and emergency services. Blue-Green Cities aim to minimise the negative impacts on these systems during times of extreme flood while maximising the positive interactions when the system is in the non-flood state.

Key barriers to effective implementation of Blue-Green infrastructure can arise if planning processes and wider urban system design and urban renewal programmes are not fully integrated.[5]

7.2 Blue-Green Infrastructure Components

Many infrastructure components and common practices may be employed when planning and developing a Blue-Green City, in line with specific local objectives, e.g. water management, delivery of multi-functional green infrastructure, biodiversity action plans. A Blue-Green City actively works with existing grey infrastructure to provide optimal management of the urban water system during a range of flood events; from no flood, to minimal flooding, to extreme rainfall events where the drainage system may be exceeded.[6]

The key functions of Blue-Green infrastructure components include water use/reuse, water treatment, detention and infiltration, conveyance, evapotranspiration, local amenity provision, and generation of a range of viable habitats for local ecosystems. In most cases, the components serve several functions.

Blue-Green infrastructure includes;

- Bioretention systems

A photograph of a bioretention system, or rain garden, in Portland, Oregon, US.

- Bioretention swales

- Swales and buffer strips

- Storage ponds, lakes and reservoirs

- Controlled storage areas, e.g. car parks, recreational areas, minor roads, playing fields, parkland and hard standing in school playgrounds and industrial areas

- Sand filters and infiltration trenches

- Permeable paving

- Rain gardens

- Stream and river restoration

- De-canalisation of river corridors and re-introduction of meanders

- Constructed wetlands

- Property level strategies to reduce surface water and manage runoff, such as water butts (or rainwater tanks in the US),

- Open green space

- Parks and gardens

- Street trees

- Pocket parks

- Vegetated ephemeral waterways

- Planted drainage assets (green roofs and green walls)

- Green outfalls

- Restored, rehabilitated and enhanced urban watercourses offering green erosion protection (see river restoration)

7.3 Benefits associated with Blue-Green Cities

A Blue-Green City contains an interconnected network of blue and green infrastructure that work in harmony to generate a range of benefits when the system is in both the flood state and non-flood state. A wide range of environmental, ecological, economic and socio-cultural benefits are directly and indirectly attributed to Blue-Green Cities. Many benefits are realised during times of no flood (green benefits), giving Blue-Green Cities a competitive edge over otherwise comparable, conventional cities. Multi-functional infrastructure is a key to generating the maximum benefits when the system is in the non-flood state. An ecosystem services approach is frequently used to determine the benefits people obtain from the environment and ecosystems. Many of the good and services provided by Blue-Green Cities have economic value, e.g. the production of clean air, water and carbon sequestration.[7]

The benefits include;

- Climate change adaptation and mitigation

- Reduction of the urban heat island effect

- Better management of stormwater and water supply, conservation of water resources through efficiency (increasing the resilience to drought)

- Carbon reduction/mitigation

- Improved air quality

- Increased biodiversity (including the reintroduction and propagation of native species)

A photograph of a stream enhancement project in Portland, Oregon, to promote wildlife habitats and increase biodiversity

- Habitat and biodiversity enhancement

- Water pollution control

- Public amenity (recreational water use, parks and recreation grounds, leisure)

- Cultural services (physical and mental health, well-being of citizens, aesthetics, spiritual)

- Community engagement

- Education

- Landscaping and quality of place

- Increased land and property values

- Labour productivity (stress reduction, attracting and retaining staff)

- Economic growth and investment

- Food production

- Healthy soils and a reduction in soil erosion and river bank retreat

- Tourism

- Reduction in the accumulation of sediment, debris and pollutants in urban watercourses

- Shading and shelter around rivers and the wider urban environment

- Economic benefits related to avoided costs from flooding

- Community cohesion and greater understanding of sustainable planning and lifestyle

A photograph of a community vegetable garden in Portland, Oregon, taken in April 2013. This illustrates some of the additional benefits of green infrastructure, e.g. community engagement, quality of place, and horticultural education

- Possible diversification of the local economy and job creation

- Strengthening ecosystem resilience

- Ecological corridors and landscape permeability (biodiversity benefits)

- Avoided impacts of flood events, including avoided damage to the economy, wildlife, buildings and infrastructure, and avoided trauma and distress (mental health impacts) associated with flooding

The multiple benefits of adopting Blue-Green infrastructure will span both the local/regional and global/international scales. The Department of Environment, Farming and Rural Affairs' (Defra) approach to flood and coastal risk management has been to seek multi-functional benefits from Flood and Coastal Erosion Risk Management (FCERM) interventions and enhance the clarity of social and environmental consequences in the decision making process. Defra note, however, that flood risk reduction benefits provided by ecosystems are not well understood and this is an area where more systematic research is needed.[8]

7.4 Case Study examples

Concepts of water sensitive cities and tools for water-centric urban design are developing in many countries.[9] During the first decade of the 21st century, Portland, Oregon, began its 'grey to green' initiative [10] and Melbourne, Australia, reached the "water cycle city" stage.[11] Few, if any UK cities have progressed beyond "the drained city" stage, with water managed for a series of single functions (including flood risk management), mostly through distribution, collection and treatment systems and drainage infrastructure that are energy intensive and which continue to degrade urban environments in general and urban watercourses, in particular. Cities in Europe (e.g. Copenhagen, Rotterdam (Waterplan),[12][13] Lodz (Blue Green Network),[14] Graz, Berlin), North America (e.g. Philadelphia, New York) and Australia (e.g. Adelaide) are also addressing the challenges of moving towards a Blue-Green City.

7.5 References

[1] Hoyer, J., Dickhaut, W., Kronawitter, L. and Weber B. 2011. Water Sensitive Urban Design. Jovis, University of Hamburg.

[2] Maksimović, Stanković, S., Xi Liu and Lalić, M. 2013. Blue Green Dream Project's Solutions for Urban Areas in the Future. Reporting for Sustainability. http://www.sciconfemc.rs/PAPERS/BLUE%20GREEN%20.pdf

[3] http://www.bluegreencities.ac.uk

[4] http://www.academia.edu/2369268/Water_purificative_landscapes_constructed_ecologies_and_contemporary_urbanism. Retrieved 4 March 2014. Missing or empty |title= (help)

[5] Novotny V., Ahern J. and Brown P. 2010. Water Centric sustainable communities: planning, retrofitting and building the next urban environment. John Wiley and Sons, New Jersey.

[6] CIRIA (2006). "Designing for exceedance in urban drainage - good practice (C635)".

[7] (PDF) http://ec.europa.eu/environment/nature/ecosystems/docs/green_infrastructure_broc.pdf. |first1= missing |last1= in Authors list (help); Missing or empty |title= (help)

[8] SWITCH, an EU-funded research programme with UNESCO-IHE as lead partner) (http://www.switchurbanwater.eu/)

[9] Howe, C. and Mitchell, C. 2012. Water Sensitive Cities. IWA Publishing, London.

[10] 'Grey to green' initiative, Portland, Oregon (http://www.portlandoregon.gov/bes/47203)

[11] Brown, R., Keath, N. and Wong, T. 2008. Transitioning to Water Sensitive Cities: Historical, Current and Future Transition States. Proceedings of the 11th International Conference on Urban Drainage, Edinburgh, Scotland (http://web.sbe.hw.ac.uk/staffprofiles/bdgsa/11th_International_Conference_on_Urban_Drainage_CD/ICUD08/pdfs/618.pdf)

[12] Rotterdam, Netherlands, Waterplan 2 (http://www.rotterdam.nl/waterplan_2_en_deelgemeentelijke_waterplannen)

[13] Rotterdam Waterplan 2013-2018 (http://www.rotterdamclimateinitiative.nl/en)

[14] Blue Green Network, Lodz, Poland (http://www.switchtraining.eu/fileadmin/template/projects/switch_training/files/Case_studies/Case_study_Lodz_preview.pdf)

-
-
-
-

7.6 External links

- Blue-Green Cities Research Project website
- Blue Green Dream, Imperial College, London
- CIRIA – Sustainable Urban Drainage Systems (SUDS)
- CRC for Water Sensitive Cities
- Engineering Natures Way - Sustainable Drainage Systems (SUDS) and flood risk management
- International Water Association (IWA) (Cities of the Future Programme)
- MARE – Managing Adaptive Responses to changing Flood Risk
- Metropolitan Glasgow Strategic Drainage Partnership (MGSDP)
- Portland Guide to Sustainable Stormwater – City of Portland, Oregon
- Stormwater Industry Association of Australia
- Sydney WSUD program
- The River Restoration Centre (RRC)

Chapter 8

Constructed wetland

Constructed wetland in an ecoological settlement in Flintenbreite near Luebeck, Germany

A **constructed wetland** (CW) is an artificial wetland created for the purpose of treating anthropogenic discharge such as municipal or industrial wastewater, stormwater runoff. It may also be created for land reclamation after mining, refineries, or other ecological disturbances such as required mitigation for natural areas lost to a development.

Constructed wetlands are engineered systems that use natural functions of vegetation, soil, and organisms to treat different water streams. Depending on the type of wastewater that has to be treated the system has to be adjusted accordingly which means that pre- or post-treatments might be necessary.

Constructed wetlands can be designed to emulate the features of natural wetlands, such as acting as a biofilter or removing sediments and pollutants such as heavy metals from the water. Some constructed wetlands may also serve as a habitat for native and migratory wildlife, although that is usually not their main purpose.

The two main types of constructed wetlands are subsurface flow and surface flow wetlands. The planted vegetation plays a role in contaminant removal but the filter bed, consisting usually of a combination of sand and gravel, has an equally important role to play.[1]

8.1 Terminology

Many terms are used to denote constructed wetlands, such as reed beds, soil infiltration beds, constructed treatment wetlands, treatment wetlands, etc. Beside "engineered" wetlands, the terms of "man-made" or "artificial" wetlands are often found as well.[1] A biofilter has some similarities with a constructed wetland, but is usually without plants.

However, the term of constructed wetlands can also be used to describe restored and recultivated land that was destroyed in the past through draining and converting into farmland, or mining.

Ponds for wastewater treatment or water purification are not considered as constructed wetlands. They are referred to as stabilization ponds or treatment ponds, respectively.

8.2 Overview

A constructed wetland is an engineered sequence of water bodies designed to filter and treat waterborne pollutants found in sewage, industrial effluent or storm water runoff. Constructed wetlands are used for wastewater treatment or for greywater treatment, and can be incorporated into an ecological sanitation approach. They can be used after a septic tank for primary treatment, in order to separate the solids from the liquid effluent. Some CW designs however do not use upfront primary treatment.

Vegetation in a wetland provides a substrate (roots, stems, and leaves) upon which microorganisms can grow as they break down organic materials. This community of microorganisms is known as the periphyton. The periphyton

Effluent from a constructed wetland for greywater treatment at an ecological housing estate in Hamburg-Allermoehe, Germany

Constructed wetland for domestic wastewater treatment in Bayawan City, the Philippines

and natural chemical processes are responsible for approximately 90 percent of pollutant removal and waste breakdown. The plants remove about seven to ten percent of pollutants, and act as a carbon source for the microbes when they decay. Different species of aquatic plants have different rates of heavy metal uptake, a consideration for plant

selection in a constructed wetland used for water treatment. Constructed wetlands are of two basic types: subsurface flow and surface flow wetlands.

Many regulatory agencies list treatment wetlands as one of their recommended "best management practices" for controlling urban runoff.

8.3   Types

The main two constructed wetlands types are:

- Subsurface flow constructed wetland - this wetland can be either with vertical flow (the effluent moves vertically, from the planted layer down through the substrate and out) or with horizontal flow (the effluent moves horizontally, parallel to the surface)

- Surface flow constructed wetland

Both types are placed in a basin with a substrate. In most cases, the bottom is lined with either a polymer geomembrane, concrete or clay (when there is appropriate clay type) in order to protect the water table and surrounding grounds. The substrate can be either gravel—generally limestone or pumice/volcanic rock, depending on local availability, sand or a mixture of various sizes of media (for vertical flow constructed wetlands).

8.3.1   Subsurface flow wetland

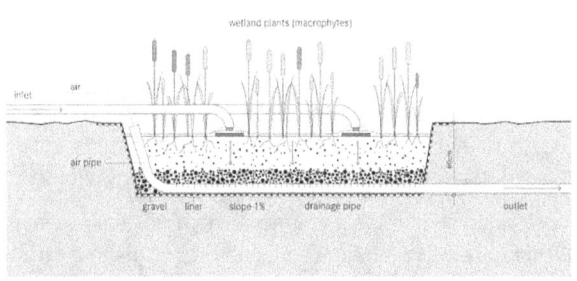

Schematic of a vertical subsurface flow constructed wetland: Effluent flows through pipes on the subsurface of the ground through the root zone to the ground.[2]

Types

Subsurface flow wetlands can be further classified as horizontal flow and vertical flow constructed wetlands. In the

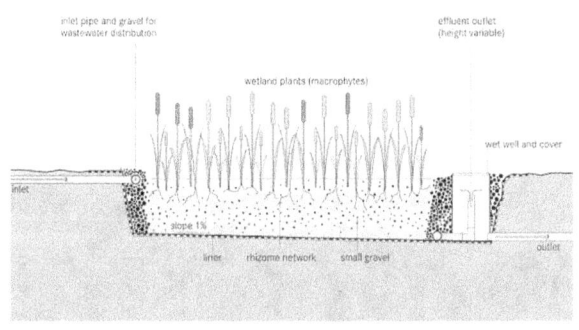

Schematic of the Horizontal Subsurface Flow Constructed Wetland: Effluent flows horizontally through the bed.[2]

vertical flow constructed wetland, the effluent moves vertically from the planted layer down through the substrate and out. In the horizontal flow CW the effluent moves horizontally, parallel to the surface. Vertical flow CWs are considered to be more efficient with less area required compared to horizontal flow CWs. However, they need to be interval-loaded and their design requires more know-how while horizontal flow CWs can receive wastewater continuously and are easier to build.[1]

The French System combines primary and secondary treatment of raw wastewater. The effluent passes various filter beds whose grain size is getting smaller (from gravel to sand).[1]

Applications

Subsurface flow wetlands can treat a variety of different wastewaters, such as household wastewater, agricultural, paper mill wastewater, mining runoff, tannery or meat processing wastes, storm water.

The quality of the effluent is determined by the design and should be customized for the intended reuse application (like irrigation or toilet flushing) or the disposal method.

Design considerations

The wastewater passes through a sand medium on which plants are rooted. A gravel medium (generally limestone or volcanic rock lavastone) can be used as well and is mainly deployed in horizontal flow systems though it does not work as efficiently as sand.[1]

Constructed subsurface flow wetlands are meant as secondary treatment systems which means that the effluent needs to first pass a primary treatment which effectively removes solids. Such a primary treatment can consist of sand and grit removal, grease trap, compost filter, septic tank,

Imhoff tank, anaerobic baffled reactor or upflow anaerobic sludge blanket (UASB) reactor.[1] The following treatment is based on different biological and physical processes like filtration, adsorption or nitrification. Most important is the biological filtration through a biofilm of aerobic or facultative bacteria. Coarse sand in the filter bed provides a surfaces for microbial growth and supports the adsorption and filtration processes. For those microorganisms the oxygen supply needs to be sufficient.

Especially in warm and dry climates the effects of evapotranspiration and precipitation are significant. In cases of water loss, a vertical flow CW is preferable to a horizontal because of an unsaturated upper layer and a shorter retention time.

The effluent can have a yellowish or brownish colour if domestic wastewater or blackwater is treated. Treated greywater usually does not tend to have a colour. Concerning pathogen levels, treated greywater meets the standards of pathogen levels for safe discharge to surface water. Treated domestic wastewater might need a tertiary treatment, depending on the intended reuse application.[1]

Plantings of reedbeds are popular in European constructed subsurface flow wetlands. Other plants are cattails (*Typha* spp.) and sedges.

Operation and maintenance

Overloading peaks should not cause performance problems while continuous overloading lead to a loss of treatment capacity through too much suspended solids, sludge or fats.

Subsurface flow wetlands require the following maintenance tasks: regular checking of the pretreatment process, of pumps, of influent loads and distribution on the filter bed.[1]

Comparisons with other types

Subsurface wetlands are less hospitable to mosquitoes compared to surface flow wetlands, as there is no water exposed to the surface. Mosquitos can be a problem in surface flow constructed wetlands. Subsurface flow systems have the advantage of requiring less land area for water treatment than surface flow. However, surface flow wetlands can be more suitable for wildlife habitat.

For urban applications the area requirement of a subsurface flow CW might be a limiting factor compared to conventional municipal wastewater treatment plants. High rate aerobic treatment processes like activated sludge plants, trickling filters, rotating discs, submerged aerated filters or membrane bioreactor plants require less space. The advantage of subsurface flow CWs compared to those technologies is their operational robustness which is particularly important in developing countries. The fact that CWs do not produce secondary sludge (sewage sludge) is another advantage as there is no need for sewage sludge treatment.[1] However, primary sludge from primary settling tanks does get produced and needs to be removed and treated.

Costs

The costs of subsurface flow CWs mainly depend on the costs of sand with which the bed has to be filled. Another factor is the cost of land.

8.3.2   Surface flow wetland

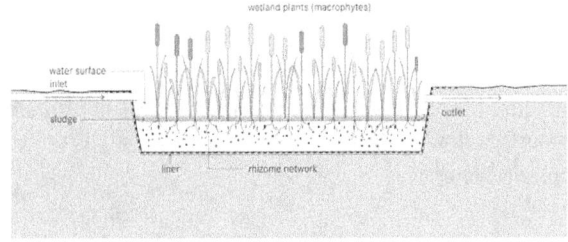

Schematic of a free-water surface constructed wetland: It aims to replicate the naturally occurring processes, where particles settle, pathogens are destroyed, and organisms and plants utilize the nutrients.

Surface flow wetlands, also known as free water surface constructed wetlands, can be used for tertiary treatment or polishing of effluent from wastewater treatment plants. They are also suitable to treat stormwater drainage.

Pathogens are destroyed by natural decay, predation from higher organisms, sedimentation and UV irradiation since the water is exposed to direct sunlight. The soil layer below the water is anaerobic but the roots of the plants release oxygen around them, this allows complex biological and chemical reactions.

Surface flow wetlands can be supported by a wide variety of soil types including bay mud and other silty clays.

Plants such as Water Hyacinth (*Eichhornia crassipes*) and *Pontederia* spp. are used worldwide (although Typha and Phragmites are highly invasive).

However, surface flow constructed wetlands may encourage mosquito breeding. They may also have high algae production that lowers the effluent quality and due to open water surface mosquitos and odours, it is more difficult to integrate them in an urban neighbourhood.

8.3.3   Hybrid systems

A combination of different types of constructed wetlands is possible to use the specific advantages of each system.[1]

Newly planted constructed wetland

Same constructed wetland, two years later

8.4   Contaminants removal

8.4.1   Overview

Physical, chemical, and biological processes combine in wetlands to remove contaminants from wastewater. An understanding of these processes is fundamental not only to designing wetland systems but to understanding the fate of chemicals once they enter the wetland. Theoretically, wastewater treatment within a constructed wetland occurs as it passes through the wetland medium and the plant rhizosphere. A thin film around each root hair is aerobic due to the leakage of oxygen from the rhizomes, roots, and rootlets.[3] Aerobic and anaerobic micro-organisms facilitate decomposition of organic matter. Microbial nitrification and subsequent denitrification releases nitrogen as gas to the atmosphere. Phosphorus is coprecipitated with iron, aluminium, and calcium compounds located in the root-bed medium.[3][4] Suspended solids filter out as they settle in the water column in surface flow wetlands or are physically filtered out by the medium within subsurface flow wetlands. Harmful bacteria and viruses are reduced by filtration and adsorption by biofilms on the gravel or sand media in subsurface flow and vertical flow systems.

8.4.2 Nitrogen removal

The dominant forms of nitrogen in wetlands that are of importance to wastewater treatment include organic nitrogen, ammonia, ammonium, nitrate, nitrite, and nitrogen gases. Total nitrogen refers to all nitrogen species. Wastewater nitrogen removal is important because of ammonia's toxicity to fish if discharged into watercourses. Excessive nitrates in drinking water is thought to cause methemoglobinemia in infants, which decreases the blood's oxygen transport ability.

Ammonia removal occurs in constructed wetlands - if they are designed to achieve biological nutrient removal - in a similar ways as in sewage treatment plants, except that no external, energy-intensive addition of air (oxygen) is needed. It is a two-step process, consisting of nitrification followed by denitrification. The nitrogen cycle is completed as follows: ammonia in the wastewater is converted to ammonium ions; the aerobic bacterium *Nitrosomonas* sp. oxidizes ammonium to nitrite; the bacterium *Nitrobacter* sp. then converts nitrite to nitrate. Under anaerobic conditions, nitrate is reduced to relatively harmless nitrogen gas that enters the atmosphere.

Nitrification

Further information: Nitrification

Nitrification is the biological conversion of organic and inorganic nitrogenous compounds from a reduced state to a more oxidized state, based on the action of two different bacteria types.[5] Nitrification is strictly an aerobic process in which the end product is nitrate (NO_3^-). The process of nitrification oxidizes ammonium (from the wastewater) to nitrite (NO_2^-), and then nitrite is oxidized to nitrate (NO_3^-).

Denitrification

Further information: Denitrification

Denitrification is the biochemical reduction of oxidized nitrogen anions, nitrate and nitrite to produce the gaseous products nitric oxide (NO), nitrous oxide (N_2O) and nitrogen gas (N_2), with concomitant oxidation of organic matter.[5] The end products, N_2O and N_2 are gases that re-enter the atmosphere.

Ammonia removal from mine water

Constructed wetlands have been used to remove ammonia and other nitrogenous compounds from contaminated mine water, including cyanide and nitrate.

8.4.3 Phosphorus removal

Phosphorus occurs naturally in both organic and inorganic forms. The analytical measure of biologically available orthophosphates is referred to as soluble reactive phosphorus (SR-P). Dissolved organic phosphorus and insoluble forms of organic and inorganic phosphorus are generally not biologically available until transformed into soluble inorganic forms.[6]

In freshwater aquatic ecosystems phosphorus is typically the major limiting nutrient. Under undisturbed natural conditions, phosphorus is in short supply. The natural scarcity of phosphorus is demonstrated by the explosive growth of algae in water receiving heavy discharges of phosphorus-rich wastes. Because phosphorus does not have an atmospheric component, unlike nitrogen, the phosphorus cycle can be characterized as closed. The removal and storage of phosphorus from wastewater can only occur within the constructed wetland itself. Phosphorus may be sequestered within a wetland system by:

1. The binding of phosphorus in organic matter as a result of incorporation into living biomass,

2. Precipitation of insoluble phosphates with ferric iron, calcium, and aluminium found in wetland soils.[6]

Biomass plants incorporation

Aquatic vegetation may play an important role in phosphorus removal and, if harvested, extend the life of a system by postponing phosphorus saturation of the sediments.[7]

Plants create a unique environment at the biofilm's attachment surface. Certain plants transport oxygen which is released at the biofilm/root interface, adding oxygen to the wetland system. Plants also increase soil or other root-bed medium hydraulic conductivity. As roots and rhizomes grow they are thought to disturb and loosen the medium, increasing its porosity, which may allow more effective fluid movement in the rhizosphere. When roots decay they leave behind ports and channels known as macropores which are effective in channeling water through the soil.

8.4.4   Metals removal

Constructed wetlands have been used extensively for the removal of dissolved metals and metalloids. Although these contaminants are prevalent in mine drainage, they are also found in stormwater, landfill leachate and other sources (e.g., leachate or FDG washwater at coal-fired power plants), for which treatment wetlands have been constructed for mines.[8]

Mine water—Acid drainage removal

Constructed wetlands can also be used for treatment of acid mine drainage from coal mines.[9]

8.5   Designs

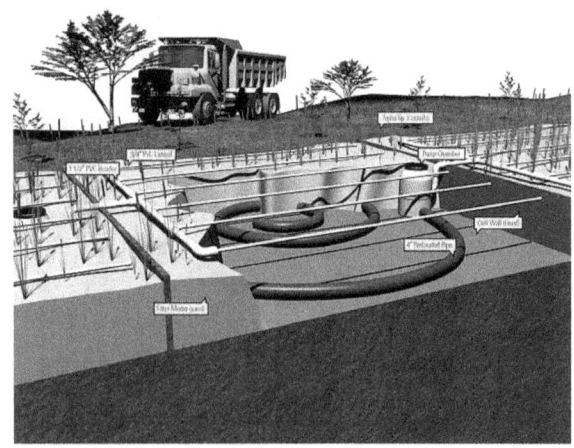

Vertical flow type of constructed wetlands

8.5.1   Design characteristics

- Surface flow CWs are characterized by the horizontal flow of wastewater across the roots of the plants. They require a relatively large area to purify water compared to subsurface flow CWs and may have increased smell and lower performance in winter.

- Subsurface flow CWs: the flow of wastewater occurs between the roots of the plants and there is no water surfacing (kept below gravel). As a result the system is more efficient, doesn't attract mosquitoes, is less odorous and less sensitive to winter conditions. Also, less area is needed to purify water—5–10 square metres (54–108 sq ft). A downside to the system are the intakes, which can clog easily, although some larger sized gravel will often bypass this problem. For large applications, they are often used in combination with vertical flow constructed wetlands. In warm climate, for organic loaded sewage, they require about 3.5 m^2 / 150 L for black and grey water combined, with an average water level of 0.50 m. In cold climate they will require the double size (7 m^2/150 L). For blackwater treatment only, they will require 2 m^2 /50 L in warm weather.

- Vertical flow CWs: these are similar to subsurface flow constructed wetlands but the flow of water is vertical instead of horizontal and the water goes through a mix of media (generally four different granulometries), it requires less space than SF but is dependent on an external energy source. Intake of oxygen into the water is better (thus bacteria activity increased), and pumping is pulsed to reduce obstructions within the intakes. The increased efficiency requires only 3 square metres (32 sq ft) of space per person, down to 1.5 square metres in hot climates.[1]

8.5.2   Plants and other organisms

See also: Organisms used in water purification

Plants

Although the majority of constructed wetland designers have long relied principally on Typhas and Phragmites, both species are extremely invasive, although effective. The field is currently evolving however towards greater biodiversity.

In North America, cattails (*Typha latifolia*) are common in constructed wetlands because of their widespread abundance, ability to grow at different water depths, ease of transport and transplantation, and broad tolerance of water composition (including pH, salinity, dissolved oxygen and contaminant concentrations). Elsewhere, Common Reed (*Phragmites australis*) are common (both in blackwater treatment but also in greywater treatment systems to purify wastewater).

A hybrid system using Flowforms in a treatment pond, in Norway.

Newly planted constructed wetland for blackwater treatment (Lima, Peru)

8.6 Costs

Since constructed wetlands are self-sustaining their lifetime costs are significantly lower than those of conventional treatment systems. Often their capital costs are also lower compared to conventional treatment systems.[10]

The large roots of this uprooted plant growing in a constructed wetlands indicate a healthy plant (Lima, Peru)

8.7 Society and Culture

8.7.1 History

Subsurface flow CWs with sand filter bed have their origin in Europe and are now used all over the world. Subsurface flow CWs with a gravel bed are mainly found in North Africa, South Africa, Asia, Australia and New Zealand.[1]

8.7.2 Examples

Australia

The Urrbrae Wetland in Australia was constructed for urban flood control and environmental education.

At the Ranger Uranium Mine, in Australia, ammonia is removed in "enhanced" natural wetlands (rather than fully engineered constructed wetlands), along with manganese, uranium and other metals.

Plants are usually indigenous in that location for ecological reasons and optimum workings.

Fish and bacteria

Locally grown bacteria and non-predatory fish can be added to surface flow constructed wetlands to eliminate or reduce pests, such as mosquitos. The bacteria are usually grown locally by submerging straw to support bacteria arriving from the surroundings.

8.8 See also

- Ecological engineering

- Floodplain restoration

- Rain garden

- Vegetative treatment system

- Water-sensitive urban design

- Wetland classification

- Wetlands Construídos (a company in Brazil)

8.9   References

[1] Hoffmann, H., Platzer, C., von Münch, E., Winker, M. (2011): Technology review of constructed wetlands - Subsurface flow constructed wetlands for greywater and domestic wastewater treatment. Deutsche Gesellschaft für Internationale Zusammenarbeit (GIZ) GmbH, Eschborn, Germany

[2] Tilley, E., Ulrich, L., Lüthi, C., Reymond, Ph., Zurbrügg, C. (2014): Compendium of Sanitation Systems and Technologies - (2nd Revised Edition). Swiss Federal Institute of Aquatic Science and Technology (Eawag), Duebendorf, Switzerland. ISBN 978-3-906484-57-0.

[3] Brix, H., Schierup, H. (1989): Danish experience with sewage treatment in constructed wetlands. In: Hammer, D.A., ed. (1989): Constructed wetlands for wastewater treatment. Lewis publishers, Chelsea, Michigan, pp. 565–573

[4] Davies, T.H., Hart, B.T. (1990): Use of aeration to promote nitrification in reed beds treating wastewater. Advanced Water Pollution Control 11: 77–84. doi:10.1016/b978-0-08-040784-5.50012-7. ISBN 9780080407845.

[5] Wetzel, R.G. (1983): Limnology. Orlando, Florida: Saunders college publishing.

[6] Mitsch, J.W., Gosselink, J.G. (1986): Wetlands. New York: Van Nostrand Reinhold Company, p. 536

[7] Guntensbergen, G.R., Stearns, F., Kadlec, J.A. (1989): Wetland vegetation. In Hammer, D.A., ed. (1989): Constructed wetlands for wastewater treatment. Lewis publishers, Chelsea, Michigan, pp. 73–88

[8] "Wetlands for Treatment of Mine Drainage". Technology.infomine.com. Retrieved 2014-01-21.

[9] Hedin, R.S., Nairn, R.W.; Kleinmann, R.L.P. (1994): Passive treatment of coal mine drainage. Information Circular (Pittsburgh, PA.: U.S. Bureau of Mines) (9389).

[10] The Interstate Technology & Regulatory Council (ITRC) (2003): Technical and Regulatory Guidance Document for Constructed Treatment Wetlands.

8.10   External links

- U.S.EPA: Constructed Wetlands resources website—United States Environmental Protection Agency

- EPA Constructed Wetlands resources—Handbook, studies and related resources

- Publications on constructed wetlands in the library of the Sustainable Sanitation Alliance

Chapter 9

Sustainable automotive air conditioning

Sustainable automotive air conditioning is the subject of a debate – nicknamed the *Cool War* – about the next-generation refrigerant in car air conditioning. The **Alliance for CO$_2$ Solutions** supports the uptake of carbon dioxide (CO$_2$) as a refrigerant in passenger cars, and the chemical industry is developing new chemical blends.[1]

The Alliance and its supporters – scientists, NGOs and business leaders – urge the car industry to replace high global warming chemical substances with the natural refrigerant carbon dioxide (CO$_2$, R744/ R-744) in car cooling and heating. This, they argue, would lead to 10% less car emissions, and knock out 1% of total greenhouse gas emissions worldwide.[2] If CO$_2$ Technology is applied in other sectors, such as commercial and industrial refrigeration, heat pumps for water heating etc., it may even save up to 3% of the world's greenhouse gases.

Opponents of the Alliance claim that CO$_2$ Technology is not cost-efficient and safe, hence seeking to postpone the global industry decision to be taken to develop new chemical blends instead.

9.1 Background

The Cool War has emanated from the decision of the European Union to phase out the current high global warming refrigerant HFC-134a in car air conditioning from January 2011 onwards.[3] To comply with the legislation, carmakers have to decide today on a new refrigerant, as they typically need 3–4 years to develop and introduce a new car platform including the air conditioning system. (It would take much less time to design merely a new air conditioning system to install in a new version of an existing model.) The current total value of the car air conditioning market is estimated to be $14.5 billion in 2007.

9.2 Arguments

9.2.1 Arguments for CO$_2$

The Alliance for CO$_2$ Solutions and its supporters agree that the refrigerant CO$_2$ is:

- More environmentally friendly with the lowest Global warming potential (GWP) of all currently used and proposed refrigerants. CO$_2$ does not deplete the ozone layer. Since the carbon dioxide used in car air conditioning is a recycled industrial waste product it becomes environmentally neutral. Overall, using a CO$_2$-based air conditioning system will reduce total car emissions by 10%, thereby sparing the planet 1% of total greenhouse gases.

- More technically ready because CO$_2$ models have been developed and tested in all climates, being now ready for mass production. They are faster to heat and cool a car, and show a superior performance in over 90% of all driving conditions.

- More cost-efficient because as a refrigerant itself, CO$_2$ is cheap and worldwide available. The servicing of CO$_2$ systems will be less costly and less complicated than that for present systems. For the consumer the total cost of ownership is lowest with CO$_2$ as it will significantly cut fuel consumption by the air conditioning device. Carmakers have to make an initial investment estimated at €20 per unit, with no additional costs once CO$_2$ Technology enters into mass production.

- Usable in Heat Pumps because at least one CO$_2$ system under development can act as a heat pump, supplying cabin heat and windshield defrosting even before the engine has warmed up.[4]

- Although the Alliance may not mention it, since CO$_2$ is so cheap and relatively harmless to the environment, the reservoirs in such systems could store additional liquid R744 to keep a vehicle cool even when the engine (or compressor) was not running.

9.2.2 Arguments against CO_2

CO_2 Technology requires the design of completely new high-pressure systems whereas so-called "drop-in solutions" (the adaptation of current systems to new substances) are potentially more cost-efficient.

The Alliance for CO_2 Solutions claims, however that the initial costs of CO_2 systems will be around €5 higher than drop-in solutions and that over a car's life cycle, CO_2 air conditioning systems will be more cost-efficient than any currently used or proposed new chemical blends. (see Arguments for CO_2). CO_2 has been classified as Safety Class A1 (low-toxic, non-flammable refrigerant) by the American Society of Heating, Refrigerating and Air-Conditioning Engineers (ASHRAE)[5] – the highest safety class possible. As the charge of CO_2 to the air conditioning systems is very small (200-400 g) there is no realistic danger for the passengers, even in case of accidental release.

9.2.3 Arguments for non-CO_2 refrigerants

- Refrigerants such as the greenpeace-developed 'Greenfreeze', based on purified butane/propane mixtures, are entirely 'natural', and due to increased efficiency over refrigerants such as R134a, allow the use of very small amounts of refrigerant to be used.

- Use of pure hydrocarbon refrigerants, which are 'backward compatible' with even early Freon (R-12) car air conditioning systems, would allow these systems to be easily converted (without modification), increasing their efficiency, and preventing further release of harmful R-134a and R-12 to the atmosphere.

9.2.4 Arguments against non-CO_2 refrigerants

Butane and propane are very flammable petroleum products; they are used as fuels for gas barbecue grills, disposable lighters, etc. Like gasoline, to which it chemically is closely related, propane has a tendency to explode if mixed with oxygen and ignited in an enclosed container.

The use of highly flammable hydrocarbon gases such as butane and propane as automotive refrigerants raises serious safety concerns. The EPA, in evaluating motor vehicle air conditioning substitutes for CFC-12 (Freon, or R-12) under its SNAP program, has classified as "Unacceptable Substitutes" other "Flammable blend[s] of hydrocarbons" by reason of "insufficient data to demonstrate safety." The EPA defines "Unacceptable" in this context as "illegal for use as a CFC-12 substitute in motor vehicle air conditioners". All of the refrigerants which EPA approved for motor vehicle use in place of CFC-12 (as of 28 September 2006) contain no more than 4% total flammable hydrocarbons (butane, isobutane, and/or isopentane).[6] Therefore, it appears unlikely, for safety reasons, that EPA will approve 'Greenfreeze' or similar hydrocarbon-based refrigerants for automotive use.

9.3 Latest and next steps

In September 2007, the German Association of the Automotive Industry (VDA) officially announced its decision to use CO_2 as the refrigerant in next-generation air conditioning. Other carmakers from Europe and the rest of the world may follow the German lead.

A working group at ACEA, the European carmakers' association, was to be drafting a common position on the issue to be adopted across the whole industry by end-2007.

However, on 9 April 2009, German public television channel ARD aired a report claiming that VDA members would be using loopholes in the law to avoid complying with the EU directive.[7]

9.4 Positions

- Deutsche Umwelthilfe - Press Release 6 September 2007

- Greenpeace Germany - News Release 6 September 2007

- Alliance for CO2 Solutions - Press Release 6 September 2007

- Alliance for CO2 Solutions - Press Release 30 July 2007

- Alliance for CO2 Solutions - Press Release 13 June 2007

- Deutsche Umwelthilfe - Press Release 13 July 2007

- German Federal Environment Agency (Umweltbundesamt) - Press Release 8 May 2007

9.5 Media coverage

- Spiegel-Online.de (06/09/2007)

- ENDS Europe Report - August edition

- "German auto industry to drop research on environmentally harmful cooling agents". International Herald Tribune. 2007-07-31. Archived from the original on 2008-04-14.

- European Voice (07/07/12)

- Euractiv (07/06/26)

- Forbes (07/06/18)

9.6 See also

- Automobile air conditioning

- EcoCute, an energy efficient electric heat pump that uses carbon dioxide as a refrigerant

9.7 References

[1] The Choice Today - New Chemical Blends or CO_2

[2] Car Air Conditioning & the Climate Change Challenge - Made Simple

[3] European Directive 2006/40/EC relating to emissions from air-conditioning systems in motor vehicles

[4] Visteon's R744 system"Feature Stories-Company-R744 Climate System". Archived from the original on 1 January 2010.

[5] ASHRAE Standard 34

[6] http://www.epa.gov/ozone/snap/refrigerants/macssubs. html#otherinfo EPA: Choosing and Using Alternative Refrigerants for Motor Vehicle Air Conditioning

[7] German TV explains car industry boycott of EU law

9.8 External links

- alliance-co2-solutions.org - Alliance for CO_2 Solutions

- R744.com Website dedicated to CO_2 Technology

- Shecco.com

- "DENSO Develops World's First CO2 Car Air Conditioner". The Auto Channel. 2002-12-04. Retrieved 2008-07-19.

Chapter 10

Creative Energy Homes

The **Creative Energy Homes**[1] (CEH) project is a showcase of innovative state-of-the-art energy-efficient homes of the future. Seven homes constructed on the University Park Campus of the University of Nottingham are being designed and constructed to various degrees of innovation and flexibility to allow the testing of different aspects of modern methods of construction including layout and form, cladding materials, roof structures, foundations, glazing materials, thermal performance, building services systems, sustainable/renewable energy technologies, lighting systems, acoustics and water supply. The project aims to stimulate sustainable design ideas and promote new ways of providing affordable, environmentally sustainable housing that are innovative in their design. The homes are fully instrumented and occupied in order to provide comprehensive post occupancy evaluation data.

10.1 The David Wilson House

The David Wilson Ecohouse

Construction of the CEH project began in 1999 with the David Wilson Ecohouse.

This is a 4-bedroom detached property which uses brick and block construction. Its main purpose was to support research into domestic-sized renewable energy systems such as micro-CHP (Combined heat and power), solar thermal, micro-wind, sun pipes, and natural ventilation devices.

The house is currently used as office space for University of Nottingham staff and continues to be used as a test bed for University research projects.

10.2 The BASF House

The BASF House

The BASF house was the 3rd house to be constructed and was completed in January 2008 taking only 25 weeks to build. The house was designed to demonstrate that it is possible to build an affordable low energy house whilst still adherering to the Passivhaus ideology. It meets the level 4 standard of the Code for Sustainable Homes.

The house has a compact floor area and relies as much as possible on passive solar design to keep costs down. The north, east and west walls are highly insulated, with the minimum number of openings compatible with acceptable day lighting standards. The southern elevation consists of a fully glazed two-layer adjustable sunspace with glazed screens that can be opened or closed to facilitate heating or cooling.

A ground-air heat exchanger, supplied by Rehau, lies to the front of the house and this consists of a network of horizontal pipes buried approximately 2m underground. At this depth the soil temperature only varies between 8-12°C throughout the year. Therefore in winter cold air can be pre-heated before being drawn into the house and in the summer the air can be cooled.

A rainwater harvesting system has also been installed, which consists of a 3500 litres tank buried under the car parking space. It collects the run-off water from the roof. This water can be used for washing clothes, watering the garden, flushing the toilet and general cleaning. Water efficiency is also an important factor of the code for sustainable homes and using a rainwater harvesting system drastically reduces the amount of mains water used per person per day.

The sun space, at the front of the house captures the solar gain from the sun, heating the air for use in the rest of the house. The south facing elevation is almost 100% glazed and the north facing elevation 30% glazed, providing improved quality of life in the home by maximising the use of natural daylight and reducing the energy required to light the home.

On the roof of the house are 3 solar thermal panels used for hot water heating. This is a larger system than is typical for this size of house, which allows storage of hot water for as long as 3 days and helps compensate for extended periods of overcast conditions.

10.3 The E.ON 2016 House

The E.ON 2016 House

The E.ON House is a replica 1930s UK semi-detached house built primarily to research the impact of various retrofitting solutions on the existing housing stock.

According to the Code for Sustainable Homes (CfSH) all the new houses built from 2016 are required to meet Level 6 requirements of the CfSH, or be carbon zero. However, it is estimated that by the year 2050, 25 million existing UK homes will still account for between 70 and 90 percent of the housing stock.[2] Around 25% of UK CO2 emissions derive from the domestic housing stock[3] so clearly there is a large opportunity to reduce UK carbon emissions by increasing the efficiency of existing homes.

The house is built to 1930s building regulations and the project developers had to receive special planning permission in order to construct the house. It has no roof or wall cavity insulation and uses single glazed windows. Hot water is provided by an outdated, inefficient non-condensing gas boiler and immersion heater. The plan is to bring this 1930s house to a top of the range home meeting Level 6 CfSH. It is hoped that the lessons learnt from this detailed research will help point the way to retrofitting homes to create more sustainable properties. The project consists of three phases:

1. During the first phase the house was monitored in its original state. There are over 100 sensors positioned around the house monitoring temperature, relative humidity, electricity consumption of individual appliances, gas consumption, water consumption, total electricity consumption and occupancy.

2. The second phase was carried out from August 2010 to June 2011. The building was upgraded with additional insulation, an improved heating system, double glazing, more efficient lighting and drought proofing. Research was carried out as to the impact of these improvements.

3. In the third phase, due to begin in late 2011, further improvements will be made incorporating the installation of a conservatory-style extension creating a buffer zone between the property and the outside elements and a sun space to maximise solar gain. Renewable energy technologies, such as Solar panel and solar hot water system will also be installed in order to supply the house with electricity and hot water.

10.4 The Tarmac Homes

Tarmac have sponsored the construction of two semi-detached eco-homes called Tarmac House. These are designed by ZEDfactory, the company responsible for the BEDZED project in south London. One home is built to Level 4 CfSH and the other to Level 6 CfSH.

The Tarmac homes

The Nottingham H.O.U.S.E

These houses demonstrate that it is possible to build a sustainable home using currently available materials and traditional methods of construction. Sustainability features of the properties include:

- Extremely well insulated walls with 250 mm of cavity wall insulation.

- A Biomass heating system situated in front of the houses.

- A Solar water heating system with roof-mounted solar panels.

- Solar century sun slates on the level 6 house helping to meet electricity demands.

10.5 The Nottingham H.O.U.S.E (Home Optimising the Use of Solar Energy)

Nottingham H.O.U.S.E is a small family "starter" home, which was built as a part of a Solar Decathlon Europe 2010 event by the students of the Department of Architecture and Built Environment. The house delivers a zero carbon sustainable design solution that is compliant with level 6 of UK's Code for Sustainable Homes. It is a prefabricated two storey house with a courtyard garden that can be used in terrace or semi-detached form.

The House was presented at the Ecobuild 2010 event and awarded with the prize for the best use of timber. The models of the house were also presented at the Shanghai Expo 2010. The prefabricated house is due to be moved on to the University Park campus of the University of Nottingham in the near future.

10.6 The Mark Group House

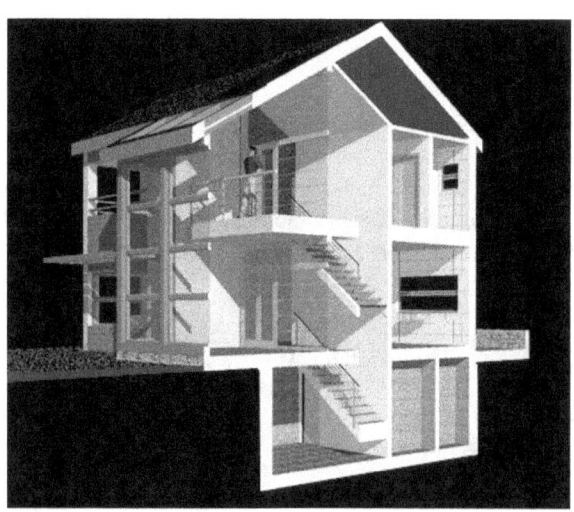

The Mark Group House

The Mark Group house is currently under construction. It is a 4-bedroom property constructed using a steel framing system with insulation cladding. Construction of the basement uses an Insulating Concrete Form (ICF) which incorporates Ground Granulated Blast Furnace Slag (GGBS) into the Portland cement.

Manufacturing Portland cement produces around 1 tonne of CO_2 per tonne of concrete whereas GGBS, as a by-product of the iron industry, produces only 0.1 tonnes of CO_2 per tonne. GGBS can be mixed with Portland cement 70%−30% to decrease its carbon footprint by around 40%.

10.7 References

[1] "Creative Energy Homes project". The University of Nottingham.

[2] "Existing Housing and Climate Change" (PDF). House of Commons. Retrieved 27 June 2011.

[3] Firth, Steven. "Investigating CO2 emissions reduction in existing urban housing using a community domestic energy model" (PDF). *Proceedings of the Eleventh International IBPSA Conference.* Retrieved 27 June 2011.

Chapter 11

Depression-focused recharge

Also referred to as depression focused groundwater recharge. In hydrology recharge implies replenishing a supply of water held within a geological formation underground. Even surface aquifers are within soil.

11.1 Recharge

If water falls uniformly over a field such that field capacity of the soil is not exceeded, then negligible water percolates to groundwater. If instead water puddles in low lying areas, the same water volume concentrated over a smaller area may exceed field capacity resulting in water that percolates down to recharge groundwater. The larger the relative contributing runoff area is, the more focused infiltration is. The recurring process of water that falls relatively uniformly over an area, flowing to groundwater selectively under surface depressions is depression focused recharge. Water tables rise under such depressions.

11.2 Depression pressure

Depression focused groundwater recharge can be very important in arid regions. More rain events are capable of contributing to groundwater supply.

Depression focused groundwater recharge also profoundly effects contaminant transport into groundwater. This is of great concern in regions with karst geological formations because water can eventually dissolve tunnels all the way to aquifers, or otherwise disconnected streams. This extreme form of preferential flow, accelerates the transport of contaminants and the erosion of such tunnels. In this way depressions intended to trap runoff water—before it flows to vulnerable water resources—can connect underground over time. Cavitation of surfaces above into the tunnels, results in pot holes or caves.

Deeper ponding exerts pressure that forces water into the ground faster. Faster flow dislodges contaminants otherwise adsorbed on soil and carries them along. This can carry pollution directly to the raised watertable below and into the groundwater supply. Thus the quality of water collecting in rapid infiltration basins is of special concern.

11.3 Pollution

Pollution in stormwater runoff collects in retention basins. Concentrating degradable contaminants can accelerate biodegradation. However, where and when water tables are high this affects appropriate design of detention ponds, retention ponds and rain gardens.

11.4 See also

- Hydrology (agriculture)
- Infiltration (hydrology)
- International trade and water
- Groundwater recharge
- Subsurface dyke

Chapter 12

Design for the Environment

US EPA's Design for the Environment Logo.

Design for the Environment Program (DfE) is a United States Environmental Protection Agency (EPA) program, created in 1992, that works to prevent pollution, and the risk pollution presents to humans and the environment.[1] The DfE program provides information regarding safer electronics, safer flame retardants, safer chemical formulations, and best environmental practices.[1] DfE employs design approaches to reduce the overall human health and environmental impact of a product, process or service, where impacts are considered across its life cycle. Different software tools have been developed to assist designers in finding optimized products or processes/services. The three main goals of DfE are:

- Promoting green cleaning and recognizing safer consumer and industrial and institutional products through safer product labeling.

- Defining Best Practices in areas ranging from auto refinishing to nail salon safety.

- Identifying safer chemicals, including life cycle considerations, through alternatives assessment.

12.1 Introduction

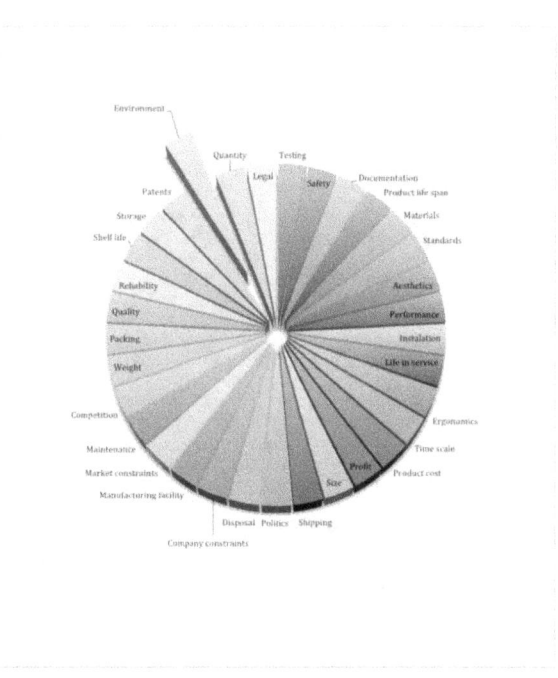

Each piece of the pie chart depicts the role that the individual processes play in the overall design and production of a product. In recent years the environment has begun to play an important role in this diagram. If any piece is missing, production may not be able to occur. This figure was adopted from Conrad Luttropp and Jessica Lagerstedt.[2]

Design for the Environment is a global movement targeting design initiatives and incorporating environmental motives to improve product design in order to minimize health and environmental impacts. The Design for the Environment (DfE) strategy aims to improve technology and design tactics to expand the scope of products. By incorporating eco-efficiency into design tactics, DfE takes into consideration the entire life-cycle of the product, while still making products usable but minimizing resource use. The key focus of DfE is to minimize the environmental-economic cost to consumers[3] while still focusing on the life-cycle framework of the product. By balancing both customer needs as well as environmental and social impacts DfE aims to "improve the product use experience both for consumers and producers, while minimally impacting the environment".

In the US, Design for the Environment is an EPA program works to "notify the public" of less harmful substitutes to certain products. They do so by labeling environmentally safe products with a DfE label to provide information on safer or alternative products to those which use harmful chemicals.

12.2 Design for Environment Practices

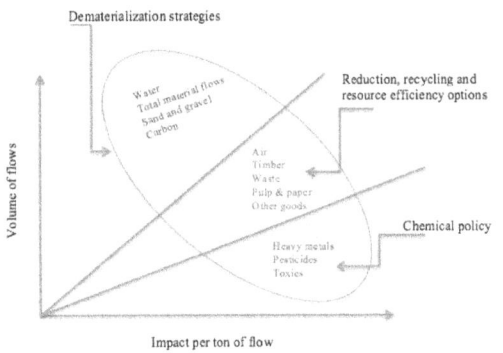

Resource consumption and mitigation strategies for product production which minimizes environmental and health impacts. This figure has been adapted from Spangenberg et al.[4]

Four main concepts that fall under the DfEt umbrella.[1]

- **Design for environmental processing and manufacturing**: This ensures that raw material extraction (mining, drilling, etc.), processing (processing reusable materials, metal melting, etc.) and manufacturing are done using materials and processes which are not dangerous to the environment or the employees working on said processes. This includes the minimization of waste and hazardous by-products, air pollution, energy expenditure and other factors.

- **Design for environmental packaging**: This ensures that the materials used in packaging are environmen-

tally friendly, which can be achieved through the reuse of shipping products, elimination of unnecessary paper and packaging products, efficient use of materials and space, use of recycled and/or recyclable materials.

- **Design for disposal or reuse**: The end-of-life of a product is very important, because some products emit dangerous chemicals into the air, ground and water after they are disposed of in a landfill. Planning for the reuse or refurbishing of a product will change the types of materials that would be used, how they could later be disassembled and reused, and the environmental impacts such materials have.

- **Design for energy efficiency**: The design of products to reduce overall energy consumption throughout the product's life.

Life cycle assessment (LCA) is employed to forecast the impacts of different (production) alternatives of the product in question, thus being able to choose the most environmentally friendly. A life cycle analysis can serve as a tool when determining the environmental impact of a product or process. Proper LCAs can help a designer compare several different products according to several categories, such as energy use, toxicity, acidification, CO_2 emissions, ozone depletion, resource depletion and many others. By comparing different products, designers can make decisions about which environmental hazard to focus on in order to make the product more environmentally friendly.

12.3 Why do firms want to design for the environment?

Modern day businesses all aim to produce goods at a low cost while maintaining quality, staying competitive in the global marketplace, and meeting consumer preferences for more environmentally friendly products. To help businesses meet these challenges, EPA encourages businesses to incorporate environmental considerations into the design process. The benefits of incorporating DfE include: cost savings, reduced business and environmental risks, expanded business and market opportunities, and to meet environmental regulations.[5]

12.4 Companies and Products

- Starbucks: Starbucks is decreasing its carbon footprint by building more energy efficient stores and facilities, conserving energy and water, and purchasing renewable energy credits. Starbucks has achieved LEED

Industry	Companies
Technology	IBM, HP, Philips, Sony, Apple, Dell
Food/beverage	Starbucks, Ice Mountain, Coca-Cola, Pepsi
Cleaning	Atlantic Chemical & Equipment Co., American Cleaning Solutions, BCD Supply, Beta Technology, Brighton USA
Automobile/	BMW, GM, Ford

Designfortheenvironment Fig3

certificates in 116 stores in 12 countries. Starbucks has even created a portable, LEED certified store in Denver. It is Starbucks' goal to reduce energy consumption by 25% and to cover 100% of its electricity with renewable energy by 2015.[6]

- Hewlett Packard: HP is working towards reducing energy used in manufacturing, developing materials that have less environmental impact, and designing easily recyclable equipment.[7]

- IBM: Their goal is to extend product life beyond just production, and to use reusable and recyclable products. This means that IBM is currently working on creating products that can be safely disposed of at the end of its product life. They are also reducing consumption of energy to minimize their carbon footprint.[8]

- Philips: For almost 20 years now, sustainable development has been a crucial part of Philips decision making and manufacturing process. Philips' goal is to produce products with their environmental responsibility in mind. Not only are they working on reducing energy during the manufacturing process, Phillips is also participating in a unique project, philanthropy through design. Since 2005, Philips has been working on and developing philanthropy through design. They collaborate with other organizations to use their expertise and innovation to help the more fragile parts of our society.[9]

Besides these large brand names there are several other consumer product companies in the DfE program this including:

- Atlantic Chemical & Equipment Co.

- American Cleaning Solutions

- BCD Supply

- Beta Technology

- Brighton USA[10]

12.5 How does a business design for the Environment?

A business can design for the environment by:[1]

- Evaluating the human health and environmental impacts of its processes and products.

- Identifying what information is needed to make human health and environment decisions

- Conducting an assessment of alternatives

- Considering cross-media impacts and the benefits of substituting chemicals

- Reducing the use and release of toxic chemicals through the innovation of cleaner technologies that use safer chemicals.

- Implementing pollution prevention, energy efficiency, and other resource conservation measures.

- Making products that can be reused and recycled

- Monitoring the environmental impacts and costs associated with each product or process

- Recognizing that although change can be rapid, in many cases a cycle of evaluation and continuous improvement is needed

12.6 Safer Product Labeling Program

The design for the environment safer product labeling program aims to protect human health and the environment. DfE works alongside consumers and producers to enhance the safety of a wide assortment of products. The program is important because it only labels products as "DfE" when they have met an environmental standard set by the EPA. According to the EPA website currently over 2,700 products carry the DfE label making it easier for consumers to identify products which utilize safer chemicals in products such as household cleaning supplies.[11] However, cleaning products are only one of many products that are eligible for the label. Other products include septic system treatments and wastewater inoculants, car care products, carpet cleaners, several types of degreasing products, laundry-care products and several others.[12] Furthermore the DfE program works alongside producers to assist them in finding safer substitutes to current chemicals used, and thus helping to minimize the companies environmental impact and

US EPA's Safer Product Label.

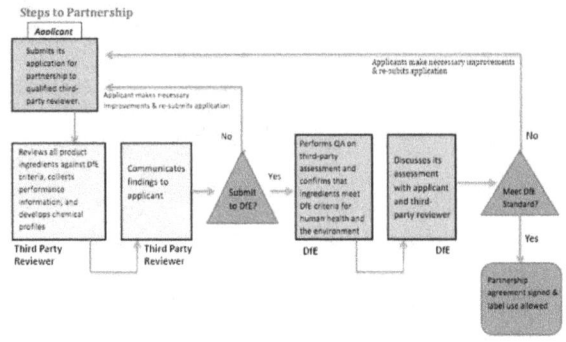

Steps to Partnership with EPA[15]

human health effects.[13] The procedure for safer product labeling consists of several steps which include selecting materials which will have the lowest environmental and health impacts, reducing the amount of material used, incorporating the "cradle-to-grave" concept of minimizing the impact of production/disposal as well as the impact during use of the product and enhancing the disposal process by using materials which can be recycled, disassembled or that will have a minimum landfill impact.[14]

12.6.1   For Consumers

EPA certified safe products are recognizable by the DfE label. The label helps the consumer identify and choose products that are safer for human health and protect the environment. The DfE logo on a product symbolizes that the DfE scientific review team has screened each ingredient contained in the product for potential human health and environmental effects and have recognized the product as being the safest option for consumers.

12.6.2   For Product Manufacturers

Manufacturers who are partners for the Design for the Environment program have the right to display the DfE logo on recognized products. These manufacturers invest in research, development and reformulation, to ensure the ingredients contained in the finished product are safe to humans and protect the environment while maintaining or improving product performance.

12.6.3   For Industrial and Institutional Purchasers

Design for the Environment is an EPA program that distinguishes safer chemical products. EPA uses rigorous criteria to ensure that Design for the Environment-labeled products are safe for human health and the environment. The Design for the Environment label makes it easy for purchasers and users to quickly identify safer chemical products that do not sacrifice quality or performance. Design for the Environment has approved more than 2,000 industrial and institutional products.[16]

12.7   Current Laws and Regulations Encouraging DfE in the Electronics Industry

12.7.1   The National Ambient Air Quality Standards (NAAQS)

The EPA has imposed the National Ambient Air Quality Standards (NAAQS) to establish an air quality standard across the U.S. The NAAQS sets standards on six main sources of pollutants, which include emissions of: ozone (0.12 ppm per 1 hour), carbon monoxide (35 ppm per 1 hour), pollutant (primary standards), particulate matter (50g/m^3 at an annual arithmetic mean), sulfur dioxide (80g/m^3 at an annual arithmetic mean), nitrogen dioxide (100g/m^3 at an annual arithmetic mean), and lead emissions (1.5g/m^3 at an annual arithmetic mean).[17]

12.7.2   Stratospheric Ozone Protection

The Stratospheric Ozone Protection is under section 602 of the Clean Air Act of 1990. This regulation aims to decrease emission of chlorofluorocarbons (CFCs) and other chemicals that are destroying the stratospheric ozone layer. The protection initiative categorizes ozone-depleting substances into two classes: Class I, and Class II.[18]

Class I substances include 20 different kinds of chemicals and have all been phased-out of production processes since 2000. Class II substances consist of the 33 different HCFCs. The EPA has already begun plans to decrease emissions in HCFCs and plan to completely phase out the class II substances by 2030.[18]

12.7.3 Reporting Requirements for Releases of Toxic Substances

A firm operating in the electronics industry in SIC Codes 20-39 that has more than 10 full-time employees and consumes more than 10,000 lbs per year of any toxic chemical lists in 40 CFR 372.65 must file a toxic release inventory.[19]

12.7.4 Other Regulations[20]

- Hazardous Air Pollutants and Maximum Achievable Control Technology (MACT) Standards

- EPA National Pollutant Discharge Elimination System

- Underground Injection Control Program

- Hazardous Waste Management

- Underground Storage Tank Management

12.8 Alternatives Assessment Program

In order to help industries choose safer chemicals for applications, DfE conducts Alternatives Assessments. This program brings together environmental organizations, industry leaders, academia, and others to evaluate the environmental and health impacts of potential alternatives to problematic chemicals. The program uses a variety of approaches to investigate safer chemistries. Life-cycle assessment can be conducted to understand the phases (e.g., production, use, and disposal) where industry can make changes to realize environmental and health benefits. DfE Hazard-based Alternatives Analyses evaluate the hazards posed by chemicals during relevant phases in the product life cycle. These approaches can be applied to identifying safer alternative chemicals for applications that now use priority chemicals of concern. The outcome of an Alternatives Assessments Partnership provides industry with the information

they need to choose safer chemicals, as well as avoid unintended consequences of switching to a poorly understood substitute.[1]

12.9 Best Practices Approach

DfE's Best Practices approach is designed to enhance the awareness of health and environmental concerns, minimize pollution, and protect workers and communities by promoting the use of safer alternative chemical products and cleaner, more efficient practices. After a chemical ingredient has been reviewed by a DfE Alternatives Assessment and no clear alternative is available, the industry is encouraged to use the Best Practices approach as formulated by DfE. Currently, there is a Best Practices approach for both the Automotive Refinishing industry and Spray Polyurethane Foam.[1]

12.10 See also

12.11 References

[1] "U.S. EPA Design for the Environment". EPA. Retrieved August 20, 2008.

[2] Luttropp, Conrad; Jessica Lagerstedt (2006). "EcoDesign and The Ten Golden Rules: generic advice for merging environmental aspects into product development". *Journal of Cleaner Production*.

[3] Luttropp, Conrad; Jessica Lagerstedt (11 November 2011). "Design and The Ten Golden Rules: generic advice for merging environmental aspects into product development". *Journal of Cleaner Production* **14**: 1396–1408. doi:10.1016/j.jclepro.2005.11.022.

[4] Spangenberg, Joachim H.; Alastair Fuad- Luke; Karen Blincoe (16 June 2010). "Design for Sustainability (DfS): the interface of sustainable production and consumption". *Journal of Cleaner Production* **18**: 1485–1493. doi:10.1016/j.jclepro.2010.06.002.

[5] "Design for the Environment Program" (PDF). Environmental Protection Agency. Retrieved 17 April 2013.

[6] Wilson, Mark. "An Experimental New Starbucks Store: Tiny, Portable, and Hyper Local". Co.design.

[7] "HP Design for Environment". Destination Green IT. Retrieved 14 April 2013.

[8] "Product Stewardship". IBM. Retrieved 14 April 2013.

[9] "Towards a Sustainable Future". Philips.

[10] "Labeled Products and Our Partners". EPA. Retrieved 14 April 2013.

[11] "What does the DfE label mean?". Environmental Protection Agency. Retrieved 14 April 2013.

[12] "Labeled Products and Our Partners". Environmental Protection Agency. Retrieved 16 April 2013.

[13] "Design for Environment Frequently Asked Questions". Environmental Protection Agency. Retrieved 14 April 2013.

[14] Young, Steven B; Jim Rollefson (Winter 2000). "Design For Environment". *Alternatives Journal*. 1 **26**: 36–37. Retrieved 14 April 2013.

[15] "How to Partner, Get the DfE Label on a Product".

[16] http://www.epa.gov/dfe/pubs/projects/formulat/ saferproductlabeling.htm.   Missing or empty |title= (help)

[17] "Clean Air Act Requirements". Environmental Protection Agency. Retrieved 17 April 2013.

[18] "Clean Air Act Requirements". Environmental Protection Agency. Retrieved 17 April 2013.

[19] "Superfund and Community Right-to-Know Requirements". Environmental Protection Agency. Retrieved 17 April 2013.

[20] "Federal Environmental Regulations Affecting the Electronics Industry". Retrieved 17 April 2013.

12.12   External links

- The European Union: The European Platform on Life Cycle Assessment

- Department Life Cycle Engineering, University of Stuttgart (English)

- Sustainable Building Alliance.org

- Sustainable Residential Design.org: Using Low-Impact Materials Resource Guide

Chapter 13

Drake Landing Solar Community

The **Drake Landing Solar Community** (DLSC) is a planned community in Okotoks, Alberta, Canada, equipped with a central solar heating system and other energy efficient technology. This heating system is the first of its kind in North America, although much larger systems have been built in northern Europe. The 52 homes in the community are heated with a solar district heating system that is charged with heat originating from solar collectors on the garage roofs and is enabled for year-round heating by underground seasonal thermal energy storage (STES).[1]

The system was designed to model a way of addressing global warming and the burning of fossil fuels. The solar energy is captured by 800 solar thermal collectors[2] located on the roofs of all 52 houses.[3] It is billed as the first solar powered subdivision in North America,[4] although its electricity and transportation needs are provided by conventional sources.

In 2012 the installation achieved a world record solar fraction of 97%; that is, providing that amount of the community's heating requirements with solar energy over a one-year time span.[5][6]

is more energy efficient, as solar collecting is more compatible with lower temperatures. This increases the total amount of heat available to each home.

In the warmer months the previously heated water is taken from the short-term storage tank to the Borehole Thermal Energy Storage (BTES). The Borehole Thermal Energy Storage unit is 144 holes located 37 m (121 ft) below the ground and stretches over an approximate area of 35 m (115 ft) in diameter. The water returns to the short-term storage tanks in the Energy Centre to be heated again in order to complete the circuit. During colder months the water from the BTES passes back to the short-term storage tank and is then directed to each home. Similar to a hot water tank, the heated water goes through a heat exchanger that blows air across the warm fan coil. Heat travels from the water to the air and is directed through the house via ductwork. When the temperature reaches that said on the thermostat, an automatic valve shuts off the heat transfer unit.[7]

13.1   How it works

There are 52 homes in this subdivision that contain an array of 800 solar thermal collectors. These solar collectors are arranged on the roofs of garages located behind the homes. During a typical summer day these collectors can generate 1.5 mega-watts of thermal power. A glycol solution (an anti-freeze solution; a mixture of water and non-toxic glycol) is heated by the sun's energy and travels through insulated piping underground through a trench system to the heat exchanger within the community's Energy Centre. This is known as the Solar Collector Loop. The glycol solution then transfers its heat to water located in the short-term storage tanks. The District Heating Loop begins with water being heated in the heat exchanger to a temperature of 40-50 °C within the Energy Centre. This lower temperature

13.2   Energy centre

The Energy Centre building is a 232 square metre (2,500 square feet) building located in the corner of the community. It is home to the short-term storage tanks and most mechanical equipment such as pumps, heat exchangers, and controls. The Solar Collector Loop, the District Heating Loop, and the Borehole Thermal Energy Storage Loop pass through the Energy Centre. Two horizontal water tanks occupy the majority of the space within the Energy Centre. These tanks are 12 ft (3.7 m) in diameter and 36 ft (11 m) in length. The remaining space within the Energy Centre houses pumps, valves, heat exchangers and other necessary equipment to operate and control the energy system. These tanks are known as Short-Term Thermal Storage (STTS).[7]

13.3 Borehole thermal energy system

The Borehole Thermal Energy System is located underground to store large quantities of heat collected in the summer to be used in the winter. It consists of 144 boreholes, which stretch to a depth of 37 m (121 ft). At the surface the pipes are joined together in groups of six to connect to the Energy Centre. The entire BTES is covered by a layer of insulation, on top of which a park is built. When the heated water is to be stored, it is pumped through the pipe series. The heat is then transferred to the surrounding soil as the water cools and returns to the Energy Centre. When the homes need heat, water flows to the centre of the BTES field and picks up the heat from the surrounding soil. The heated water then goes to the short-term energy tank in the Energy Centre and is pumped through the District Heating Loop to the homes.[7]

13.4 Sponsors and partners

This project was conceived by Natural Resources Canada's CanmetENERGY in partnership with governmental organizations and Canadian industries. Of the $7 million needed for this project this was the breakdown of funds:

- $2 million from federal government agencies.

- $2.9 million from the Federation of Canadian Municipalities and Green Municipal Investment Fund.

- $625,000 from the Alberta Government.[8]

13.5 Community members

Homeowners were willing to pay for these energy efficient homes because it ensured high quality construction. Until the solar heating system began working, ATCO Gas (an Alberta-based natural gas distribution company) fixed heating costs at $60 per month for the homeowners at the Drake Landing Solar Community. With rising fuel costs, this was a powerful incentive for homeowners to support the DLSC project. Even if the project had failed, ATCO Gas would have replaced the special hot-water furnaces with traditional natural gas ones. There was limited risk to the homeowners and this encouraged them to support the project.[9]

13.6 Local sustainability

The 52 homes in Drake Landing Solar Community are certified to Natural Resource Canada's R-2000 Standard as well as the Built Green™ Alberta Gold Standard.

13.7 Costs and financing

- Each house sold for an average of $380,000.

- Homeowners are receiving an average of $60 per month solar utility bill for heating.

- $7 million for the initial start up of the Drake Landing Solar Community project.

- If this project were repeated it would cost $4 million, as approximately $3 million was for one-time research and development.

- Optimal community size would be 200-300 homes to realize the economies of scale. The number of systems would remain the same but the number of boreholes would just need to increase.[10]

13.8 International effects

A group of researchers from South Korea visited Drake Landing Solar Community in April 2012 to study the geothermal heating technology and how it can be applied to communities in South Korea, particularly ahead of the 2018 Winter Olympics in Pyeongchang. The main focus of this research trip was to learn about the economics and reliability of the technology.[11]

13.9 Current status

On October 5, 2012 the DLSC set a new world record by covering 97% of space heating needs with solar thermal energy.[12]

13.10 See also

- List of energy storage projects

13.11 References

[1] "Drake Landing Solar Community". Retrieved 2008-02-10.

[2] Climate Change Central. "Case Study: Drake Landing". Retrieved 2007-02-09.

[3] Natural Resources Canada. "Unique Community a Model for a Greener, Healthier Canada". Archived from the original on 2007-11-06. Retrieved 2008-02-09.

[4] "North America's First Solar Powered Subdivision - Drake Landing". Town of Okotoks. Archived from the original on 2008-01-03. Retrieved 2008-02-09.

[5] "Canadian Solar Community Sets New World Record for Energy Efficiency and Innovation". *Natural Resources Canada*. 5 October 2012.

[6] Wong, B.; Thornton, J. (2013). "Integrating Solar & Heat Pumps" (PDF). *Presentation at Renewable Heat Workshop*. Retrieved 31 January 2013.

[7] "Drake Landing Solar Community". *dlsc.ca*.

[8] "CanmetENERGY" (PDF). *nrcan.gc.ca*.

[9] http://qspace.library.queensu.ca/bitstream/1974/1696/1/ Wamboldt_Jason_M_200901_Master.pdf

[10] "CanmetENERGY" (PDF). *nrcan.gc.ca*.

[11] "Korean researchers learn from Drake Landing". *Okotoks Western Wheel*.

[12] "Startpage - 404 - SolarServer". *solarserver.com*.

13.12 External links

- Drake Landing Solar Community Retrieved on September 30, 2009

- Drake Landing Solar Community: Workings Retrieved on September 30, 2009

Coordinates: 50°43′51″N 113°57′01″W / 50.73095°N 113.95029°W

Chapter 14

Eco-cities

An **eco-city** is a city built off the principles of living within the means of the environment. The ultimate goal of many eco-cities is to eliminate all carbon waste, to produce energy entirely through renewable sources, and to incorporate the environment into the city; however, eco-cities also have the intentions of stimulating economic growth, reducing poverty, organizing cities to have higher population densities, and therefore higher efficiency, and improving health.

14.1 History

14.1.1 Origins

The concept of the "eco-city" was born out of one of the first organizations focused on eco-city development, "Urban Ecology." The group was founded by Richard Register in Berkeley, California in 1975,[1] and was founded with the idea of reconstructing cities to be in balance with nature.[2] They worked to plant trees along the main streets, built solar greenhouses, and worked within the Berkeley legal system to pass environmentally friendly policies and encourage public transportation. Urban Ecology then took the movement another step further with the creation of The Urban Ecologist, a journal they started publishing in 1987.

14.1.2 International Eco-City Conference

Urban Ecology further advanced the movement when they hosted the first International Eco-City Conference in Berkeley, California in 1990.[3] The Conference focused on urban sustainability problems and encouraged the over 700 participants to submit proposals on how to best reform cities to work within environmental means. In 1992 Richard Register founded the organization Ecocity Builders which has acted as convener of the conference series ever since. Eco-City Conferences have been held in Adelaide, Australia; Yoff, Senegal; Curitiba, Brazil; Shenzhen, China; Bangalore, India; San Francisco, United States; Istanbul,

Turkey; Montreal, Canada; Nantes, France and Abu Dhabi (2015).[4]

14.2 Eco-city criteria

There are currently no set criteria for what is considered an "eco-city," although several sets of criteria have been suggested, encompassing the economic, social, and environmental qualities an eco-city should satisfy. The ideal "eco-city" has been described as a city that fulfils the following requirements:[2][5]

An example of a green roof project

- Operates on a self-contained economy, resources needed are found locally

- Has completely carbon-neutral and renewable energy production

- Has a well-planned city layout and public transportation system that makes the priority methods of transportation as follows possible: walking first, then cycling, and then public transportation.

- Resource conservation—maximizing efficiency of water and energy resources, constructing a waste management system that can recycle waste and reuse it, creating a zero-waste system

- Restores environmentally damaged urban areas

- Ensures decent and affordable housing for all socio-economic and ethnic groups and improve jobs opportunities for disadvantaged groups, such as women, minorities, and the disabled

- Supports local agriculture and produce

- Promotes voluntary simplicity in lifestyle choices, decreasing material consumption, and increasing awareness of environmental and sustainability issues

In addition to these initial requirements, the city design must be able to grow and evolve as the population grows and the needs of the population change.[6] This is especially important when taking into consideration infrastructure designs, such as for water systems, power lines, etc. These must be built in such a way that they are easy to modernize (as opposed to the dominant current strategy of placing them underground, and therefore making them highly inaccessible).

Each individual eco-city development has also set its own requirements to ensure their city is environmentally sustainable; these criteria range from zero-waste and zero-carbon emissions, such as in the Sino-Singapore Tianjin Eco-city project and the Abu Dhabi Masdar City project, to simple urban revitalization and green roof garden projects in Augustenborg, Malmö, Sweden.[7][8][9][10]

Using a different set of criteria, the International Eco-Cities Initiative recently identified as many as 178 significant eco-city initiatives at different stages of planning and implementation around the world.[11] To be included in this census, initiatives needed to be at least district-wide in their scale, to cover a variety of sectors, and to have official policy status. Although such schemes display great variety in their ambitions, scale, and conceptual underpinnings, since the late 2000s there has been an international proliferation of frameworks of urban sustainability indicators and processes designed to be implemented across different contexts.[12] This may suggest that a process of de facto eco-city 'standardisation' is underway.

14.3 Practical achievements

14.3.1 Economic impact

One of the major and most noticeable economic impacts of the movement towards becoming an eco-city is the notable increase in productivity across existing industries as well as the introduction of new industries, thus creating jobs.

First, the movement away from carbon-producing energy sources to more renewable energy sources, such as wind, water and solar power, provides local economies with new, thriving industries. The creation of these industries, in turn, births an increase in the demand for labor; thus, not only does total employment increase, but an increase in wages also mimics increasing employment.[13]

Moreover, one of the main priorities of a sustainable city is to reduce its ecological footprint by reducing total carbon emissions, which, economically speaking means increasing productivity. Merely increasing the rate of productivity in an industry reduces costs, both monetary and environmental; that is, as an industry becomes more productive, it can more efficiently allocate and use both its physical and human capital, reducing the time it takes to make the same amount of goods which also allows for a higher wage (because employees are doing more) and a lesser environmental impact (because using less energy and resources to produce the same amount).[13]

In all, although the initial movement towards becoming a sustainable city may be quite costly for a smaller, poorer city, the benefits of such movement are plentiful in the long-run economic model. Moreover, as more and more countries move towards becoming more sustainable, the technologies required to initiate this movement will become more readily accessible and cheaper; therefore, many rich, developed nations should put themselves forth as an example of what other cities should model themselves like, thus sparking the innovation towards a future of sustainable technology.

14.3.2 Environmental standards

Although local environmental standards may differ across eco-cities, each city nonetheless has its own appropriate and practical goals and expectations that have provided the foundation for their recognition as a sustainable city. Differences in these goals and expectations are to be expected, however, due to the limitations of technology and local financing.

The primary goal for all sustainable cities is to significantly decrease total carbon emissions as quickly as possible in order to work towards becoming a carbon-free city; that is, sustainable cities work to move towards an economy based solely on renewable energy. Actions towards carbon-reductions can be seen on both the corporate and individual levels: many industries are working towards cleaner production, but individuals are also moving away from environmentally costly forms of transportation to more sustainable methods, such as public transportation or biking. On this note, another common environmental goal is to increase and make more efficient the public transportation systems.

Many sustainable cities also work towards becoming more densely populated (urban density); having its citizens living closer to energy production means less environmental costs of transporting said energy to citizen households. Additionally, citizens living closer to the city-center also mean that transportation to work is significantly reduced.[14]

Often a city's primary goal is to increase environmental education in hopes of achieving better citizen involvement and cooperation. By making the private sector more aware of how its behavior affects the environment, a reduction in carbon emissions becomes more of a reality.

In terms of international standards, however, we can look to the International Finance Corporation (IFC). The IFC has a long history of implementing environmental and social standards in localized economies, and its primary mission is to promote sustainable development across the globe, primarily in developing countries. One of its plans to accomplish this goal is to encourage international cooperation in order to accelerate and promote sustainable growth across nations.[15]

Overall, the most important aspect of setting an environmental goal is making it plausible. Many cities across the globe set goals that, although they may be super-sustainable, are not entirely possible. These exaggerated goals include too much sustainable development for a small time period or an expectation that is simply too expensive. The globe needs to work together to make steps towards a sustainable future that are possible and execute them well, ultimately resulting in an overall spiral towards complete global sustainability.

14.3.3 Social

Poverty reduction

The development of eco-cities has aided in reducing poverty in various locations via job creation in environmentally friendly business sectors. By promoting social equity based on meeting the needs of local populations, eco-cities create sustainable business models that encourage local investment and the subsequent expansion of the job market. Johannesburg, South Africa serves as a prime example of the manner in which adopting eco-city standards can aid in reducing urban poverty. According to the United Nations Environment Program, the "EcoCity [program] has mobilized the disadvantaged and unemployed people of Ivory Park (part of the city of Johannesburg) to form cooperatives to grow and buy food, to recycle, to repair bicycles, to build homes, to use and promote green energy so-

lutions, to become eco-tourism guides and more than 300 jobs have been created" between 1991 and 2001.[16] By creating small local businesses, residents of eco-cities create self-sufficient small enterprises that, as an aggregate, greatly alleviate the scarcity of quality employment and create economic opportunities that continuously aid in poverty reduction. These ecologically sound small-scale practices are additionally less sensitive to economic shocks, allowing for enduring economic sustainability in eco-cities. In addition to creating green jobs, eco-cities promote the deployment of green methods of saving money, such as investing in ecologically sustainable local infrastructure, carpooling, and reducing consumption of water and energy, to decrease the financial burden on the poor.[17]

Population distribution

Increasing proportions of the world population are now located in cities. As a result, eco-city models place substantial attention on mitigating and reducing the environmental damage caused by growing urban populations.[18] Because urbanization does not appear to be slowing, eco-cities aim to increase urban density while integrating "green infrastructure" or "green spaces" into the urban landscape. Eco-cities promote compact use of land by people for residential and commercial purposes. In this way, increasing urban density reduces the strain on the environment by centralizing and, thereby, reducing resource consumption. For example, the 2006 plans for the Chinese eco-city of Dongtan employed this strategy.[19] At the time, the city planned to divide its residential and commercial land into three compact districts divided by farms, parks, lakes, and pagodas. Additionally, residents would live in ecologically designed apartment buildings six to eight stories high but appropriately spaced apart as to avoid heat island effects. Although no construction of Dongtan has happened yet, these principles are generally applicable to all eco-cities.[20]

Furthermore, increased urban density reduces urban sprawl, thus, decreasing dependence on cars. According to Kenworthy, urban density is accountable for 84% of the variance in car travel.[21] Because of the compact urban layout of eco-cities, residents are able to easily navigate their surrounding environment on foot, by bicycle, or through use of public transportation. As a result, eco-cities avoid much of the negative effects of car pollution.

Additionally, by centralizing the population within a given area, eco-cities increase the amount of land that can be used for parks and urban agriculture. As such, eco-cities increase food security and promote ecological preservation within urban areas. Urban agriculture allows for "production of fresh food and vegetables, reduction on transportation load and enrichment of environmental quality" (Lim 2010).[22]

Public health

Eco-cities aid in creating healthier urban populations through the implementation of sustainable practices that improve environmental standards and, as a result, decrease the strain on public health. By employing practices that aim to reduce air pollution, eco-city standards have an indirect effect on decreasing rates of respiratory disease within urban areas. According to the World Health Organization, urban outdoor air pollution is responsible for over 1.3 million deaths worldwide per year. Additionally, "the mortality in cities with high levels of pollution exceeds that observed in relatively cleaner cities by 15–20%."[23] Through the implementation of "clean" practices, eco-cities greatly assist in decreasing the disease burden placed on urban residents by decreasing the risk factors associated with cardiovascular and respiratory diseases as well as various forms of cancer.

Additionally, the "greenspaces" that constitute the infrastructure of eco-cities provide a unique method of reducing air pollution and promoting clean air. Urban foliage naturally cleans the air by absorbing carbon monoxide, nitrogen dioxide, and sulfur dioxide. Green spaces also absorb airborne particulates and reduce heat, allowing for improved levels of public health.[24]

The decreased dependency on cars encouraged by the compact, walkable layout of eco-cities will also help combat obesity and other chronic diseases by encouraging frequent physical activity. The World Health Organization estimates that physical inactivity leads to 3.2 million deaths per year. 2.6 million of these deaths are centralized in low and middle-income countries.[23] Thus, by reducing urban sprawl, eco-cities may help decrease rates of coronary heart disease and stroke, diabetes, hypertension, colon cancer, breast cancer, osteoporosis, and depression. Furthermore, by decreasing the concentration of cars within city limits, eco-cities are also able to reduce the number of preventable deaths among the working age population. It is estimated that traffic accidents kill 1.2 million people per year and is the leading cause of death among people under the age of 25 (WHO, 2011).[23]

Increased access to affordable vegetation via urban agriculture also permits the improvement of public health conditions by making healthy foods more available and affordable. Lye and Chen note, "An eco-city must not become or be perceived as an enclave for only the rich and powerful but must welcome and be accessible to people from various walks of life."[22] By investing in urban agriculture, eco-cities can help eliminate the prominent issue of food deserts in urban poor areas. Expanded access to vegetables will in hand aid in decreasing rates of obesity, cancer, cardiovascular disease, diabetes and other chronic illness, especially among low-income residents.[23]

14.3.4 Technology and urban layout

Transportation

By decreasing urban sprawl, eco-cities decrease the residential and commercial dependence on automobiles. Concurrently, improved public transportation further decreases the demand for cars. The development of metro station and light rail transit systems provide mass transit not only within sectors of a city but between cities.[20] Furthermore, many eco-cities are employing expanded "clean" bus routes in order to decrease the emissions from single household vehicles. Critics note, however, that the high price of "clean" diesel, CNG/LNG, hybrid electric buses, and super capacitor-powered buses may not prove "economically and operationally viable" (World Bank, 2009).[20]

Urbanism

Jakriborg in Sweden, started in the late 1990s as a new urbanist eco-friendly new town near Malmö

Eco-cities may also seek to create sustainable urban environments with long-lasting structures, buildings and a great liveability for its inhabitants. The most clearly defined form of walkable urbanism is known as the *Charter of New Urbanism*. It is an approach for successfully reducing environmental impacts by altering the built environment to create and preserve smart cities which support sustainable transport. Residents in compact urban neighborhoods drive fewer miles, and have significantly lower environmental impacts across a range of measures, compared with those living in sprawling suburbs.[25] The concept of Circular flow land use management has also been introduced in Europe to promote sustainable land use patterns that strive for compact cities and a reduction of greenfield land take by urban sprawl.

In sustainable architecture the recent movement of New

Classical Architecture promotes a sustainable approach towards construction, that appreciates and develops smart growth, walkability, architectural tradition and classical design.[26][27] This in contrast to modernist and globally uniform architecture, as well as opposing solitary housing estates and suburban sprawl.[28] Both trends started in the 1980s.

Landscape

Eco-cities primarily employ green roofs, vertical landscaping, and bridge links as methods of decreasing the environmental impact of land use. Constructing green roofs and investing in vertical landscaping create natural insulation for residential and commercial properties as well as allows for rainfall collection. Additionally, green roofs and vertical landscaping lower urban temperatures and help prevent the heat island effect. Bridge links allow for development of a walkable city without disrupting the soil to run utility lines by connecting buildings with above ground walkways.

Energy

London skyscraper with integrated wind turbines

Eco-cities look to employ renewable energy sources, such as wind turbines, solar panels, and biogas, to reduce emissions. Wind turbines present the opportunity of being able to provide both localized districts within eco-cities and the larger region as a whole with emission-free renewable energy that can additionally supplement existing power sources. Furthermore, by designing buildings with natural ventilation systems, eco-cities reduce the need for air conditioning, thus, drastically decreasing commercial and residential energy use. The energy generated can come from large scale energy production systems such as solar farms which supply many homes and businesses or from individual buildings energying at least in part their own energy from solar

photovoltaic or small scale wind turbines or biomass. Many eco-cities additionally look to deploy solar thermal energy. By installing solar collectors, developers will be able to provide hot water for space heating and individual and community needs while reducing dependence on gas fueled boilers. While solar thermal energy appears to be a more efficient source of renewable energy, many urban planners also view photovoltaics as a viable source of energy. Photovoltaics directly convert solar energy into electricity; however, the extensive costs associated with developing this technology on the city-scale may limit its use when compared to its potential payback. Biogas technology is also deployed as a source of renewable energy as the organic material from wastewater is converted into fuel.[29]

Water

Eco-cities aim to decrease water consumption by employing technologies that reduce the amount of water that is needed for irrigation and sewage flow while also preventing blackwater and greywater runoff from entering ground water sources. Developers suggest installing low flow fixtures, rainwater harvesting systems, and sustainable urban drainage systems to meet eco-city standards. Additionally, advanced irrigation systems (xeriscaping) aid in maintaining green infrastructure while decreasing green space consumption of water for irrigation.

14.4   Leading eco-cities

14.4.1   Curitiba, Brazil

The city of Curitiba, Brazil proactively began to address the challenges of sustainable urban development in 1966 with a master plan that outlined future integration between urban development, transportation and public health.[30]

This plan has been realized in modern Curitiba, which is defined by linear stretches of urban development surrounded by green space and low-density residential areas.[30] The city was designed for the mobility of people, not the mobility of cars. The city's bus system is highly developed, with high-capacity busses and dedicated lanes, it effectively reaches about 90% of the population.[30] This bus system is utilized by 45% of the population, which has caused private automobile use to drop to 22%.[30] Despite this decline, to prevent congestion central areas of the city have been closed to cars. These road closures have led to dynamic economic growth for local shops and the development of community space for pedestrians.[30]

The resulting public health and education gains from this initiative have also been substantial. Curitiba maintains the

Linear urban development in Curitiba contrasted with surrounding residential development

lowest air pollution rates in Brazil and over 300,000 trees in the city helps reduce natural flooding.[30] Curitiba has also dedicated resources to environmental education in primary school, which has translated into environmentally conscious citizens. Over 70% of city residents participate in recycling programs which fuels the city's progressive waste processing system.[30]

Curitiba has maintained a consistent vision of the future and worked to attain it by through careful urban planning that takes into account transportation, while also encouraging environmental initiatives and public health. In 2010, Curitiba recognized for their achievement with the Globe Sustainable City Award due to "their understanding of sustainable city development – both regarding policy and implementation."[31]

14.4.2 Auroville, India

Auroville was founded in 1968 with the intention of realizing human unity, and is now home to approximately 2,000 individuals from over 45 nations around the world. Its focus is its vibrant community culture and its expertise in renewable energy systems, habitat restoration, ecology skills, mindfulness practices, and holistic education.

14.4.3 Freiburg, Germany

The city of Freiburg, Germany, whose sustainable policies date all the way back to the 1970s, has constructed itself as a sustainable city by actively committing to its target areas of energy, transportation, and to its three pillars for sustainable development: energy saving, new technology, and renewable energy sources.[32] One of the largest motivators for success can be accredited to citizen's engagement; in

the 1970s opposition to local nuclear power led to the creation of a campaign for sustainable solutions for the energy needs of the city. A network of environmentalists, research organizations, and businesses was established, helping the agenda of a sustainable city push forward.[32]

Taking advantage of Freiburg's location, educated and active residents, and political priorities invested in the environment and economy has led Freiburg to be considered a Solar Capital.[33] Along with high solar electricity rates,[32] Freiburg hosts such innovations as the world's first football stadium with its own solar power plant and the world's first self-sustaining solar energy building.[33] In terms of both ecology and economy, Freiburg has been extremely successful in the fields of research and marketing of renewable energy. The Freiburg science network and solar industry embraces many research institutions, like the Fraunhofer Institute for Solar Energy Systems ISE, Europe's largest solar research institute.

In addition to solar initiatives, over the last four decades Freiburg has made improvements to their transportation systems.[34] Freiburg has over 500 km of bicycle paths and more than 5000 bicycle parking spaces as well as car-free centers, 30 kph zones, a region wide bus service, and tram lines.[32]

Long before it was taken seriously Freiburg was resolving to cut carbon dioxide emissions. In 1966, the city resolved to lower carbon dioxide by 25 percent by 2010. Although they did not reach their initial goal by 2010, they are continually extending their goals. By 2030, they resolved to cut carbon dioxide emissions by 40 percent and be climate neutral by 2050.[33]

Freiburg also focuses initiatives on waste management. Paper products are composed to 80 percent recycled materials. Financial incentive programs, like discounts for collective waste disposal and people who compost, are used to increase waste avoidance. Since 2005, Freiburg's non-recyclable waste has been incinerated and the heat energy released is converted to supply electricity to almost 25,000 households in the city.[33]

Freiburg is a green city. 43 percent of borough area is woodland. In 2001, the Freiburg Woodland Convention was adopted and since 2009, the city officially supported the Freiburg Convention on the Protection of Ancient Woodland. For over 20 years Freiburg has worked to maintain their public parks with principles that work with nature: they no longer use pesticides, grass is mown less, and almost 50,000 trees line streets and parks.[33]

14.4.4 Stockholm, Sweden

Stockholm in Sweden has been an environmentally focused city that is redeveloping itself to become an eco-city through efficient urban planning and resource use. Stockholm has established six environmental goals, called Vision 2030, that act as the foundation of this initiative. These goals include development of efficient transportation, sustainable energy, land, and water use, waste treatment improvements, and safe building and product materials.[30] Beyond Vision 2030, Stockholm is planning to be fossil fuel free by 2050.[30]

In terms of urban planning, Stockholm currently requires mandatory reuse of land before urban sprawl can continue.[30] This policy has led to complete revitalization of run-down and abandoned industrial areas that have been transformed into modern, efficient and integrated residential and business communities.[30] The Hammarby Sjostad district of Stockholm is the primary example of this practice, as this resurrected industrial area has become twice as energy efficient as the rest of the city after an environmentally focused redevelopment.[30]

These gains are measured by the environmental load profile of the area, a life-cycle assessment tool developed by the City of Stockholm, the Royal Institute of Technology, and a consultancy firm.[30] This unique measure allows for environmental performance analyses, on both the small and large scale, in terms of environmental costs and benefits.[30] This comprehensive measure has allowed Stockholm to quantify their environmental progress and could be applied as a decision-making tool in other cities or districts to aid their environmental efforts.[30]

Stockholm has pursued green development and optimization of urban systems and achieved results. These efforts were recognized in 2010 by European Union, which deemed Stockholm the European Green Capital for "leading the way towards environmentally friendly urban living."[35]

14.4.5 Adelaide, Australia

Urban forests

In Adelaide, South Australia (a city of 1.3 million people) Premier Mike Rann (2002 to 2011) launched an urban forest initiative in 2003 to plant 3 million native trees and shrubs by 2014 on 300 project sites across the metro area. The projects range from large habitat restoration projects to local biodiversity projects. Thousands of Adelaide citizens have participated in community planting days. Sites include parks, reserves, transport corridors, schools, water courses and coastline. Only trees native to the local area are planted to ensure genetic integrity. Premier Rann said the project aimed to beautify and cool the city and make it more liveable; improve air and water quality and reduce Adelaide's

greenhouse gas emissions by 600,000 tonnes of C02 a year. He said it was also about creating and conserving habitat for wildlife and preventing species loss.[36]

Solar power

The Rann government also launched an initiative for Adelaide to lead Australia in the take-up of solar power. In addition to Australia's first 'feed-in' tariff to stimulate the purchase of solar panels for domestic roofs, the government committed millions of dollars to place arrays of solar panels on the roofs of public buildings such as the Museum, Art Gallery, Parliament, Adelaide Airport, 200 schools and Australia's biggest rooftop array on the roof of Adelaide Showgrounds' convention hall which was registered as a power station.

Wind power

South Australia went from zero wind power in 2002 to wind power making up 26% of its electricity generation by October 2011. In 5 years to 2011 there was a 15% drop in emissions, despite strong economic growth.[37]

Waste recycling

For Adelaide the South Australian government also embraced a Zero Waste recycling strategy, achieving a recycling rate of nearly 80% by 2011 with 4.3 million tonnes of materials diverted from landfill to recycling. On a per capita basis this was the best result in Australia, the equivalent of preventing more than a million tonnes of C02 entering the atmosphere. In the 1970s container deposit legislation was introduced. Consumers are paid a 10 cent rebate on each bottle/can/container they return to recycling. In 2009 nonreusable plastic bags used in supermarket checkouts were banned by the Rann Government preventing 400 million plastic bags per year entering the litter stream. In 2010 Zero Waste SA was commended by a UN Habitat Report entitled 'Solid Waste Management in the World Cities'.[38]

14.5 Challenges

Despite the sustainability, efficiency and other established benefits of ecocities, actual implementation can be difficult to attain. Existing infrastructure, both in terms of the physical city layout and existing local bureaucracy, are major, often insurmountable, obstacles to large-scale sustainable development.[18] The high cost of the technological integration necessary for eco-city development is a major challenge, as many cities either can't afford, or are not willing to take on, the extra costs.[30]

Challenges associated with planning and managing sustainable programs are also large. Cities that want to become more sustainable are faced with retrofitting existing structures and concurrent management of sustainable urban expansion and development. The costs and infrastructure needed to manage these large scale, two-pronged projects are great, and beyond the capabilities of most cities.[30] In addition, many cities around the world are currently struggling to maintain the status quo, with budgetary issues, high rates of poverty, transportation inefficiencies, and rapid population growth encouraging reactive, coping policy.[30] While there are many examples worldwide, the development of ecocities is still limited due to the vast challenges and high costs associated with sustainability.

14.6 References

[1] "Urban Ecology". Retrieved 21 November 2011.

[2] Roseland, Mark (1997). "Dimensions of the Eco-city". *Cities* **14** (4): 197–202.

[3] Devuyst, Dimitri (2001). *How green is the city?*. New York: Columbia University Press.

[4] "Ecocity Builders". Retrieved 21 November 2011.

[5] Harvey, Fiona. "Green vision: the search for the ideal eco-city". Financial Times. Retrieved 21 November 2011.

[6] Graedel, Thomas. "Industrial Ecology and the Ecocity". National Academy of Engineering. Retrieved 21 November 2011.

[7] Caprotti, F. (2014) 'Critical research on eco-cities? A walk through the Sino-Singapore Tianjin Eco-City'. *Cities* 36: 10-36.

[8] Yoneda, Yuka. "Tianjin Eco City is a Futuristic Green Landscape for 350,000 Residents Read more: Tianjin Eco City is a Futuristic Green Landscape for 350,000 Residents | Inhabitat - Green Design Will Save the World". Inhabitat. Retrieved 21 November 2011.

[9] Palca, Joe. "Abu Dhabi Aims to Build First Carbon-Neutral City". NPR. Retrieved 21 November 2011.

[10] "Ekostaden Augustenborg". World Habitat Awards. Retrieved 21 November 2011.

[11] Joss, S., Tomozeiu, D. and Cowley, R., 2011. "Eco-Cities - a global survey: eco-city profiles", University of Westminster (ISBN 978-0-9570527-1-0). Available from: https://www.westminster.ac.uk/ecocities/publications

[12] Joss, S., Cowley, R., de Jong, M., Müller, B., Park, B-S., Rees, W., Roseland, M., and Rydin, Y. (2015). Tomorrow's City Today: Prospects for Standardising Sustainable Urban Development. London: University of Westminster. (ISBN 978-0-9570527-5-8) Available from: http://www.westminster.ac.uk/ecocities-leverhulme

[13] Gerber, James (2010). *International Economics*. Prentice Hall. pp. 167–175. ISBN 978-0-13-510015-8.

[14] "Eco-Cities". Good Planet. Retrieved 20 November 2011.

[15] "International Finance Corporation's Policy on Social and Environmental Policy". International Finance Corporation. Retrieved 20 November 2011.

[16] "Eco-City, Johannesburg" (PDF). United Nations Environment Program. Retrieved 17 November 2011.

[17] "CSR Best Practices- Eco-City: Johannesburg, South Africa". Article 13. Retrieved 17 November 2011.

[18] "Eco2 Cities" (PDF). World Bank. Retrieved 16 November 2011.

[19] Pearce, Fred. "Eco-cities Special: A Shanghai Surprise" (PDF). New Scientist Tech. Retrieved 17 November 2011.

[20] "SIno-Singapore Tianjin Eco-City: A Case Study of an Emerging Eco-City in China" (PDF). World Bank. Retrieved 17 November 2011.

[21] Kenworthy, Jeffrey. "The eco-city: ten key transport and planning dimensions for sustainable city development" (PDF). Retrieved 17 November 2011.

[22] Fook, Lye Liang (2010). *Towards a Liveable and Sustainable Urban Environment: Eco-cities in Asia*. Singapore: World Scientific. ISBN 978-981-4287-76-0.

[23] "Air quality and health". World Health Organization. Retrieved 18 November 2011.

[24] "Towards a Green Economy" (PDF). United Nations Environment Program. Retrieved 17 November 2011.

[25] Ewing, R "Growing Cooler - the Evidence on Urban Development and Climate Change". Retrieved on: 2009-03-16.

[26] Charter of the New Urbanism

[27] "Beauty, Humanism, Continuity between Past and Future". Traditional Architecture Group. Retrieved 23 March 2014.

[28] Issue Brief: Smart-Growth: Building Livable Communities. American Institute of Architects. Retrieved on 2014-03-23.

[29] "Purac launches new biogas technology for Eco-city in China". Lackeby Water Group. Retrieved 18 November 2011.

[30] Hiroaki Suzuki; Arish Dastur; Sebastian Moffatt; Nane Yabuki; Hinako Maruyama (2010). *Eco2 Cities: Ecological Cities as Economic Cities*. World Bank Publications. p. 170.

[31] "Globe Award". Retrieved 20 November 2011.

[32] "Freiburg in a pathway towards a sustainable city" (PDF). Gaia Consulting. Retrieved 20 November 2011.

[33] "Green City-Freiburg" (PDF). Retrieved 20 November 2011.

[34] "Germany to test first smart road in Europe".

[35] "European Green Capital". City of Stockholm. Retrieved
     20 November 2011.

[36] http://www.milliontrees.com.au

[37] http://www.theclimategroup.com

[38] http://www.zerowaste.sa.gov.au

Chapter 15

Eco-municipality

"Eco-municipality" has a specific meaning. For a more general discussion of the sustainability of cities, see Sustainable city.

An **eco-municipality**, (also known as an **eco-town**) is a local government area that has adopted ecological and social justice values in its charter. The development of eco-municipalities stems from changing systems in Sweden, where more than seventy municipal governments have accepted varying principles of sustainability in their operations as well as community-wide decision making processes.[1] The purpose of these policies is to increase the overall sustainability of the community.

The distinction between an eco-municipality and other sustainable development projects (such as green building and alternative energy) is the focus on community involvement and social transformation in a public agency as well as the use of a holistic systems approach. An eco-municipality is one that recognizes that issues of sustainability are key to all decisions made by government. [2]

15.1 History

In 1983 the Övertorneå community of Sweden first adopted an Eco-municipality framework followed by a formal organization in 1995 (SEKOM).

15.2 Framework

In becoming an eco-municipality, cities or towns typically adopt a resolution, based on the Natural Step framework (or Framework for Strategic Sustainable Development (FSSD)), which sets the following objectives:

- Reduce dependence upon fossil fuels.

- Reduce dependence upon synthetic chemicals.

- Reduce encroachment upon nature.

- Better meet human needs fairly and efficiently.[3]

15.3 Municipalities adopting framework

Communities in North America, Europe and Africa ranging in size from villages of 300 to cities of 700,000 have become eco-municipalities. In Sweden, over sixty municipalities have officially become eco-municipalities. They have formed a national association of eco-municipalities to assist one another and work to influence national policy. Whistler, BC, was awarded first place in a United Nations-endorsed international competition for sustainable communities. Its long-term sustainability plan, Whistler 2020, is based on the Natural Step framework.[4]

In Wisconsin, there is a growing eco-municipality movement which began in the Chequamegon Bay region. As of November 2007, twelve local communities had formally adopted eco-municipality resolutions. The resolutions state the community's intention to become an eco-municipality, endorsing the Natural Step sustainability principles and framework as a guide.[5]

15.4 See also

- Ecovillage

- Eco-towns

- Green municipalism

- Sustainable city

15.5   Notes

[1] Miranda Spencer (September 22, 2005) Building Sustainable Cities: Scandinavia's "Eco-Municipalities" Show the Way.sustainablebusiness.com. Retrieved on: November 5, 2007.

[2] Torbjorn Lahti and Sarah James (May 17, 2005) The Eco-municipality Model for Sustainable Community Change: A systems approach to creating sustainable communities. Retrieved on: November 5, 2007

[3] Alliance for Sustainability Ashland, WI and Duluth, MN become Sustainable Cities. Retrieved on: February 10, 2008.

[4] Sustain Dane Eco-municipalities: Where Are They?. Retrieved on: February 10, 2008.

[5] 1,000 Friends of Wisconsin. Eco-municipalities: A Model for Sustainable Communities in Wisconsin. Retrieved on: February 10, 2008.

15.6   References

- James, S. and T. Lahti (2004). *The Natural Step for Communities: How Cities and Towns can Change to Sustainable Practices.* Gabriola Island, British Columbia: New Society Publishers. ISBN 0-86571-491-6

15.7   External links

- National Association of Swedish Eco-municipalities Website of SEKOM

- Sustainable Sweden Association Website

- The Natural Step Case study TNS case study on North American Eco-municipality Network

- Sarah James Associates Consulting firm working in the field.

- Wisconsin Chapter of the American Planning Association Eco Municipalities links

- 1,000 Friends of Wisconsin page on Eco-Municipalities

- The American Association of Planners policy guide on sustainability.

- Sustain Dane Website on Dane (US)

- Sustainable Lawrence Website on Lawrence (Canada)

- Piscataqua Sustainability Website on Piscataqua (Canada)

Chapter 16

Ecodistrict

An **ecodistrict** or eco-district is a neologism associating the terms "district" and "eco" as an abbreviation of ecological. It designates an urban planning aiming to integrate objectives of "sustainable development" and reduce the ecological footprint of the project. This notion insists on the consideration of the whole environmental issues by attributing them ambitious levels of requirements.

16.1   Examples

Ecodistricts can be found in metropolises such as :

- Stockholm (Hammarby Sjöstad) (Sweden)
- Hanover (Germany)
- Freiburg im Breisgau (Vauban, Freiburg) (Germany)
- Malmö(BO01) (Sweden)
- London (BedZED) (United Kingdom)
- Grenoble (De Bonne and Blanche Monier) (France)
- Dongtan (China)
- EVA Lanxmeer (Netherlands)
- Amersterdam Noord (Netherlands)[1]
- Jono district low-carbon project (KitaKyushu, Japan)
- Frequel-Fontarabie (Paris, France)
- Energy Hub Project— Tweewaters Leuven (Belgium)

16.2   References

[1] Amsterdam Noord

16.3   See also

- Sustainable city
- Urban ecology
- Green building
- Vertical farming
- Urban agriculture
- Ecovillage
- Ecological footprint
- Ecological debt
- Sustainable transport
- Bicycle City
- Transition town
- Sustainable design
- Peri-urbanisation

Chapter 17

Ecological design

See also: Sustainable design and Environmental design

Ecological design is defined by Sim Van der Ryn and Stuart Cowan as "any form of design that minimizes environmentally destructive impacts by integrating itself with living processes."[1] Ecological design is an integrative ecologically responsible design discipline.

It helps connect scattered efforts in green architecture, sustainable agriculture, ecological engineering, ecological restoration and other fields. The "eco" prefix was used to ninety sciences including eco-city, eco-management, eco-technique, eco-tecture. It was used by John Button in 1998 at the first time. The inchoate developing nature of ecological design was referred to the "adding in "of environmental factor to the design process, but later it was focused on the details of eco-design practice such as product system or individual product or industry as a whole.[2] By including life cycle models through energy and materials flow, ecological design was related to the new interdisciplinary subject of industrial ecology. Industrial ecology meant a conceptual tool emulating models derived from natural ecosystem and a frame work for conceptualizing environmental and technical issues.

Living organisms exist in various systems of balanced symbiotic relationships. The ecological movement of the late twentieth-century is based on understanding that disruptions in these relationships has led to serious breakdown of natural ecosystems. In human history, technological means have resulted in growth of human populations through fire, implements and weapons. This dramatic increase in explosive population contributed the introduction of mechanical energies in machine production and there have been improvements in mechanized agriculture, manufactured chemical fertilizers and general health measures. Although the earlier invention inclined energy adjusting the ecological balance, the latest population growth after industrial revolution led to change ecology abnormally.[3]

17.1 Ecological design issues and the role of designers

Since the Industrial Revolution, many propositions in the design field were raised with unsustainable design principles. The architect-designer Victor Papanek suggested that industrial design has murdered by creating new species of permanent garbage and by choosing materials and processes that pollute the air.[4] For these issues, R. Buckminster Fuller, who was invited as University Professor at Southern Illinois University in Carbondale in 1960s, demonstrated how design could play a central role in identifying major world problems between 1965 and 1975. That included following contents:[5]

- Review and analysis of world energy resources

- Defining more efficient uses of natural resources such as metals

- Integrating machine tools into efficient systems of industrial production

In the 1992 conference, 'The Agenda 21: The Earth Summit Strategy to Save Our Planet", a proposition was put forward that our world is on a path of energy production and consumption that cannot be sustained. The report drew attention to Individuals and groups around the world who have a set of principles to develop strategies for change that might be effective in world economics and trade policies, and the design professions will play a role in it. Namely, those meant that design profession becomes not what new products to make, but how to reinvent design culture likely to be realized. He noted designers firstly have to realize that design has historically been a dependent, contingent practice rather than one based on necessity. The design theorist, Clive Dilnot noted design becomes once again a means of ordering the world rather than merely of shaping products.[6] As a broader approach, the conference of 'Agenda 21: The Earth Summit Strategy to Save Our Planet'

1992, emphasized that designers should challenge for facing human problems. These problems were mentioned to six themes: quality of life, efficient use of natural resources, protecting the global commons, managing human settlements, the use of chemicals and the management of human industrial waste, and fostering sustainable economic growth on a global scale.[7]

17.2 History

- 1971 Ian McHarg, in his book "Design with Nature", popularized a system of analyzing the layers of a site in order to compile a complete understanding of the qualitative attributes of a place. McHarg gave every qualitative aspect of the site a layer, such as the history, hydrology, topography, vegetation, etc. This system became the foundation of today's Geographic Information Systems (GIS), a ubiquitous tool used in the practice of ecological landscape design.

- 1978 Permaculture. Bill Mollison and David Holmgren coin the phrase for a system of designing regenerative human ecosystems. (Founded in the work of Fukuoka, Yeoman, Smith, etc..

- 1994 David Orr, in his book "Earth in Mind: On Education, Environment, and the Human Prospect", compiled a series of essays on "ecolgocial design intelligence" and its power to create healthy, durable, resilient, just, and prosperous communities.

- 1994 Canadian biologists John Todd (biologist) and Nancy Jack Todd, in their book "From Eco-Cities to Living Machines" describe the precepts of ecological design.

- 2004 Fritjof Capra, in his book "The Hidden Connections: A Science for Sustainable Living", wrote this primer on the science of living systems and considers the application of new thinking by life scientists to our understanding of social organization.

- 2004 K. Ausebel compiled compelling personal stories of the world's most innovative ecological designers in "Nature's Operating Instructions."

17.3 Influence

There are some clothing companies that are using several ecological design methods to change the future of the textile industry into a more environmentally friendly one. Recycling used clothing to minimize the use of resources, using biodegradable textile materials to reduce the impact on the environment, and using plant dyes instead of poisonous chemicals to improve the appearance of fabric.[8]

17.4 See also

- Circles of Sustainability

- Ecodesign

- Ecological restoration

- Energy-efficient landscape design

- Environmental design

- Environmental graphic design

- Green roof

- Permaculture

- Principles of Intelligent Urbanism

- Sustainable design

- Sustainable landscape architecture

- Terreform ONE

- Transition Design

17.5 Notes and references

[1] Van der Ryn S, Cowan S(1996). "Ecological Design". Island Press, p.18

[2] Anne-Marie Willis (1991), "An international Eco Design" conference

[3] John McHale (1969), "An Ecological Overview", in The Future of the Future, New York; George Braziller, pp.66-74

[4] Victor Papanek (1972), "Design for the Real World: Human Ecological and social change", Chicago: Academy Edition, ix.

[5] Victor Margolin (1997), "Design for a Sustainable World", Design Issues, vol14, 2. pp. 85

[6] Clive Dilnot (1982), "Design as a Society Significant Activity: An Introduction", Design studies 3:2. pp.144

[7] Victor Margolin (1988), "Design for a Sustainable World", Design Issues, vol14,2. pp. 91

[8] Taieb, Amine Hadj et al. (2010). "Sensitising Children to Ecological Issues through Textile Eco-Design". International Journal of Art & Design Education, vol. 29, 3. p313-320

17.6   Further reading

- *From Bauhaus to Ecohouse: A History of Ecological Design.* By Peder Anker, Published by Louisiana State University Press, 2010. ISBN 0-8071-3551-8.

- *Ecological Design.* By Sim Van der Ryn, Stuart Cowan, Published by Island Press, 2007. ISBN 978-1-59726-141-8 (2nd ed., 1st, 1996)

- *Ignorance and Surprise: Science, Society, and Ecological Design.* By Matthias Gross, Published by MIT Press, 2010. ISBN 0-262-01348-7

17.7   External links

- Sustainable Design & Development Resource Guide

Chapter 18

Ecosa Institute

The **Ecosa Institute for Ecological Design** is a design school located in Prescott, Arizona. Since 2000 Ecosa has advocated a radical departure from the traditional approach to teaching design. Ecosa uses nature as a model for their design curriculum, and asks students to design with nature as their client. Their core program offering is an Ecological Design Certificate program, in which students learn to use design and ecological awareness to explore the exciting challenge of creating a healthy, just and sustainable world. Students are challenged to envision and design for a world where human activities are in balance with the natural world.

18.1 Ecosa Institute of Ecological Design

Ecosa Institute advocates a radical departure from the traditional approach to teaching design. As an in-depth overview to sustainable and ecological design, their certificate program explores the many ways in which design can solve the environmental challenges of the 21st century. Architecture, urban planning, landscape architecture, industrial design, and other design-thinking disciplines are unique problem-solving tools that have the potential to create a healthy, just and sustainable world.

The report of the American Institute of Architects Committee on the Environment has said "In many ways, what students get in one semester is a more holistic understanding of sustainability than they could presently get at any traditional design school."

18.2 History

The vision for Ecosa was formulated during the 1980s and 90's by English architect and educator Antony Brown. His dedication to issues of sustainability and ecological design developed after joining Paolo Soleri's Cosanti Foundation and working with the Italian architect on his conceptual designs for a new vision of urban settlements. Brown worked on the resulting urban prototype, Arcosanti, as architect-in-residence supervising both design work and construction. During his time studying with Soleri and teaching the philosophy of the arcology concept, Brown began to cultivate his own vision of an ecological future and the new approach to design education he saw as necessary to achieve it.

Brown left the Arcosanti project and began to explore his ideas through a series of classes he developed and taught at Prescott College. This opportunity to experiment with teaching methods convinced him that experiential education was the best way to reach students and to personalize learning. At Prescott College, Brown tried turning students who were environmentalists into designers, but later realized teaching designers to become environmentalists may be more effective in reaching his goal. Brown's goal was not to tack on sustainable design to a conventional curriculum, but to restructure the underlying ethos of architectural education and bring a new sensitivity to the practice of architecture.

In 1996 Brown formally founded the Ecosa Institute in Prescott, Arizona, and in 1998 the organization was granted 501(c)(3) status. The Ecosa Institute offered its first semester in sustainable design in 2000. Ecosa Institute merged with Prescott College in 2012.

18.3 Curriculum

Ecosa's curriculum is unlike those at traditional colleges and universities in that sustainable and ecological design at its core. It is an intensive, multifaceted mix of lectures, field trips, student presentations, readings, discussions, studio time and meetings with community groups and clients. New learning is then applied to real world design projects, giving a practical use for theoretic information. Final presentations on larger projects occur at the end of each semester, while smaller project presentations occur throughout the semesters. On completion of the Certificate

Program, students have a total overview of the issues, solutions and promise of ecological design. Additional information about the programs at Ecosa Institute can be found at www.ecosa.org.

18.4   Guest speakers

Each semester an adjunct faculty consisting of educators from colleges and universities, professionals in the architecture, design, and construction industries and experts from nationally recognized organizations join Ecosa as visiting faculty and lecturers. These individuals, regarded as experts in their field, teach specific segments of the curriculum. Regular Ecosa guest lectures include design-build architect and University of Washington professor Steve Badanes, Pliny Fisk III, co-director of the Center for Maximum Potential Building Systems , Nate Cormier of SVR Design , father of arcology Paolo Soleri, sustainable architect and author Sim Van der Ryn, and architects Will Bruder and Eddie Jones .

18.5   References

- New York Times
- Architecture Week
- Daily Courier
- Metropolis Magazine
- Arcosanti
- Treehugger

18.6   External links

- Ecosa website

Chapter 19

Energy-efficient landscaping

Energy-efficient landscaping is a type of landscaping designed for the purpose of conserving energy. There is a distinction between the embedded energy of materials and constructing the landscape, and the energy consumed by the maintenance and operations of a landscape.

Design techniques include:

- Planting trees for the purpose of providing shade, which reduces cooling costs.

- Planting or building windbreaks to slow winds near buildings, which reduces heat loss.

- Wall sheltering, where shrubbery or vines are used to create a windbreak directly against a wall.

- Earth sheltering and positioning buildings to take advantage of natural landforms as windbreaks.

- Green roofs that cool buildings with extra thermal mass and evapotranspiration.

- Reducing the heat island effect with pervious paving, high albedo paving, shade, and minimizing paved areas.

- Site lighting with full cut off fixtures, light level sensors, and high efficiency fixtures

Energy-efficient landscaping techniques include using local materials, on-site composting and chipping to reduce greenwaste hauling, hand tools instead of gasoline-powered, and also may involve using drought-resistant plantings in arid areas, buying stock from local growers to avoid energy in transportation, and similar techniques.

19.1   See also

- Natural Materials
- Green building

- Building material

- Energy conservation

- Keyline design

- Drought tolerance

- Drought-tolerant plants

- Roof garden

- Water conservation

- Xeriscaping

- Climate-friendly gardening

- Sustainable gardening

- Sustainable landscape architecture

- Sustainable architecture

- Landscape design

- Landscape architecture

- Garden design

- Landscape planning

19.2   External links

- The Green Building Sourcebook has a section on landscaping

- ecoLogical Home Ideas Magazine for green home building/remodeling

- Stopwaste.org

- Sustainable Residential Design: Increasing Energy Efficiency Resource Guide

19.3   References

Chapter 20

Environmental impact assessment

Environmental impact assessment (EIA) is the formal process used to predict the environmental consequences (positive or negative) of a plan, policy, program, or project prior to the decision to move forward with the proposed action. Formal impact assessments may be governed by rules of administrative procedure regarding public participation and documentation of decision making, and may be subject to judicial review. An impact assessment may propose measures to adjust impacts to acceptable levels or to investigate new technological solutions.

The purpose of the assessment is to ensure that decision makers consider the environmental impacts when deciding whether or not to proceed with a project. The International Association for Impact Assessment (IAIA) defines an environmental impact assessment as "the process of identifying, predicting, evaluating and mitigating the biophysical, social, and other relevant effects of development proposals prior to major decisions being taken and commitments made."[1] EIAs are unique in that they do not require adherence to a predetermined environmental outcome, but rather they require decision makers to account for environmental values in their decisions and to justify those decisions in light of detailed environmental studies and public comments on the potential environmental impacts.[2]

Engineering and consulting companies work hand in hand as contractors for mining, energy, oil&gas companies executing EIAs. Companies operating globally such as Royal HaskoningDHV, Golder Associates, Amec Foster Wheeler, Schlumberger Water Services (an Schlumberger company) are an example of a much bigger pool of expertise globally. These contractors are the ones not only in charge of preparing an EIA study but most importantly getting these studies approved by each country government offices prior to the execution of a project. Each country will also have its own local contractors offering the same kind of service hence breaking out monopolies by increasing the supply of EIAs execution consultants.[3]

20.1 History

Environmental impact assessments commenced in the 1960s, as part of increasing environmental awareness.[notes 1] EIAs involved a technical evaluation intended to contribute to more objective decision making. In the United States, environmental impact assessments obtained formal status in 1969, with enactment of the National Environmental Policy Act. EIAs have been used increasingly around the world. The number of "Environmental Assessments" filed every year "has vastly overtaken the number of more rigorous Environmental Impact Statements (EIS)."[4] An Environmental Assessment is a "mini-EIS designed to provide sufficient information to allow the agency to decide whether the preparation of a full-blown Environmental Impact Statement (EIS) is necessary."[5][6] EIA is an activity that is done to find out the impact that would be done before development will occur.

20.2 Methods

General and industry specific assessment methods are available including:

- *Industrial products* - Product environmental life cycle analysis (LCA) is used for identifying and measuring the impact of industrial products on the environment. These EIAs consider activities related to extraction of raw materials, ancillary materials, equipment; production, use, disposal and ancillary equipment.[7]

- *Genetically modified plants* - Specific methods available to perform EIAs of genetically modified organisms include GMP-RAM and INOVA.[8]

- *Fuzzy logic* - EIA methods need measurement data to estimate values of impact indicators. However, many of the environment impacts cannot be quantified, e.g. landscape quality, lifestyle quality and social acceptance. Instead information from similar EIAs, expert judgment and community sentiment are employed. Approximate reasoning methods known as fuzzy logic can be used.[9] A fuzzy arithmetic approach has also been proposed [10] and implemented using a software tool (TDEIA). More information can be found at ARAI web site.

20.3 Follow-up

At the end of the project, an audit evaluates the accuracy of the EIA by comparing actual to predicted impacts. The objective is to make future EIAs more valid and effective. Two primary considerations are:

- *Scientific* - to examine the accuracy of predictions and explain errors

- *Management* - to assess the success of mitigation in reducing impacts

Audits can be performed either as a rigorous assessment of the null hypothesis or with a simpler approach comparing what actually occurred against the predictions in the EIA document.[11]

After an EIA, the precautionary and polluter pays principles may be applied to decide whether to reject, modify or require strict liability or insurance coverage to a project, based on predicted harms.

The Hydropower Sustainability Assessment Protocol is a sector specific method for checking the quality of Environmental and Social assessments and management plans.

20.4 Around the world

20.4.1 Australia

The history of EIA in Australia could be linked to the enactment of the U.S. National Environment Policy Act (NEPA) in 1969, which made the preparation of environmental impact statements a requirement. In Australia, one might say that the EIA procedures were introduced at a State Level prior to that of the Commonwealth (Federal), with a majority of the states having divergent views to the Commonwealth. One of the pioneering states was New South Wales, whose State Pollution Control Commission issued EIA guidelines in 1974. At a Commonwealth (Federal) level, this was followed by passing of the Environment Protection (Impact of Proposals) Act in 1974. The Environment Protection and Biodiversity Conservation Act 1999 (EPBC) superseded the Environment Protection (Impact of Proposals) Act 1974 and is the current central piece for EIA in Australia on a Commonwealth (Federal) level. An important point to note is that this Commonwealth Act does not affect the validity of the States and Territories environmental and development assessments and approvals; rather the EPBC runs as a parallel to the State/Territory Systems.[12] Overlap between federal and state requirements is addressed via bilateral agreements or one off accreditation of state processes, as provided for in the EPBC Act.

The Commonwealth Level

The EPBC Act provides a legal framework to protect and manage nationally and internationally important flora, fauna, ecological communities and heritage places-defined in the EPBC Act as matters of 'national environmental significance'. Following are the eight matters of 'national environmental significance' to which the EPBC ACT applies:[13]

- World Heritage sites

- National Heritage places

- RAMSAR wetlands of international significance

- Listed threatened species and ecological communities

- Migratory species protected under international agreements

- The Commonwealth marine environment

- Nuclear actions (including uranium mining)

- National Heritage.

In addition to this, the EPBC Act aims at providing a streamlined national assessment and approval process for activities. These activities could be by the Commonwealth, or its agents, anywhere in the world or activities on Commonwealth land; and activities that are listed as having a 'significant impact' on matters of 'national environment significance'.[13]

The EPBC Act comes into play when a person (a 'proponent') wants an action (often called a 'proposal' or 'project') assessed for environmental impacts under the EPBC Act, he or she must refer the project to the Department of Environment, Water, Heritage and the Arts (Australia). This 'referral' is then released to the public, as well as relevant state, territory and Commonwealth ministers, for comment on whether the project is likely to have a significant impact on matters of national environmental significance.[13] The

Department of Environment, Water, Heritage and the Arts assess the process and makes recommendation to the minister or the delegate for the feasibility. The final discretion on the decision remains of the minister, which is not solely based on matters of 'national environmental significance' but also the consideration of social and economic impact of the project.[13]

The Australian Government environment minister cannot intervene in a proposal if it has no significant impact on one of the eight matters of 'national environmental significance' despite the fact that there may be other undesirable environmental impacts.[13] This is primarily due to the division of powers between the States and the Federal government and due to which the Australian Government environment minister cannot overturn a state decision.

There are strict civil and criminal penalties for the breach of EPBC Act. Depending on the kind of breach, civil penalty (maximum) may go up to $550,000 for an individual and $5.5 million for a body corporate, or for criminal penalty (maximum) of seven years imprisonment and/or penalty of $46,200.[13]

The State and Territory Level

Australian Capital Territory (ACT)   EIA provisions within Ministerial Authorities in the ACT are found in the Chapters 7 and 8 of the *Planning and Development Act 2007* (ACT). EIA in ACT was previously administered with the help of Part 4 of the Land (Planning and Environment) Act 1991 (Land Act) and Territory Plan (plan for land-use).[12] Note that some EIA may occur in the ACT on Commonwealth land under the EPBC Act (Cth). Further provisions of the *Australian Capital Territory (Planning and Land Management) Act 1988* (Cth) may also be applicable particularly to national land and "designated areas".

New South Wales (NSW)   In New South Wales, the Environment Planning Assessment Act 1979 (EPA) establishes three pathways for EIA. The first is under Part 5.1 of the EPAA, which provides for EIA of 'State Significant Infrastructure' projects. (From June 2011, this Part replaced Part 3A, which previously covered EIA of major projects). The second is under Part 4 of the Act dealing with development control. If a project does not require approval under Part 3A or Part 4 it is then potentially captured by the third pathway, Part 5 dealing with environment impact assessment.[12]

Northern Territory (NT)   The EIA process in Northern Territory is chiefly administered under the Environmental Assessment Act (EAA).[14] Although EAA is the primary

tool for EIA in Northern Territory, there are further provisions for proposals in the Inquiries Act 1985 (NT).[12]

Queensland (QLD)   There are four main EIA processes in Queensland.[15] Firstly, under the Integrated Planning Act 1997 (IPA) for development projects other than mining. Secondly, under the Environmental Protection Act 1994 (EP Act) for some mining and petroleum activities. Thirdly, under the State Development and Public Works Organization Act 1971 (State Development Act) for 'significant projects'. Finally, Environment Protection and Biodiversity Conservation Act 1999 (Cth) for 'controlled actions'.[15]

South Australia (SA)   The local governing tool for EIA in South Australia is the Development Act 1993. There are three levels of assessment possible under the Act in the form of an environment impact statement (EIS), a public environmental report (PER) or a Development Report (DR).[12]

Tasmania (TAS)   In Tasmania, an integrated system of legislation is used to govern development and approval process, this system is a mixture of the Environmental Management and Pollution Control Act 1994 (EMPCA), Land Use Planning and Approvals Act 1993 (LUPAA), State Policies and Projects Act 1993 (SPPA), and Resource Management and Planning Appeals Tribunal Act 1993.[12]

Victoria (VIC)   The EIA process in Victoria is intertwined with the Environment Effects Act 1978 and the Ministerial Guidelines for Assessment of Environmental Effects (made under the s. 10 of the EE Act).[16]

Western Australia (WA)   The Environmental Protection Act 1986 (Part 4) provides the legislative framework for the EIA process in Western Australia.[17] The EPA Act oversees the planning and development proposals and assesses their likely impacts on the environment.

20.4.2   Canada

In *Friends of the Oldman River Society v. Canada (Minister of Transportation)*,(SCC 1992) La Forest J of the Supreme Court of Canada described environmental impact assessment in terms of the proper scope of federal jurisdiction with respect to environments matters,

> "Environmental impact assessment is, in its simplest form, a planning tool that is now generally regarded as an integral component of

sound decision-making."[18]

Supreme Court Justice La Forest cited (Cotton, Emond & 1981 245), "The basic concepts behind environmental assessment are simply stated: (1) early identification and evaluation of all potential environmental consequences of a proposed undertaking; (2) decision making that both guarantees the adequacy of this process and reconciles, to the greatest extent possible, the proponent's development desires with environmental protection and preservation."[19]

La Forest referred to (Jeffrey 1989, 1.2,1.4) and (Emond 1978, p. 5) who described "...environmental assessments as a planning tool with both an information-gathering and a decision-making component" that provide "...an objective basis for granting or denying approval for a proposed development."[20][21]

Justice La Forest addressed his concerns about the implications of Bill C-45 regarding public navigation rights on lakes and rivers that would contradict previous cases.(La Forest & 1973 178-80)[22]

The Canadian Environmental Assessment Act 2012 (CEAA 2012)[23] "and its regulations establish the legislative basis for the federal practice of environmental assessment in most regions of Canada."[24][25][26] CEAA 2012 came into force July 6, 2012 and replaces the former *Canadian Environmental Assessment Act* (1995). EA is defined as a planning tool to identify, understand, assess and mitigate, where possible, the environmental effects of a project.

"The purposes of this Act are: (a) to protect the components of the environment that are within the legislative authority of Parliament from significant adverse environmental effects caused by a designated project; (b) to ensure that designated projects that require the exercise of a power or performance of a duty or function by a federal authority under any Act of Parliament other than this Act to be carried out, are considered in a careful and precautionary manner to avoid significant adverse environmental effects; (c) to promote cooperation and coordinated action between federal and provincial governments with respect to environmental assessments; (d) to promote communication and cooperation with aboriginal peoples with respect to environmental assessments; (e) to ensure that opportunities are provided for meaningful public participation during an environmental assessment; (f) to ensure that an environmental assessment is completed in a timely manner; (g) to ensure that projects, as defined in section 66, that are to be carried out on federal lands, or those that are outside Canada and that are to be carried out or financially supported by a federal authority, are considered in a careful and precautionary manner to avoid significant adverse environmental effects; (h) to encourage federal authorities to take actions that promote

sustainable development in order to achieve or maintain a healthy environment and a healthy economy; and (i) to encourage the study of the cumulative effects of physical activities in a region and the consideration of those study results in environmental assessments."[27]

Canadian Environmental Assessment Act

Opposition

Environmental Lawyer Dianne Saxe argued that the CEAA 2012 "allows the federal government to create mandatory timelines for assessments of even the largest and most important projects, regardless of public opposition." (Saxe 2012)[28]

"Now that federal environmental assessments are gone, the federal government will only assess very large, very important projects. But it's going to do them in a hurry."

Dianne Saxe[28]

On 3 August 2012 the Canadian Environmental Assessment Agency nine "designated projects" with their timelines: Enbridge Northern Gateway Pipeline Joint Review Panel (JRP) 18 months; Marathon Platinum Group Metals and Copper Mine Project (JRP): 13 months; Site C Clean Energy Project (JRP) 8.5 months; Deep Geologic Repository Project (JRP) 17 months; Enbridge Northern Gateway Project (JRP) 18 months; Jackpine Mine Expansion Project (JRP) 11.5 months; Pierre River Mine Project: 8 months; New Prosperity Gold-Copper Mine Project (JRP) 7.5 months; Frontier Oil Sands Mine Project (JRP)8.5 months; EnCana/Cenovus Shallow Gas Infill Project (JRP) 5 months.[29]

Saxe compares these timelines with environmental assessments for the Mackenzie Valley Pipeline. Thomas R. Berger, Royal Commissioner of the Mackenzie Valley Pipeline Inquiry (9 May 1977), worked extremely hard to ensure that industrial development on Aboriginal people's land resulted in benefits to those indigenous people.[30]

On 22 April 2013, Official Opposition Environment critic Megan Leslie issued a statement claiming that the federal government's recent changes to "fish habitat protection, the Navigable Waters Protection Act and the Canadian Environmental Assessment Act", along with gutting existing laws and making cuts to science and research, "will be disastrous, not only for the environment, but also for Canadians' health and economic prosperity."[31] On 26 September 2012, Leslie argued that with the changes to the Canadian Environmental Assessment Act that came into effect 6 July 2012, "seismic testing, dams, wind farms and power plants" no longer required any federal environmental assessment.

She also claimed that because the CEAA 2012—which she claimed was rushed through Parliament—dismantled the CEAA 1995, the Oshawa ethanol plant project would no longer have a full federal environmental assessment.[32] Mr. Peter Kent (Minister of the Environment) explained that the CEAA 2012 "provides for the Government of Canada and the Environmental Assessment Agency to focus on the large and most significant projects that are being proposed across the country." The 2,000 to 3,000-plus smaller screenings that were in effect under CEAA 1995 became the "responsibility of lower levels of government but are still subject to the same strict federal environmental laws."[32] Anne Minh-Thu Quach, MP for Beauharnois—Salaberry, QC, argued that the mammoth budget bill dismantled 50 years of environmental protection without consulting Canadians about the "colossal changes they are making to environmental assessments." She claimed that the federal government is entering into "limited consultations, by invitation only, months after the damage was done."[32]

20.4.3 China

The Environmental Impact Assessment Law (EIA Law) requires that an environmental impact assessment be completed prior to project construction. However, if a developer completely ignores this requirement and builds a project without submitting an environmental impact statement, the only penalty is that the environmental protection bureau (EPB) may require the developer to do a make-up environmental assessment. If the developer does not complete this make-up assessment within the designated time, only then is the EPB authorized to fine the developer. Even so, the possible fine is capped at a maximum of about US$25,000, a fraction of the overall cost of most major projects. The lack of more stringent enforcement mechanisms has resulted in a significant percentage of projects not completing legally required environmental impact assessments prior to construction.[33]

China's State Environmental Protection Administration (SEPA) used the legislation to halt 30 projects in 2004, including three hydro-power plants under the Three Gorges Project Company. Although one month later (Note as a point of reference, that the typical EIA for a major project in the USA takes one to two years.), most of the 30 halted projects resumed their construction, reportedly having passed the environmental assessment, the fact that these key projects' construction was ever suspended was notable.

A joint investigation by SEPA and the Ministry of Land and Resources in 2004 showed that 30-40% of the mining construction projects went through the procedure of environment impact assessment as required, while in some areas only 6-7% did so. This partly explains why China has witnessed so many mining accidents in recent years.

SEPA alone cannot guarantee the full enforcement of environmental laws and regulations, observed Professor Wang Canfa, director of the centre to help environmental victims at China University of Political Science and Law. In fact, according to Wang, the rate of China's environmental laws and regulations that are actually enforced is estimated at barely 10%.[34]

20.4.4 Egypt

Environmental Impact Assessment (EIA) EIA is implemented in Egypt under the umbrella of the Ministry of state for environmental affairs. The Egyptian Environmental Affairs Agency (EEAA) is responsible for the EIA services.

In June 1997, the responsibility of Egypt's first full-time Minister of State for Environmental Affairs was assigned as stated in the Presidential Decree no.275/1997. From thereon, the new ministry has focused, in close collaboration with the national and international development partners, on defining environmental policies, setting priorities and implementing initiatives within a context of sustainable development.

According to the Law 4/1994 for the Protection of the Environment, the Egyptian Environmental Affairs Agency (EEAA) was restructured with the new mandate to substitute the institution initially established in 1982. At the central level, EEAA represents the executive arm of the Ministry.

The purpose of EIA is to ensure the protection and conservation of the environment and natural resources including human health aspects against uncontrolled development. The long-term objective is to ensure a sustainable economic development that meets present needs without compromising future generations ability to meet their own needs. EIA is an important tool in the integrated environmental management approach.

EIA must be performed for new establishments or projects and for expansions or renovations of existing establishments according to the Law for the Environment. [35]

20.4.5 EU

There is a wide range of instruments in the Environmental policy of the European Union. Among them the European Union has established a mix of mandatory and discretionary procedures to assess environmental impacts.[36] European Union Directive (85/337/EEC) on Environmental Impact Assessments (known as the *EIA Directive*) [37] was first introduced in 1985 and was amended in 1997. The directive was amended again in 2003, following EU signature of the 1998 Aarhus Convention, and once more in 2009. The initial Directive of 1985 and its three amendments have been codified in Directive 2011/92/EU of 13 December 2011.[38] In 2001, the issue was enlarged to the assessment of plans and programmes by the so-called *Strategic Environmental Assessment (SEA) Directive* (2001/42/EC), which is now in force.[36] Under the EU directive, an EIA must provide certain information to comply.[39] There are seven key areas that are required:

1. Description of the project

   - Description of actual project and site description

   - Break the project down into its key components, i.e. construction, operations, decommissioning

   - For each component list all of the sources of environmental disturbance

   - For each component all the inputs and outputs must be listed, e.g., air pollution, noise, hydrology

2. Alternatives that have been considered

   - Examine alternatives that have been considered

   - Example: in a biomass power station, will the fuel be sourced locally or nationally?

3. Description of the environment

   - List of all aspects of the environment that may be affected by the development

   - Example: populations, fauna, flora, air, soil, water, humans, landscape, cultural heritage

   - This section is best carried out with the help of local experts, e.g. the RSPB in the UK

4. Description of the significant effects on the environment

   - The word significant is crucial here as the definition can vary

   - 'Significant' must be defined

   - The most frequent method used here is use of the Leopold matrix

   - The matrix is a tool used in the systematic examination of potential interactions

   - Example: in a windfarm development a significant impact may be collisions with birds

5. Mitigation

   - This is where EIA is most useful

   - Once section 4 is complete, it is obvious where impacts are greatest

   - Using this information ways to avoid negative impacts should be developed

   - Best working with the developer with this section as they know the project best

   - Using the windfarm example again construction could be out of bird nesting seasons

6. Non-technical summary (EIS)

   - The EIA is in the public domain and be used in the decision making process

   - It is important that the information is available to the public

   - This section is a summary that does not include jargon or complicated diagrams

   - It should be understood by the informed layperson

7. Lack of know-how/technical difficulties

   - This section is to advise any areas of weakness in knowledge

   - It can be used to focus areas of future research

   - Some developers see the EIA as a starting block for poor environmental management

Annexed projects

All projects are either classified as Annex 1 or Annex 2 projects. Those lying in Annex 1 are large scale developments such as motorways, chemical works, bridges, powerstations etc. These always require an EIA under the Environmental Impact Assessment Directive (85,337,EEC as amended). Annex 2 projects are smaller in scale than those referred to in Annex 1. Member States must determine whether these project shall be made subject to an assessment subject to a set of criteria set out in Annex 3 of codified Directive 2011/92/EU.

The Netherlands

EIA was implemented in Dutch legislation on September 1, 1987. The categories of projects that require an EIA are summarised in Dutch legislation, the Wet milieubeheer. The use of thresholds for activities makes sure that EIA is obligatory for those activities that may have considerable impacts on the environment.

For projects and plans that fit these criteria, an EIA report is required. The EIA report defines a.o. the proposed initiative, it makes clear the impact of that initiative on the environment and compares this with the impact of possible alternatives with less a negative impact.[40]

20.4.6 Hong Kong

EIA in Hong Kong, since 1998, is regulated by the *Environmental Impact Assessment Ordinance 1997*.

The original proposal to construct the Lok Ma Chau Spur Line overground across the Long Valley failed to get through EIA, and the Kowloon–Canton Railway Corporation had to change its plan and build the railway underground. In April 2011, the EIA of the Hong Kong section of the Hong Kong-Zhuhai-Macau Bridge was found to have breached the ordinance, and was declared unlawful. The appeal by the government was allowed in September 2011. However, it was estimated that this EIA court case had increased the construction cost of the Hong Kong section of the bridge by HK$6.5 billion in money-of-the-day prices.[41]

20.4.7 India

The Ministry of Environment and Forests (MoEF) of India has been in a great effort in Environmental Impact Assessment in India. The main laws in action are the Water Act(1974), the Indian Wildlife (Protection) Act (1972), the Air (Prevention and Control of Pollution) Act (1981) and the Environment (Protection) Act (1986),Biological Diversity Act(2002).[42] The responsible body for this is the Central Pollution Control Board. Environmental Impact Assessment (EIA) studies need a significant amount of primary and secondary environmental data. Primary data are those collected in the field to define the status of the environment (like air quality data, water quality data etc.). Secondary data are those collected over the years that can be used to understand the existing environmental scenario of the study area. The environmental impact assessment (EIA) studies are conducted over a short period of time and therefore the understanding of the environmental trends, based on a few months of primary data, has limitations. Ideally, the primary data must be considered along with the secondary data for complete understanding of the existing environmental status of the area. In many EIA studies, the secondary data needs could be as high as 80% of the total data requirement. EIC is the repository of one stop secondary data source for environmental impact assessment in India.

The Environmental Impact Assessment (EIA) experience in India indicates that the lack of timely availability of reliable and authentic environmental data has been a major bottle neck in achieving the full benefits of EIA. The environment being a multi-disciplinary subject, a multitude of agencies are involved in collection of environmental data. However, no single organization in India tracks available data from these agencies and makes it available in one place in a form required by environmental impact assessment practitioners. Further, environmental data is not available in enhanced forms that improve the quality of the EIA. This makes it harder and more time-consuming to generate environmental impact assessments and receive timely environmental clearances from regulators. With this background, the Environmental Information Centre (EIC) has been set up to serve as a professionally managed clearing house of environmental information that can be used by MoEF, project proponents, consultants, NGOs and other stakeholders involved in the process of environmental impact assessment in India. EIC caters to the need of creating and disseminating of organized environmental data for various developmental initiatives all over the country.

EIC stores data in GIS format and makes it available to all environmental impact assessment studies and to EIA stakeholders in a cost effective and timely manner. So that we can manage that in different proportions such as remedy measures etc.,

20.4.8 Korea, South

Recycling culture and policy Ministry of Environment

20.4.9 Malaysia

In Malaysia, Section 34A, Environmental Quality Act, 1974[43] requires developments that have significant impact to the environment are required to conduct the Environmental impact assessment.

20.4.10 Nepal

In Nepal, EIA has been integrated in major development projects since the early 1980s. In the planning history of Nepal, the sixth plan (1980–85), for the first time, recognized the need for EIA with the establishment of Environmental Impact Study Project (EISP) under the Department of Soil Conservation in 1982 to develop necessary instruments for integration of EIA in infrastructure development projects. However, the government of Nepal enunciated environment conservation related policies in the seventh plan (NPC, 1985–1990). To enforce this policy and make necessary arrangements, a series of guidelines were developed, thereby incorporating the elements of environmental factors right from the project formulation stage of the development plans and projects and to avoid or minimize adverse effects on the ecological system. In addition, it has also emphasized that EIAs of industry, tourism, water resources, transportation, urbanization, agriculture, forest and other developmental projects be conducted.

In Nepal, the government's Environmental Impact Assessment Guideline of 1993 inspired the enactment of the Environment Protection Act (EPA) of 1997 and the Environment Protection Rules (EPR) of 1997 (EPA and EPR have been enforced since 24 and 26 June 1997 respectively in Nepal) to internalizing the environmental assessment system. The process institutionalized the EIA process in development proposals and enactment, which makes the integration of IEE and EIA legally binding to the prescribed projects. The projects, requiring EIA or IEE, are included in Schedules 1 and 2 of the EPR, 1997 (GoN/MoLJPA 1997). Progresses were made in the Environmental protection issue during the 8th five-year plan (1992–1997). The following development in Environmental protection were achieved during that time:

- Formulation of Environmental Protection Act 1997

- Establishment of Ministry of Environment

- Development of National Environmental Policies and Action Plan, EIA guidelines developed

- Consideration of environmental concerns in hydropower projects

- Development of industrial, irrigation and agricultural policies that undertook environmental concerns

Source: Bhatta R. and Khanal S. 2010.African Journal of Environmental Science and Technology Vol. 4(9), pp. 586–594

20.4.11   New Zealand

In New Zealand, EIA is usually referred to as *Assessment of Environmental Effects* (AEE). The first use of EIA's dates back to a Cabinet minute passed in 1974 called Environmental Protection and Enhancement Procedures. This had no legal force and only related to the activities of government departments. When the Resource Management Act was passed in 1991, an EIA was required as part of a resource consent application. Section 88 of the Act specifies that the AEE must include "such detail as corresponds with the scale and significance of the effects that the activity may have on the environment". While there is no duty to consult any person when making a resource consent application (Sections 36A and Schedule 4), proof of consultation is almost certain required by local councils when they decide whether or not to publicly notify the consent application under Section 93.

20.4.12   Russian Federation

As of 2004, the state authority responsible for conducting the State EIA in Russia has been split between two Federal bodies: 1) Federal service for monitoring the use of natural resources – a part of the Russian Ministry for Natural Resources and Environment and 2) Federal Service for Ecological, Technological and Nuclear Control. The two main pieces of environmental legislation in Russia are: The Federal Law 'On Ecological Expertise, 1995 and the 'Regulations on Assessment of Impact from Intended Business and Other Activity on Environment in the Russian Federation, 2000.[44]

Federal Service for monitoring the use of natural resources

In 2006, the parliament committee on ecology in conjunction with the Ministry for Natural Resources and Environment, created a working group to prepare a number of amendments to existing legislation to cover such topics as stringent project documentation for building of potentially environmentally damaging objects as well as building of projects on the territory of protected areas. There has been some success in this area, as evidenced from abandonment of plans to construct a gas pipe-line through the only remaining habitat of the critically endangered Amur leopard in the Russian Far East.

Federal Service for Ecological, Technological and Nuclear Control

The government's decision to hand over control over several important procedures, including state EIA in the field of all types of energy projects, to the Federal Service for Ecological, Technological and Nuclear Control had caused a major controversy and criticism from environmental groups that blamed the government for giving nuclear power industry control over the state EIA.

Not surprisingly the main problem concerning State EIA in Russia is the clear differentiation of jurisdiction between the two above-mentioned Federal bodies.

20.4.13   Sri Lanka

Environmental Impact Assessments

One popular approach to assist in smart growth in democratic countries is for law-makers to require prospective developers to prepare environmental impact assessments of their plans as a condition for state and/or local governments to go for Environmental Impact Assessments.

These reports often indicate how significant impacts the development generates can be mitigated, usually at developer expense. These assessments are frequently controversial. Conservationists, neighborhood advocacy groups and NIMBYs are often skeptical about such impact reports, even when prepared by independent agencies and approved by decision makers rather than promoters. Conversely, developers sometimes strongly resist requirements to implement the mitigation measures required by the local government, as they may be quite costly.

The importance of the Environmental Impact Assessment as an effective tool for the purpose of integrating environmental considerations with development planning is highly recognized in Sri Lanka. The application of this technique is considered as a means of ensuring that the likely effects of new development projects on the environment are fully understood and taken into account before development is allowed to proceed. The importance of this management tool to foresee potential environmental impacts and problems caused by proposed projects and its use as a mean to make project more suitable to the environment are highly appreciated.

20.4.14   United States

Main article: National Environmental Policy Act

The National Environmental Policy Act of 1969 (NEPA), enacted in 1970, established a policy of environmental impact assessment for federal agency actions, federally funded activities or federally permitted/licensed activities that in the U. S. is termed "environmental review" or simply "the NEPA process."[45] The law also created the Council on Environmental Quality, which promulgated regulations to codify the law's requirements.[46] Under United States environmental law an Environmental Assessment (EA) is compiled to determine the need for an *Environmental Impact Statement* (EIS). Federal or federalized actions expected to subject or be subject to significant environmental impacts will publish a Notice of Intent to Prepare an EIS as soon as significance is known. Certain actions of federal agencies must be preceded by the NEPA process. Contrary to a widespread misconception, NEPA does not prohibit the federal government or its licensees/permittees from harming the environment, nor does it specify any penalty if an environmental impact assessment turns out to be inaccurate, intentionally or otherwise. NEPA requires that plausible statements as to the prospective impacts be disclosed in advance. The purpose of NEPA process is to ensure that the decision maker is fully informed of the environmental aspects and consequences prior to making the final decision.

Environmental assessment

An **environmental assessment** (EA) is an environmental analysis prepared pursuant to the National Environmental Policy Act to determine whether a federal action would significantly affect the environment and thus require a more detailed *Environmental Impact Statement (EIS)*. The certified release of an Environmental Assessment results in either a *Finding of No Significant Impact (FONSI)* or an EIS.

The Council on Environmental Quality (CEQ), which oversees the administration of NEPA, issued regulations for implementing the NEPA in 1979. Eccleston reports that the NEPA regulations barely mention preparation of EAs. This is because the EA was originally intended to be a simple document used in relatively rare instances where an agency was not sure if the potential significance of an action would be sufficient to trigger preparation of an EIS. But today, because EISs are so much longer and complicated to prepare, federal agencies are going to great effort to avoid preparing EISs by using EAs, even in cases where the use of EAs may be inappropriate. The ratio of EAs that are being issued compared to EISs is about 100 to 1.[47]

Likewise, even the preparation of an accurate EA is viewed today as an onerous burden by many entities responsible for the environmental review of a proposal. Federal agencies have responded by streamlining their regulations that implement NEPA environmental review, by defining categories of projects that by their well understood nature may be safely excluded from review under NEPA, and by drawing up lists of project types that have negligible material impact upon the environment and can thus be exempted.

Content   The Environmental Assessment is a concise public document prepared by the federal action agency that serves to:

1. briefly provide sufficient evidence and analysis for determining whether to prepare an EIS or a Finding of No Significant Impact (FONSI)

2. Demonstrate compliance with the act when no EIS is required

3. facilitate the preparation of an EIS when a FONSI cannot be demonstrated

The Environmental Assessment includes a brief discussion of the purpose and need of the proposal and of its alternatives as required by NEPA 102(2)(E), and of the human environmental impacts resulting from and occurring to the proposed actions and alternatives considered practicable, plus a listing of studies conducted and agencies and stakeholders consulted to reach these conclusions. The action agency must approve an EA before it is made available to the public. The EA is made public through notices of availability by local, state, or regional clearing houses, often triggered by the purchase of a public notice advertisement in a newspaper of general circulation in the proposed activity area.

Structure   The structure of a generic Environmental Assessment is as follows:

1. Summary

2. Introduction

   - Background

   - Purpose and Need for Action

   - Proposed Action

   - Decision Framework

   - Public Involvement

   - Issues

3. Alternatives, including the Proposed Action

   - Alternatives

   - Mitigation Common to All Alternatives

   - Comparison of Alternatives

4. Environmental Consequences

5. Consultation and Coordination

Procedure   The EA becomes a draft public document when notice of it is published, usually in a newspaper of general circulation in the area affected by the proposal. There is a 15-day review period required for an Environmental Assessment (30 days if exceptional circumstances) while the document is made available for public commentary, and a similar time for any objection to improper process. Commenting on the Draft EA is typically done in writing or email, submitted to the lead action agency as published in the notice of availability. An EA does not require a public hearing for verbal comments. Following the mandated public comment period, the lead action agency responds to any comments, and certifies either a FONSI or a Notice of Intent (NOI) to prepare an EIS in its public environmental review record. The preparation of an EIS then generates a similar but more lengthy, involved and expensive process.

Environmental impact statement

Main article: Environmental impact statement

The adequacy of an environmental impact statement (EIS) can be challenged in federal court. Major proposed projects have been blocked because of an agency's failure to prepare an acceptable EIS. One prominent example was the Westway landfill and highway development in and along the Hudson River in New York City.[48] Another prominent case involved the Sierra Club suing the Nevada Department of Transportation over its denial of the club's request to issue a supplemental EIS addressing air emissions of particulate matter and hazardous air pollutants in the case of widening U.S. Route 95 through Las Vegas.[49] The case reached the United States Court of Appeals for the Ninth Circuit, which led to construction on the highway being halted until the court's final decision. The case was settled prior to the court's final decision.

Several state governments that have adopted "little NEPAs," state laws imposing EIS requirements for particular state actions. Some those state laws such as the California Environmental Quality Act refer to the required environmental impact study as an **environmental impact report**.[50]

These variety of state requirements are yielding voluminous data not just upon impacts of individual projects, but also to elucidate scientific areas that had not been sufficiently researched. For example, in a seemingly routine *Environmental Impact Report* for the city of Monterey, California, information came to light that led to the official federal endangered species listing of Hickman's potentilla, a rare coastal wildflower.

20.5   Transboundary application

Environmental threats do not respect national borders. International pollution can have detrimental effects on the atmosphere, oceans, rivers, aquifers, farmland, the weather and biodiversity. Global climate change is transnational. Specific pollution threats include acid rain, radioactive contamination, debris in outer space, stratospheric ozone depletion and toxic oil spills. The Chernobyl disaster, precipitated by a nuclear accident on April 26, 1986, is a stark reminder of the devastating effects of transboundary nuclear pollution.[51]

Environmental protection is inherently a cross-border issue and has led to the creation of transnational regulation via multilateral and bilateral treaties. The United Nations Conference on the Human Environment (UNCHE or Stockholm Conference) held in Stockholm in 1972 and the United Nations Conference on the Environment and Development (UNCED or Rio Summit, Rio Conference, or Earth Summit) held in Rio de Janeiro in 1992 were key in the creation of about 1,000 international instruments that include at least some provisions related to the environment and its protection.[52]

The United Nations Economic Commission for Europe's Convention on Environmental Impact Assessment in a Transboundary Context was negotiated to provide an international legal framework for transboundary EIA.[53]

However, as there is no universal legislature or administration with a comprehensive mandate, most international treaties exist parallel to one another and are further developed without the benefit of consideration being given to potential conflicts with other agreements. There is also the issue of international enforcement.[54] This has led to duplications and failures, in part due to an inability to enforce agreements. An example is the failure of many international fisheries regimes to restrict harvesting practises.[55]

20.6 Criticism

As per Jay *et al.*, EIA is used as a decision aiding tool rather than decision making tool. There is growing dissent about them as their influence on decisions is limited. Improved training for practitioners, guidance on best practice and continuing research have all been proposed.[56]

EIAs have been criticized for excessively limiting their scope in space and time. No accepted procedure exists for determining such boundaries. The boundary refers to 'the spatial and temporal boundary of the proposal's effects'. This boundary is determined by the applicant and the lead assessor, but in practice, almost all EIAs address only direct and immediate on-site effects.[57]

Development causes both direct and indirect effects. Consumption of goods and services, production, use and disposal of building materials and machinery, additional land use for activities of manufacturing and services, mining and refining, etc., all have environmental impacts. The indirect effects of development can be much higher than the direct effects examined by an EIA. Proposals such as airports or shipyards cause wide-ranging national and international effects, which should be covered in EIAs.[58]

Broadening the scope of EIA can benefit the conservation of threatened species. Instead of concentrating on the project site, some EIAs employed a habitat-based approach that focused on much broader relationships among humans and the environment. As a result, alternatives that reduce the negative effects to the population of whole species, rather than local subpopulations, can be assessed.[59]

Thissen and Agusdinata [60] have argued that little atten-

tion is given to the systematic identification and assessment of uncertainties in environmental studies which is critical in situations where uncertainty cannot be easily reduced by doing more research. In line with this, Maier et al.[61] have concluded on the need to consider uncertainty at all stages of the decision-making process. In such a way decisions can be made with confidence or known uncertainty. These proposals are justified on data that shows that environmental assessments fail to predict accurately the impacts observed. Tenney et al.[62] and Wood et al.[63] have reported evidence of the intrinsic uncertainty attached to EIAs predictions from a number of case studies worldwide. The gathered evidence consisted of comparisons between predictions in EIAs and the impacts measured during, or following project implementation. In explaining this trend, Tenney et al.[62] have highlighted major causes such as project changes, modelling errors, errors in data and assumptions taken and bias introduced by people in the projects analyzed. Cardenas [64] provides a comprehensive review on the issues of uncertainty in environmental impact assessments.

20.7 See also

- Environmental good

- Environmental impact design

- Environmental indicator

- Environmental policy of the European Union

- Equator Principles

- Healthy development measurement tool

- Hydropower Sustainability Assessment Protocol

- Leopold matrix

- List of international environmental agreements

- Natural landscape

- Phase I Environmental Site Assessment

- Social Impact Assessment

- Strategic Environmental Assessment

- United Nations Environment Programme

- Sustainability appraisal

20.8 References

[1] "Principle of Environmental Impact Assessment Best Practice." International Association for Impact Assessment. 1999.

[2] Holder, J., (2004), Environmental Assessment: The Regulation of Decision Making, Oxford University Press, New York; For a comparative discussion of the elements of various domestic EIA systems, see Christopher Wood Environmental Impact Assessment: A Comparative Review (2 ed, Prentice Hall, Harlow, 2002).

[3] "EIA engineering and consulting companies". Digiscend.com. 2015. Retrieved 2015-08-01.

[4] Clark & Canter 1997, p. 199.

[5] Rychlak & Case 2010, p. 111-120.

[6] Kershner 2011.

[7] Daniel, S., Tsoulfas, G., Pappis, C., & Rachaniotis, N. (2004) Aggregating and evaluating the results of different Environmental Impact Assessment methods Ecological indicators 4:125-138

[8] Hitzschky, K., & Silviera, J. (2009) A proposed impact assessment method for genetically modified plants (As-GMP method) Environmental Impact Assessment review 29: 348-368

[9] Peche, R., & Rodriguez, E., (2009) Environmental impact Assessment procedure: A new approach based on Fuzzy logic Environmental Impact Assessment review 29:275-283

[10] Duarte O. (2000) Técnicas Difusas en la Evaluación de Impacto Ambiental. Ph.D. Thesis, Universidad de Granada

[11] Wilson, L., (1998), A Practical Method for Environmental Impact Assessment Audits Environ Impact Assess Rev 18: 59-71

[12] Elliott, M. & Thomas, I. (2009), "Environment Impact Assessment in Australia: Theory and Practice, 5th Edn, Federation Press, Sydney"

[13] *The Environment Protection and Biodiversity Conservation Act*, Australia: The Department of the Environment, Water, Heritage and the Arts, retrieved 9 September 2010

[14] The Northern Territory Government, viewed 10 September 2010,

[15] The Environment Defenders, viewed 10 September 2010, EDO factsheet

[16] The Law Handbook, viewed 9 September 2010,

[17] The Government of Western Australia, viewed 9 September 2010,

[18] SCC 1992.

[19] Roger, Cotton; Emond, D. Paul (1981). John Swaigen, ed. *Environmental Impact Assessment*. Environmental Rights in Canada. Toronto, Ontario: Butterworths.

[20] Jeffery, Michael I. (1989). *Environmental Approvals in Canada*. Toronto, Ontario: Butterworths.

[21] Emond, D. P. (1978). *Environmental Assessment Law in Canada*. Toronto, Ontario: Emond-Montgomery Ltd.

[22] La Forest,, Gérard V. (1973). *Water Law in Canada*. Ottawa: Information Canada.

[23] {{{year}}} SCC {{{num}}} (18 July 2012)

[24] 2012 SCC {{{num}}}

[25] {{{year}}} SCC {{{num}}}

[26] CEAA (2012). "What is the Canadian Environmental Assessment Act, 2012?". Basics of Environmental Assessment. Canadian Environmental Assessment Agency.

[27] CEAA 2012.

[28] Saxe, =Dianne (9 August 2012). "Federal environmental assessments will be rushed". Environmental Law and Litigation.

[29] Saxe 2012.

[30] Host:Allan McFee (9 May 1977). "The Berger Report is released". *As It Happens*http://ms.radio-canada.ca/archives_new/2002/en/wma/berger19770509er1.wma |transcripturl= missing title (help). CBC Radio. CBC Radio 1. Retrieved 9 December 2011.

[31] "Statement by Official Opposition Environment critic Megan Leslie on Earth Day". NDP. 22 April 2013.

[32] Megan Leslie ({{{year}}}). *Question period*

[33] Wang, Alex (2007-02-05). "Environmental protection in China: the role of law".

[34] Gu, Lin (2005-09-29). "China Improves Enforcement of Environmental Laws". China Features.

[35] "About MSEA". EEAA. Retrieved 2013-01-03.

[36] Watson, Michael (November 13–15, 2003). "Environmental Impact Assessment and European Community Law". XIV International Conference "Danube-River of Cooperation".

[37] Council Directive 85/337/EEC on the Assessment of the Effects of Certain Public and Private Projects on the Environment (1985-06-27) from Eur-Lex

[38] DIRECTIVE 2011/92/EU OF THE EUROPEAN PARLIAMENT AND OF THE COUNCIL of 13 December 2011 on the assessment of the effects of certain public and private projects on the environment from Eur-Lex

[39] Directive 2001/42/EC of the European Parliament and of the Council (2001-06-27) from Eur-Lex

[40] List of criteria for EIA/SEA

[41] "LCQ1: Judicial review case regarding the Environmental Impact Assessment reports of the Hong Kong-Zhuhai-Macao Bridge". *Press Releases*. www.info.gov.hk. October 26, 2011. Retrieved October 27, 2011.

[42] Shibani Ghosh, "Demystifying the Environmental Clearance Process in India", *NUJS LAW REVIEW*, January 2, 2015

[43] http://www.doe.gov.my/portal/wp-content/uploads/2010/07/Appendix_1.pdf

[44] Department of Environmental Protection, Russia

[45] United States. National Environmental Policy Act, P.L. 91-190, 83 Stat. 852, 42 U.S.C. § 4321 *et seq.* Approved January 1, 1970.

[46] U.S. Council on Environmental Quality. "NEPA and Agency Planning." *Code of Federal Regulations*, 40 C.F.R. 1501.

[47] Eccleston, Charles; Doub, J. Peyton (2012). *Preparing NEPA Environmental Assessments: A User's Guide to Best Professional Practices*. CRC Press. ISBN 9781439808825.

[48] *Sierra Club v. United States Army Corps of Engineers*, 701 F.2d 1011, 18 ERC 1748. (2d Cir., 02/25/1983)

[49] Ritter, John (2003-06-03). "Lawsuit pits risks and roads". USA Today.

[50] "Sive,D. & Chertok,M., "Little NEPAs" and Environmental Impact Assessment Procedures" (PDF). Retrieved 2013-01-03.

[51] Sands, P., (1989), The Environment, Community and International Law, Harvard International Law Lournal, 393, p402

[52] Weiss, E., (1999), Understanding Compliance with International Environmental Agreements: The Bakers Dozen Myths, Univ Richmond L.R. 32, 1555

[53] "Convention on Environmental Impact Assessment in a Transboundary Context (Espoo, 1991)". Unece.org. Retrieved 2013-01-03.

[54] Wolfrum, R., & Matz, N., (2003), Conflicts in International Environmental Law, Max-Planck-Institut für Ausländisches Öffentliches Recht und Völkerrech

[55] Young, O., (1999), The Effectiveness of International Environmental Regimes, MIT Press

[56] Jay, S.; Jones, C.; Slinn, P.; Wood, C. (2007). "Environmental Impact Assessment: Retrospect and Prospect". *Environmental Impact Assessment Review* (Elsevier) **27** (4): 289–300. doi:10.1016/j.eiar.2006.12.001.

[57] Lenzen, M.; Murray, S.; Korte, B.; Dey, C. (2003). "Environmental impact assessment including indirect effects—a case study using input-output analysis". *Environmental Impact Assessment Review* (Elsevier) **23** (3): 263–282. doi:10.1016/S0195-9255(02)00104-X.

[58] Shepherd, A.; Ortolano, L. (1996). "Strategic environmental assessment for sustainable urban development". *Environmental Impact Assessment Review* (Elsevier) **16** (4-6): 321–335. doi:10.1016/S0195-9255(96)00071-6.

[59] Fernandes, João P. (2000). "EIA procedure, Landscape ecology and conservation management—Evaluation of alternatives in a highway EIA process". *Environmental Impact Assessment Review* (Elsevier) **20** (6): 665–680. doi:10.1016/S0195-9255(00)00060-3.

[60] Thissen, WIH; Agusdinata, DB (2008). "Handling deep uncertainties in impact assessment". *In proceedings of the 28th annual conference of the IAIA May 5–9, 2008, Perth, Australia*. (IAIA).

[61] Maier, HR; Ascough, JC; Wattenbach, M; Renschler, CS; Labiosa, WB (2008). "Uncertainty in Environmental Decision Making: Issues, Challenges and Future Directions". *Publications from USDA-ARS/UNL Faculty. Paper 399*. (USDA-ARS/UNL Faculty).

[62] Tenney, A; Kværner, J; Gjerstad, KI (2006). "Uncertainty in environmental impact assessment predictions: the need for better communication and more transparency". *Impact Assessment and Project Appraisal* **24** (1): 45–56.

[63] Wood, C; , Dipper, B; Jones, C (2000). "Auditing the assessments of the environmental impacts of planning projects". *Journal of Environmental Planning and Management* **43** (1): 23–47.

[64] Cardenas, IC. "Coping with uncertainty in environmental impact assessments. A multi-field synthesis of literature". *Under review by a refereed journal*.

20.8.1 Sources

- Clark, Ray; Canter, Larry, eds. (1997). *Environmental Policy and NEPA: Past, Present and Future*. Boca Raton, Florida: St. Lucie Press.

- Rychlak, Ronald J.; Case, David W. (2010). *Environmental Law: Oceana's Legal Almanac Series*. New York: Oxford University Press. pp. 111–120.

- Kershner, Jim (27 August 2011). "NEPA, the National Environmental Policy Act" (9903).

- *Friends of the Oldman River Society v. Canada (Minister of Transport)* 1992 CanLII 110, [1992] 1 SCR 3 (23 January 1992), Supreme Court of Canada (Canada)

20.9 Notes

[1] The United States Environmental Protection Agency was established on 2 December 1970, in response to elevated concern about environmental pollution.

20.10 Further reading

- Carroll, B. and Turpin T. (2009). *Environmental impact assessment handbook,* 2nd ed. Thomas Telford Ltd, ISBN 978-0-7277-3509-6

- Glasson, J; Therivel, R; Chadwick A. (2005). *Introduction to Environmental Impact Assessment.* London: Routledge

- Hanna, K. (2009). *Environmental Impact Assessment: Practice and Participation,* 2nd ed. Oxford

- Petts, J. (ed.), *Handbook of Environmental Impact Assessment,* Vols 1 & 2. Oxford, UK: Blackwell. ISBN 0-632-04772-0

- Ruddy, T. F.; Hilty, L. M. (2008). "Impact assessment and policy learning in the European Commission". *Environmental Impact Assessment Review* **28** (2–3): 90. doi:10.1016/j.eiar.2007.05.001.

20.11 External links

- European Commission - EIA website

- European Commission-funded project on Impact Assessment Tools

- Environmental Impact Assessment at the University of Sydney

- International Association for Impact Assessment (IAIA)

- Netherlands Commission for Environmental Assessment

- UNU Open Educational Resource on EIA: A Course Module, Wiki and Instructional Guide

- ELM EIA Law Matrix ~ Environmental Law Alliance Worldwide

-

Chapter 21

Erosion control

Terraces, conservation tillage, and conservation buffers save soil and improve water quality on this Iowa farm.

Erosion control is the practice of preventing or controlling wind or water erosion in agriculture, land development, coastal areas, river banks and construction. Effective erosion controls are important techniques in preventing water pollution, soil loss, wildlife habitat loss and human property loss.

Hydroseeding in UK

21.1 Usage

Erosion controls are used in natural areas, agricultural settings or urban environments. In urban areas erosion controls are often part of stormwater runoff management programs required by local governments. The controls often involve the creation of a physical barrier, such as vegetation or rock, to absorb some of the energy of the wind or water that is causing the erosion. On construction sites they are often implemented in conjunction with sediment controls such as sediment basins and silt fences.

Bank erosion is a natural process: without it, rivers would not meander and change course. However, land management patterns that change the hydrograph and/or vegetation cover can act to increase or decrease channel migration rates. In many places, whether or not the banks are unstable due to human activities, people try to keep a river in a single place. This can be done for environmental reclamation or to prevent a river from changing course into land that is being used by people. One way that this is done is by placing riprap or gabions along the bank.

21.2 Examples

Examples of erosion control methods include:

- buffer strip
- cellular confinement systems [1]
- crop rotation
- conservation tillage
- contour bunding
- contour plowing
- cover crops
- fiber rolls
- gabions
- hydroseeding
- level spreaders
- mulching
- perennial crops
- plasticulture
- polyacrylamide (as a coagulant)
- reforestation
- riparian strip
- riprap
- strip farming
- sand fence
- vegetated waterway (bioswale)
- terracing
- wattle (construction)
- windbreaks [2]

21.3 Mathematical modeling

Since the 1920s and 1930s[3] scientists have been creating mathematical models for understanding the mechanisms of soil erosion and resulting sediment surface runoff, including an early paper by Albert Einstein applying Baer's law.[4] These models have addressed both gully and sheet erosion. Earliest models were a simple set of linked equations which could be employed by manual calculation. By the 1970s the models had expanded to complex computer models addressing nonpoint source pollution with thousands of lines of computer code.[5] The more complex models were able to address nuances in micrometerology, soil particle size distributions and micro-terrain variation.

21.4 See also

- Bridge scour
- Burned area emergency response
- Certified Professional in Erosion and Sediment Control
- Coastal management
- Dust Bowl
- Natural Resources Conservation Service (United States)
- Universal Soil Loss Equation
- Vetiver System

21.5 Notes

[1] State of California Department of Transportation, Division of Environmental Analysis, Stormwater Program. Sacramento, CA."Cellular Confinement System Research." 2006.

[2] Tennessee Department of Environment and Conservation. Nashville, TN."Tennessee Erosion and Sediment Control Handbook." 2002.

[3] Robert E. Horton. 1933

[4] Albert Einstein. 1926

[5] C. Michael Hogan, Leda Patmore, Gary Latshaw, Harry Seidman et al. 1973

21.6 References

- Albert Einstein. 1926. *Die Ursache der Mäanderbildung der Flußläufe und des sogenannten Baerschen Gesetzes*, Die Naturwissenschaften, 11, S. 223–224

- C. Michael Hogan, Leda Patmore, Gary Latshaw, Harry Seidman et al. 1973. *Computer modeling of pesticide transport in soil for five instrumented watersheds*, U.S. Environmental Protection Agency Southeast Water laboratory, Athens, Ga. by ESL Inc., Sunnyvale, California

- Robert E. Horton. 1933. *The Horton Papers*

- U.S. Natural Resources Conservation Service (NRCS). Washington, DC. "National Conservation Practice Standards." National Handbook of Conservation Practices. Accessed 2009-03-28.

21.7   External links

- "Saving Runaway Farm Land", November 1930, Popular Mechanics One of the first articles on the problem of soil erosion control

- Erosion Control Technology Council - a trade organization that mission is to educate and standardize the erosion control industry

- International Erosion Control Association - Professional Association, Publications, Training

- WatchYourDirt.com - Erosion Control Educational Video Resource

- Soil Bioengineering and Biotechnical Slope Stabilization - Erosion Control subsection of a website on Riparian Habitat Restoration

- Vetiver Network International - Soil and water management method

- Groenkreatief Markvoort Holten - Groenkreatief Markvoort Holten, Netherlands

Chapter 22

Grassed waterway

Grassed waterway in Velm, Belgium, during a sunny day

A **grassed waterway** consists in a 2-metre (6.6 ft) to 48-metre-wide (157 ft) native grassland strip of green belt. It is generally installed in the thalweg, the deepest continuous line along a valley or watercourse, of a cultivated dry valley in order to control erosion. A study carried out on a grassed waterway during 8 years in Bavaria showed that it can lead to several other types of positive impacts, e.g. on biodiversity.[1]

Grassed waterway in Velm, Belgium, after a thunderstorm

22.2 Runoff and erosion mitigation

Runoff generated on cropland during storms or long winter rains concentrates in the thalweg where it can lead to rill or gully erosion.

Rills and gullies further concentrate runoff and speed up its transfer, which can worsen damage occurring downstream. This can result in a muddy flood.

In this context, a grassed waterway allows increasing soil cohesion and roughness. It also prevents the formation of rills and gullies. Furthermore, it can slow down runoff and allow its re-infiltration during long winter rains. In contrast,

22.1 Distinctions

Confusion between "grassed waterways" and "vegetative filter strips" should be avoided. The latter are generally narrower (only a few metres wide) and rather installed along rivers as well as along or within cultivated fields. However, buffer strip can be a synonym, with shrubs and trees added to the plant component, as does a riparian zone.

its infiltration capacity is generally not sufficient to reinfil-
trate runoff produced by heavy spring and summer storms.
It can therefore be useful to combine it with extra measures,
like the installation of earthen dams across the grassed waterway, in order to buffer runoff temporarily.[2]

22.3   External links

- (Dutch) Water Agency of the Melsterbeek river, in
  Belgium (agency that had a pioneer role to implement
  erosion control measures in central Belgium)

- (English) PhD thesis on muddy flood problems and solutions in central Belgium

22.4   References

[1] Fiener P., Auerswald K. (2003). Concept and effects of a
    multi-purpose grassed waterway. Soil Use and Management
    19, 65-72.

[2] Evrard, O., Vandaele, K., van Wesemael, B., Bielders, C.L,
    2008. A grassed waterway and earthen dams to control
    muddy floods from a cultivated catchment of the Belgian
    loess belt. Geomorphology 100, 419-428.

Chapter 23

Green furniture

Green Furniture, often symbolized by a tree, are products that use materials from sustainable forests, have low toxic material levels, locally manufactured and are durable enough to last. It should lend itself to easy repair, disassembly, and recycling. Products certified by MBDC's C2C (Cradle 2 Cradle) product regimen are a perfect example, like certified office chairs from Herman Miller, Mebelluks (Russia) and Steelcase. These product can be easily taken apart, sorted into their constituent parts, and recycled at the end of their useful lives.

Chapter 24

Groundwater recharge

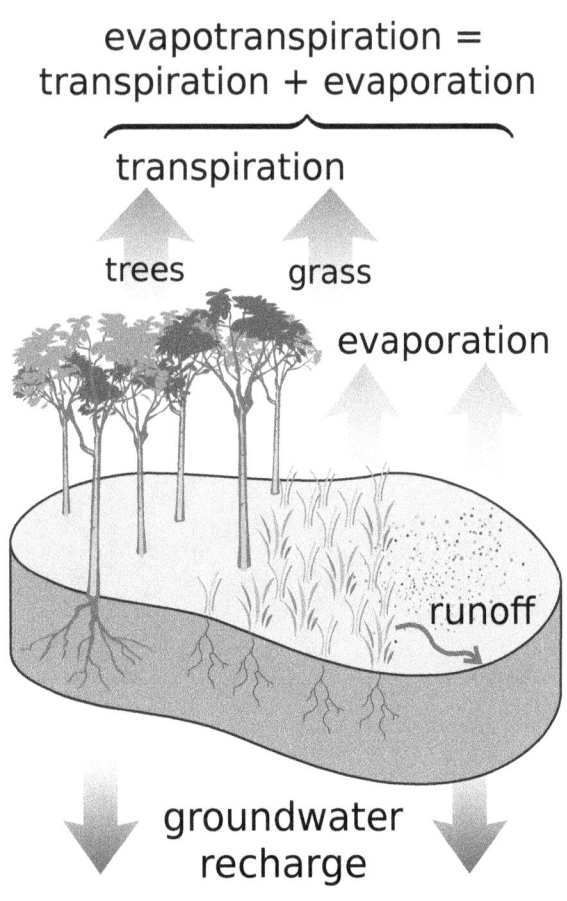

evapotranspiration =
transpiration + evaporation

transpiration

trees   grass

evaporation

runoff

groundwater
recharge

Water balance

Groundwater recharge or **deep drainage** or **deep percolation** is a hydrologic process where water moves downward from surface water to groundwater. This process usually occurs in the vadose zone below plant roots and is often expressed as a flux to the water table surface. Recharge occurs both naturally (through the water cycle) and through anthropogenic processes (i.e., "artificial groundwater recharge"), where rainwater and or reclaimed water is routed to the subsurface.

24.1 Processes

Groundwater is recharged naturally by rain and snow melt and to a smaller extent by surface water (rivers and lakes). Recharge may be impeded somewhat by human activities including paving, development, or logging. These activities can result in loss of topsoil resulting in reduced water infiltration, enhanced surface runoff and reduction in recharge. Use of groundwaters, especially for irrigation, may also lower the water tables. Groundwater recharge is an important process for sustainable groundwater management, since the volume-rate abstracted from an aquifer in the long term should be less than or equal to the volume-rate that is recharged.

Recharge can help move excess salts that accumulate in the root zone to deeper soil layers, or into the groundwater system. Tree roots increase water saturation into groundwater reducing water runoff.[1] Flooding temporarily increases river bed permeability by moving clay soils downstream, and this increases aquifer recharge.[2]

Artificial groundwater recharge is becoming increasingly important in India, where over-pumping of groundwater by farmers has led to underground resources becoming depleted. In 2007, on the recommendations of the International Water Management Institute, the Indian government allocated Rs 1800 crore (US$400million) to fund dug-well recharge projects (a dug-well is a wide, shallow well, often lined with concrete) in 100 districts within seven states where water stored in hard-rock aquifers had been over-exploited. Another environmental issue is the disposal of waste through the water flux such as dairy farms, industrial, and urban runoff.

24.1.1 Wetlands

Wetlands help maintain the level of the water table and exert control on the hydraulic head (O'Brien 1988; Winter 1988). This provides force for groundwater recharge and discharge to other waters as well. The extent of groundwater recharge by a wetland is dependent upon soil, vegetation, site, perimeter to volume ratio, and water table gradient (Carter and Novitzki 1988; Weller 1981). Groundwater recharge occurs through mineral soils found primarily around the edges of wetlands (Verry and Timmons 1982) The soil under most wetlands is relatively impermeable. A high perimeter to volume ratio, such as in small wetlands, means that the surface area through which water can infiltrate into the groundwater is high (Weller 1981). Groundwater recharge is typical in small wetlands such as prairie potholes, which can contribute significantly to recharge of regional groundwater resources (Weller 1981). Researchers have discovered groundwater recharge of up to 20% of wetland volume per season (Weller 1981).

24.2 Estimation methods

Rates of groundwater recharge are difficult to quantify, since other related processes, such as evaporation, transpiration (or evapotranspiration) and infiltration processes must first be measured or estimated to determine the balance.

24.2.1 Physical

Physical methods use the principles of soil physics to estimate recharge. The *direct* physical methods are those that attempt to actually measure the volume of water passing below the root zone. *Indirect* physical methods rely on the measurement or estimation of soil physical parameters, which along with soil physical principles, can be used to estimate the potential or actual recharge. After months without rain the level of the rivers under humid climate is low and represents solely drained groundwater. Thus the recharge can be calculated from this base flow if the catchment area is known.

24.2.2 Chemical

Chemical methods utilize the presence of relatively inert water-soluble substances, such as an isotopic tracer or chloride,[3] moving through the soil, as deep drainage occurs.

24.2.3 Numerical models

Recharge can be estimated using numerical methods, using such codes as HELP, UNSAT-H, SHAW, WEAP, and MIKE SHE. The 1D-program HYDRUS1D is available online. These codes generally use climate and soil data to ar-

rive at a recharge estimate, and use Richards equation in some form to model groundwater flow in the vadose zone.

24.3 See also

- Aquifer storage and recovery
- Contour trenching
- Depression focused recharge
- Drainage
- Infiltration (hydrology)
- Hydrology (agriculture)
- Soil salinity control by subsurface drainage
- Watertable control
- Category: Water conservation

24.4 References

[1] "Urban Trees Enhance Water Infiltration". *Fisher, Madeline*. The American Society of Agronomy. November 17, 2008. Retrieved October 31, 2012.

[2] "Major floods recharge aquifers". University of New South Wales Science. January 24, 2011. Retrieved October 31, 2012.

[3] Allison, G.B.; Hughes, M.W. (1978). "The use of environmental chloride and tritium to estimate total recharge to an unconfined aquifer". *Australian Journal of Soil Research* **16** (2): 181–195. doi:10.1071/SR9780181.

- Allison, G.B.; Gee, G.W.; Tyler, S.W. (1994). "Vadose-zone techniques for estimating groundwater recharge in arid and semiarid regions". *Soil Science Society of America Journal* **58**: 6–14. doi:10.2136/sssaj1994.03615995005800010002x. OSTI 7113326.

- Bond, W.J. (1998). *Soil Physical Methods for Estimating Recharge*. Melbourne: CSIRO Publishing.

24.5 Further reading

- LaMoreaux, Philip E., & Tanner, Judy T, ed. (2001). *Springs and bottled water of the world: Ancient history, source, occurrence, quality and use*. Berlin, Heidelberg, New York: Springer-Verlag. ISBN 3-540-61841-4. Retrieved 13 July 2010. Provides a good overview of hydrogeological processes, including groundwater recharge.

- Pierre D. Glynn & L. Niel Plummer (March 2005).   "Geochemistry and the understanding of ground-water systems" (PDF). *Hydrogeology Journal* **13** (1): 263–287.   Bibcode:2005HydJ...13..263G. doi:10.1007/s10040-004-0429-y.   Retrieved 4 July 2010.

Chapter 25

Hydropower Sustainability Assessment Protocol

The **Hydropower Sustainability Assessment Protocol** is a tool that promotes and guides more sustainable hydropower projects. It is a methodology used to measure the performance of a hydropower project across more than twenty environmental, social, technical and economic topics.

Hydropower projects can have both a positive and a negative environmental and social impact. This is because the construction of a dam, power plant and reservoir creates certain social and physical changes in the area it effects. The protocol provides a common language to allow governments, civil society, financial institutions and the hydropower sector to talk about and evaluate sustainability issues.

Assessments are based on objective evidence and the results are presented in a standardised way, to show how new projects are being developed or how existing facilities are performing.

It has been designed to work on projects and facilities anywhere in the world.[1][2]

25.1 Application

25.1.1 Purpose

The Protocol is used by different hydropower stakeholders for different reasons.[3]

Popular uses include:

- Independent review of sustainability issues

- Guiding improvement of sustainability practice

- Comparison with international best practice

- Communication with stakeholders

- Facilitating access to finance

- Preparing clients to meet bank requirements

- Reducing investment risk

25.1.2 Users

Crédit Agricole, Societe Generale, Standard Chartered, Citi, and UBS now refer to the Protocol in their sector guidance.[4][5][6][7][8]

The World Bank has analysed the value of the Protocol for use by their clients, concluding that it is a useful tool for guiding the development of sustainable hydropower in developing countries.[9]

The International Institute for Environment and Development has reviewed social and environmental safeguards for large dam projects, concluding that the Protocol currently offers the best available 'measuring stick' for the World Commission on Dams provisions.[10]

25.1.3 Process

A Protocol assessment takes place over a one week period at the project site and provides a rapid sustainability check.

A Protocol assessment does not replace an environmental and social impact assessment (ESIA), which takes place over a much longer period of time as a mandatory regulatory requirement. A Protocol assessment will, amongst other things, check the scope and quality of the ESIA which has been done.

To ensure high quality, all commercial use of the Protocol is carried out by accredited assessors. These assessors have significant experience of the hydropower sector or relevant sustainability issues, and have passed a rigorous accreditation course.[11]

25.2 Scope

25.2.1 Tools

The Protocol can be used at any stage of hydropower development, from the early planning stages through to operation. Each project stage is assessed using a different tool:

- The early stage tool, a screening tool for potential hydropower projects

- The preparation tool, which covers planning and design, management plans and commitments.

- The implementation tool, used through the construction phase.

- The operation tool, used on working projects.

25.2.2 Topics

The Protocol covers a range of topics that need to be understood to assess the overall sustainability of a hydropower project.

The name and range of topics changes slightly at different project stages.

Social aspects

- Communications and consultation
- Project benefits
- Project affected communities and livelihoods
- Resettlement
- Indigenous peoples
- Labour and working conditions
- Public health
- Cultural heritage

Environmental aspects

- Environmental and social assessment and management
- Biodiversity and invasive species
- Erosion and sedimentation
- Water quality
- Waste, noise and air quality

- Reservoir planning / preparation and filling / management
- Downstream flow regimes

Business aspects

- Governance
- Procurement
- Integrated project management
- Financial viability
- Economic viability

Technical aspects

- Demonstrated need and strategic fit
- Siting and design
- Hydrological resource
- Asset reliability and efficiency
- Infrastructure safety

The Protocol also includes 'cross-cutting issues' such as climate change, gender and human rights, which feature in multiple topics.

25.2.3 Criteria

For each sustainability topic, performance is assessed against a range of criteria at two levels: basic good practice and proven best practice.

Table 2: Criteria requirement at different levels.

25.3 History

A multi-stakeholder forum developed the Protocol between 2008 and 2010.[12]

The following key group were represented: social and environmental NGOs, governments of developed and developing countries, financial institutions, development banks, and the hydropower industry.

The forum jointly reviewed, enhanced and built consensus on what a sustainable hydropower project should look like.

Policies taken into account included the World Commission on Dams' Criteria and Guidelines, World Bank Safeguard

Policies, IFC Performance Standards, and the Equator Principles.

A draft of the Protocol was released in 2009, which was trialled in 16 countries across six continents and subjected to further consultation involving 1,933 individual stakeholders from 28 countries.

The final version was produced in 2010.[13]

The diversity of the forum was important to ensure that the Protocol became globally applicable and universally accepted. Diversity also ensured that the multiple perspectives and stakeholder interests surrounding a hydropower project were incorporated into the document.

25.4 Governance

The Protocol is governed by a multi-stakeholder body, the Hydropower Sustainability Assessment Council (HSA Council).[14]

The mission of the Council is to ensure multi-stakeholder input and confidence in the Protocol's content and application.

All individuals and organisations engaged in hydropower are welcome and encouraged to join the Council. This approach to governance ensures that all stakeholder voices are heard in the shaping of the use of the Protocol and its future development.

The Council consists of a series of Chambers, each representing a different segment of hydropower stakeholders. Each chamber elects a chair and alternate chair for a two-year term. The chamber chairs come together regularly to form the decision-making Protocol Governance Committee. The current Governance Committee (elected May 2015) is shown in the table below.

25.5 References

[1] "Hydropower Sustainability - Home". *www.hydrosustainability.org*. Retrieved 2015-10-01.

[2] "Hydropower Sustainability - Protocol Basics". *www.hydrosustainability.org*. Retrieved 2015 10 01.

[3] "Hydropower Sustainability - How it can help". *www.hydrosustainability.org*. Retrieved 2015-10-01.

[4] http://mediacommun.ca-cib.com/sitegenic/medias/DOC/13870/2012-12-politique-sectorielle-rse-energie-hydroelectricite-eng.pdf

[5] http://www.societegenerale.com/sites/default/files/documents/Document%20RSE/Finance%20responsable/Dams%20&%20Hydropower%20Sector%20Policy.pdf

[6] https://www.sc.com/en/resources/global-en/pdf/sustainabilty/Dams_and_Hydropower_Position_Statement.pdf

[7] http://www.citigroup.com/citi/environment/data/1160840_Sector_Brief_HydroPower.pdf

[8] https://www.ubs.com/content/dam/ubs/global/about_ubs/corporate_responsibility/UBS-ESR-framework.pdf

[9] "The Protocol for World Bank clients". *documents.worldbank.org*. Retrieved 2015-10-01.

[10] "A review of social and environmental safeguards for large dam projects". *pubs.iied.org*. Retrieved 2015-10-01.

[11] "Hydropower Sustainability - Accredited Assessors". *www.hydrosustainability.org*. Retrieved 2015-10-01.

[12] "Hydropower Sustainability - History". *www.hydrosustainability.org*. Retrieved 2015-10-01.

[13] "Hydropower Sustainability - Download Protocol Document". *www.hydrosustainability.org*. Retrieved 2015-09-30.

[14] http://www.hydrosustainability.org/Governance.aspx

Chapter 26

Integrated Modification Methodology

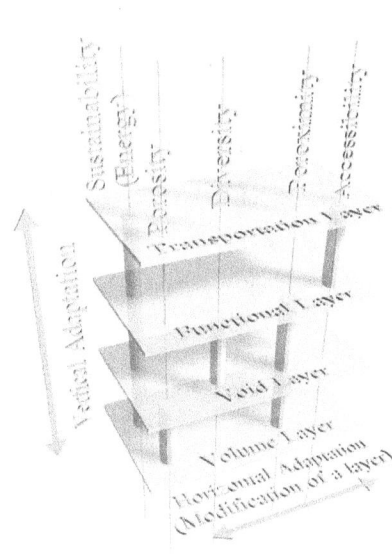

A complex adaptive system (CAS) is composed of four or more subsystems; it is considered to be a superposition of products of the sub-systems' states. Superimposition is a process of integration of two or more sub-systems. Once the subsystems interact, their states are no longer independent.

Integrated Modification Methodology (IMM)[1] is a design methodology based on a specific process with the main goal of improving the urban energy performance, through the modification of its constituents and optimization of the architecture of their ligands. According to this view, the city, considered as a complex adaptive system (CAS),[2] is not solely a mere aggregation of disconnected energy consumers and the total energy consumption of the city is different from the sum of all of the buildings' consumption. This considerable gap between the total energy consumption of the city and the sum of all consumers is concealed from the urban morphology and urban form of the city. IMM is a multi-stage, iterative process, applied to urban components, for improving their environmental and energy performances, is fundamentally holistic,

multi-layer, multi-scale; it investigates the relationships between urban morphology[3] and energy consumption by focusing mostly on the 'subsystems' characterized by physical characters and arrangement. In this methodology, a city consists of the superimposition of an enormous number of interrelated components, categorized in different layers or 'subsystems', which through their inner arrangement and the architecture of their ligands provide a certain physical and provisional arrangement. The constituents of the CAS adapt themselves to react to the newly imposed constraints, in order to improve upon the entire system's performance. The complex adaptive system is composed of heterogeneous elements, linked together either directly or indirectly, and the final system performance emerges from all of the elements as a whole. This adaptation occurs within or on members of a single subsystem, known as horizontal adaptation, and between the different subsystems, termed Vertical Adaptation.[4] In other words, the adaptation of existing members in a subsystem, or horizontal adaptation, as a response to the newly imposed conditions and constraints, changes the subsystem's performance, which will be the cause of the entire system's transformation over time.

26.1 Background

Over half of the greenhouse gas emissions are created in and by cities; the majority of the population lives and works in cities, where up to 80% of energy is consumed. The onward march of the population growth rate has been reaching a dramatic measure and has created a series of questions regarding to the overall sustainability of the ecosystem. In fact, this unrestrained trend, which may lead the actual world population to 9.2 billion by 2050, and to 13.2 billion people by 2080, is an urban succession, which impacts directly or indirectly on other phenomena, such as:

- Urbanization growth.

- Planet's deforestation with loss of wildlife habitat, as well as other natural resources.

- Progressive expansion of the land occupation, for agricultural and dwelling purposes.

- CO_2 emissions increment.

- Air, Land and water quality deterioration.

In this scenario, it is clear how urban areas, as well as their urban design, play a key role in the definition of a long-term strategy for a sustainable development, despite other ephemeral remedies. Reconsidering the location where cities should be located and designed could reduce the CO_2 emissions and energy demand, accordingly. In fact, the ultimate goal of our methodology is to identify useful principles and tools, aiming to direct the increasing world urbanization towards more sustainable long-term models, which are characterized by better energy performance and, consequently, better balance that would be achieved between the available resources and the required consumption The enthralling report of the US Energy Information Administration reveals that as the world energy consumption increased from 472 quadrillion Btu in 2006, it will increase further to 552 quadrillion Btu in 2015, and 678 quadrillion Btu in 2030, which a total increase of 44 per cent over the projection period. Also in all EU Member States, gross inland consumption of primary energy had increased throughout the period from 1999 to 2009, except for the United Kingdom (-10.1%). Additionally, the World Bank report illustrates that the modern patterns of city growth are increasingly land-intensive. "Average urban densities (that is, the number of inhabitants per square kilometre of built-up area) have been declining for the past two centuries. As transportation continues to improve, the tendency is for cities to use up more and more land per person. The built-up area of cities with populations of 100,000 or more presently occupy a total of about 400,000 km2, half of it in the developing world. Cities in developing countries have many more people, but occupy less space per inhabitant. In both developing and industrialized countries, the average density of cities has been declining quickly: at an annual rate of 1.7 per cent over the last decade in developing countries and 2.2 per cent in industrialized countries". The significance impacts of urban form on energy performances of city, as well as emitted pollution, has been illustrated by different researches. The major part of mentioned studies explicates that the reduction of the residential density matches the steady increase in the amount of energy needs. In this scenario, it is clear how urban areas as well as urban design play a key role in the definition of a long term strategy for a sustainable development, despite the other ephemeral remedies: "Although cities embody environmental damage, namely, increasing emissions due to transportation, energy consumption and other factors, policymakers and experts increasingly recognize the potential value of cities for long-term sustainability, after

all, the majority of energy is consumed in cities. Therefore, sustainability is an urban issue". Consequently, new demands have risen, and fundamental questions to which the city has to deal with, such as: How can the city contribute to overall urban sustainability? Can urban design contribute with an appropriate approach to climate mitigation and emission reduction? Is the urban form correlated with these issues? And eventually, how can the urban transformation be performed, in order to achieve a sustainable urban form? And moreover, how can a city address both its competitive status, development and its ecological stewardship? IMM theory considers the city as a complex adaptive system. Furthermore, it sketches out the relationships between urban morphology and energy consumption, providing some new basic design principles to re-shape urban assessment, as well as designing new sustainable neighbourhoods as an integrated part of the city. Morphology plays an essential role for any energy-saving policy, urban efficiency, liveability and, generally, sustainable urban environments to succeed. It is necessary to adopt new principles and new urban design methodologies. One of the main objectives of the research is to find, thanks to a holistic and multidisciplinary approach, new methodologies that can help to shape a better comprehension of the different performances of different urban assessment; then, to apply to the new design principles in order to improve the system's performance. A complex system, to put it in a nutshell, is an arrangement of interconnected heterogeneous elements that, as a whole, shows one or more performances, and the final result of the whole system is utterly different from every individual constituent's performance.

26.2 Theory

As mentioned above, the IMM is a design methodology with the aim to improve the performance of the CAS; the main characteristics of the IMM are based on three fundamental approaches: holistic, multi-layer and multi-scale. The complex adaptive system is composed of heterogeneous elements, linked together either directly or indirectly, and the final system performance emerges from all of the elements as a whole. This adaptation occurs within or on members of a single subsystem, hereafter known as horizontal adaptation, and between the different subsystems, hereafter termed vertical adaptation. In other words, the adaptation of existing members in a subsystem, or horizontal adaptation, as a response to the newly imposed conditions and constraints, changes the subsystem's performance, which will be the cause of the entire system's transformation over time. One can sharpen the performances of the entire complex system, utilizing the adaptive behaviors of the CAS, both horizontal and vertical. The entire complex sys-

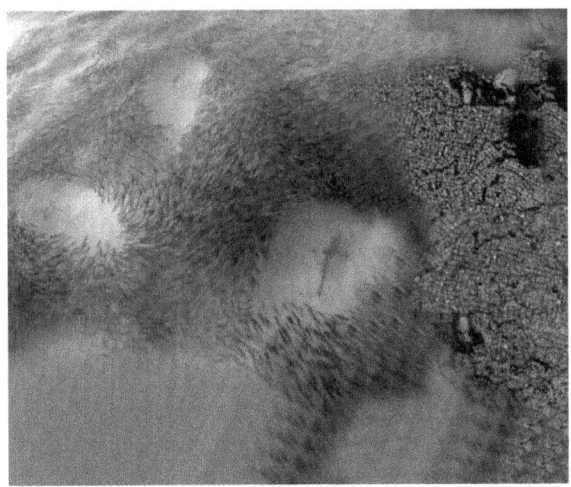

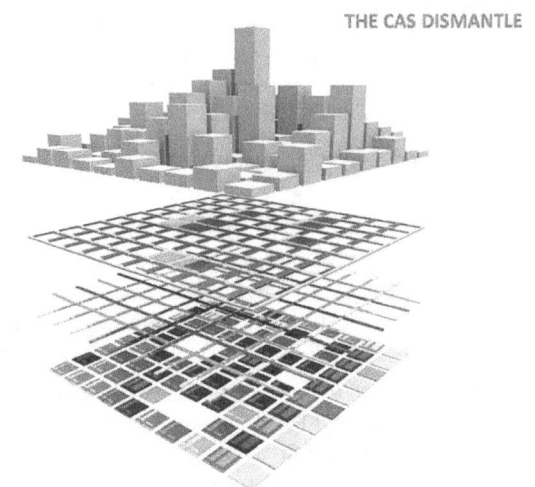

Complex Adaptive System Flocking low : There are many rules that bond the agents and systems together; while every agent follows its own interest, the relationship between agents are defined by these norms. Members constantly adapt and change their functions, behavior and performances to stay under these fundamental laws. These are known as flocking laws. Final system product emerges through these relations and interactions.

Urban subsystem. The CAS's components or subsystems (layers), which affect the urban morphology: Volume Layer; Void Layer; Functional Layer; Transportation Layer

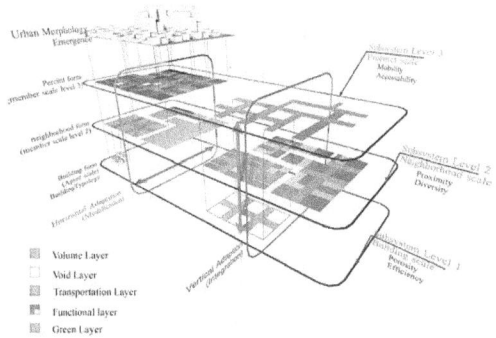

Horizontal and Vertical modification in IMM

tem will be transformed by the mentioned symbiotic adaptive behaviors between the elements and subsystems, modification and integration, over time. By boosting the performance of one subsystem through the assistance of the transformation of another subsystem, one creates a collaborative relation, which ultimately leads to transformation of the complex system in an optimal way. To reiterate, modification happens when the members of one layer are optimized, in order to improve their own layer's performances. On the other hand, integration is a symbiotic relation between different layers, for better performance, which ultimately improves the entire system's performance. Due to the fact that the IMM investigates on the relationships between urban morphology and energy consumption,[5] the theory focus mostly on the 'Subsystems' characterized by physical characters and arrangement:

- Urban Volumes (built-up mass layer)

- Urban Voids (open spaces, streets, etc.)

- Functional (land use layer)

- Transportation and Mobility Layer

They are structurally organized and linked together in a provisional physical structure, outlining a distinctive and specific morphology. Actually it is the architecture of their ligands, which provides a certain physical and provisional arrangement of the CAS every time different. Moreover the CAS is also a single energy entity; accordingly with this

assumption, a more efficient and sustainable urban form emerges through modification of its elements and integration of its subsystems over time. In IMM holistic methodology, the final system performance results from the whole elements; moreover, the city reshapes itself through a dynamic and on-going adaptation process of its constituents. The IMM process highlights the transformation of mid-scale areas, which is a determined area and acts as a bridge between the local scale and global scale. However, the limit of this area has to be mapped, as the intervention and project site by the designers and planners. The main criterion to confine the intervention's border is based on the wide-ranging contextual features, such as morphological aspects, social and functional layers. In IMM theory the modification of CAS elements, which causes the final transformation of the system, occurs in different scales. Equally, the urban interventions are operated in different scales. As the modifications of CAS are classified in the local, intermediate and global scale, any intervention effect has to be considered in the three mentioned scales. The intermediate interventions bridge the gap between local and global scales.[6] Hence, the IMM (Integrated Modification Methodology) acted to transform locally a neighbourhood in order to start a reaction involving the entire urban context and changing the CAS structurally.

26.3 A Phasing process

1	1a	Horizontal Investigation	Dismantling the system to investigate	Actual CAS Arrangement	Investigation/ Observation & Measurement
	1b	Vertical investigation	The actual value of Key Categories		
	1c	Actual performance of the system based on 12 indicators		Actual CAS performances	
2	2a	Detection of the transformation's Horizontal and Vertical *Catalysts and Reactants*.		Catalysts selection and Reactants ordering	Assumption and Interpretation/ Formulation
	2b	Assumption of the 12 IMM Ordering principles		DOP Arrangement	
3	3a	Horizontal Modification	The *catalyst* drives the local transformation; changing the structure of the layers/Ligands	Catalyzers' Modification and chain reaction	Modification Intervention & Design
	3b	Vertical Modification	Local transformation acts globally, changing the entire system's configuration		
4	4a	Performance of the new CAS based on 12 indicators		New CAS performances	Retrofitting
	4b	Local modification/optimization is a process involving again the first level of superimposition for improving locally their performance. Local optimization works using selected tools/features: - Volume/Voids Solar Gain; Wind Tunnel; - Volume/Function = Level of mixed use - Function/Voids - Function distribution - Transports/Voids = Number of intersection - Transports/Function - Service area control. - Transport/Volume= Catchment area control		Local modification of the new CAS	Optimization
	4c	Universal indicators		Comparison	

Different phases of the IMM

The IMM methodology is based on a multi-stage process composed by different but full integrated four phases, respectively:

- Phase 1. Investigation/Analysis.

- Phase 2. Interpretation/Assumption.

- Phase 3. Modification, Transformation.

- Phase 4. Retrofitting and Optimization.

26.3.1 Phase 1. Investigation/Analysis.

This phase investigates the actual configuration of an Urban System (CAS) considered in a provisional state and effects of an endless transformation process. This phase is devoted to investigation of the relationship between urban morphology and energy consumption of the CAS, which involve its own subsystems and their correlation, which affect the urban form as well as energy consumption. Actually the comprehension of the configuration of the involved Subsystems

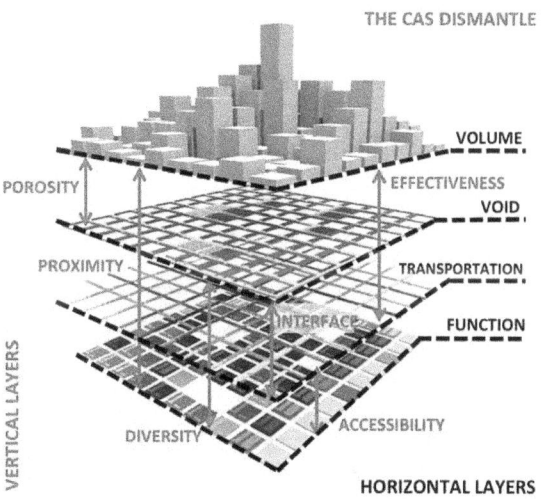

City as Complex Adaptive System. The disassembling process of the CAS's components by Horizontal Investigation. Analyses of physical assessment of the subsystems (layers), which affect the urban morphology: Volume Layer; Void Layer; Functional Layer; Transportation Layer. The CAS's ligands (Key Categories) by Vertical Investigation. The comprehensive configuration of the CAS is mostly described by the correlation between the different subsystems

and their Links play a significant role in the IMM final result. Furthermore the current structure of the system can be considered as just a temporary configuration produced by a preceding process of integration of two or Sub-systems, called Superimposition process. Once the subsystems interact, their states are no longer independent and they start working, depending on the condition, such as Catalyst or Reactants. Through to the investigation phase the designer activates disassembling procedure of the CAS (Horizontal investigation) into its mains physical components or subsystems, such as: Voids, Volumes (Built Spaces), Functions, and Transportations. Each subsystem will be firstly described on its own, in order to describe its individual structure and characteristics respectively on a Morphological, Typological and Technological point of view.

Then the correlations/links or the architecture of the ligands between the subsystems will be analyzed in a more specific way, through a more detailed investigation named Vertical investigation which works through special features named Key Categories are respectively: Porosity, Proximity, Diversity, Interface, Accessibility and Efficiency. The main outcomes of this Investigation's phase are:

- Comprehension of the physical arrangement of the CAS

- Appraisal of the role and value of the Key categories

- Evaluation of the current energy performance of the CAS

DIFFERENT PHASES OF THE IMM INVESTIGATION PROCEDURE
(The formulas are still under the evaluation by the authors)

Horizontal Investigation		
Volume	Built volume density, Dwelling density, Human density	$V_l = V_{built} / Area$
Void	Open space area	$V_d = V_{open} / Area$
Function	Job density, Number of legal entities in the intervention area	$F_n = J_{number} / Area$
Transportation	Number of carried out urban trips	N_{tr}

Vertical Investigation Key Categories		
Porosity	Factuality of urban voids [4]	$P_s = cat^{-1} \sum [1 - (n_i x_i A^i)]^2$
Proximity	Number of key functions within walking distance area from the dwellings	$P_x = N_f N_{dw}^{-1}$
Diversity	Diversity of subdivision use [4]	$D_s = cat^{-1} \sum [1 - (S_i / S_i^{bj})]^2$
Interface	Cycomatic complexity of pedestrian [4] (L: number of Links, Number of Nodes)	$\mu = L - N + 1$
Accessibility	Number of available jobs reachable in 20 min, Number of available public transportation mode in the area	N_{Acc}
Efficiency	The number of public transportation trips and the total number of trips	$E_f = N_{ptr} N_{tt}^{-1}$

Procedure for the (CAS) system's dismantling

One of the most important goal is this stage is the evaluation of the current energy performance of the CAS. Hence 12 Indicators will be used for achieving this result, and then the same indicators will be used in the CAS Retrofitting process (Step 4a. Second measurement) necessary for the final evaluation of the system performance, after the transformation design process. It is important to emphasize that the 12 Indicators are also connected with a series of design principles, named Design Ordering Principles (DOP), tools used to arrange later the structure of the CAS.

26.3.2   Phase 2. Interpretation/Assumption

The second moment of the IMM process called Interpretation/Assumption is halfway between investigation and design steps and it is essentially dedicated to establish a Supposition/Hypothesis, like a possible way for modifying structurally the CAS in order to achieve its improvement in terms of quality and energy performance. The consideration on how to fulfill the initial intentions and simultaneously to reach the final goal plays the main role in this Assumption phase. As mentioned, the configuration of the CAS emerges through local modification and integration of the system's components; therefore, the effect of the Local modification (on selected layers) plays a great role in the entire system's performances, changing the final CAS global configuration.

First measurement (Actual CAS Energy Performances)		
Compactness	**1. Ground Use:** a) Urban Built density* b) Compactness factor* c) Number of building per hectare d) Weighted Urban Proliferation WUP* index of urban sprawl	BD = $\sum$ Floor of every storey of Building / Ground level surf C = Surface / (Volume) ^ (2/3) N = Ha WUP = UP x w_1 (DIS) - w_2(UD)
	2. Population and energy: a) Consumption per capita. b) Rate of energy coming from renewable sources* c) Renewable energy percentage in transport	KWh year per capita $T_{ENR} = T_{renewable en. prod.} / T_{total en. prod.}$ (%) $T_{ENt} = T_{renewable tr.} / T_{total tr.}$ (%)
Complexity	**3. Walkability:** a) Number of key function in a walking distance from residential buildings b) Car free or minimal car traffic streets c) High quality Street paths	Living within 300 m from key services (in-number of inhabitants or ha) Length km $T_{Length} = T_{HQ Street paths} / T_{total street path}$ (%)
	4. Uses of the space: a) Ratio between numbers of residents and activities* b) Housing diversity* c) Ratio of place dedicated to Innovation and Knowledge*	C = $\sum$ Residing / $\sum$ Activities I.B.L. = $\sum$ Low income housing / $\sum$ Total housing (%) A&K = $\sum$ Num Sur of IK activities / $\sum$ total min. class (%)
	5. Open Spaces: a) Ratio of green open spaces b) Extent and number of parks, Number of trees per ha. c) Surface of and number of Public space, paved (scaled) surfaces	T = $T_{Green urban area}$ / $T_{total ground area}$ (%) N = ha Ha & N
	6. Urban biodiversity: a) Proportion of your municipality is comprised of natural or semi-natural areas. b) Number of native species (plants, animals) d) Number of Park area available to citizens. e) % vegetation cover in the urban core in the overall area	% Number Number %
Connectivity	**7. Cycling:** a) Length of biking roads (km)	Km per capita
	8. Transportation and Mobility: a) Private passenger transport energy use per person. b) Passenger number in public transport. c) Inhabitants leaving within 300m from public transport d) Length of roads per capita. Road Ratio* e) Vehicles distance travelled (VDT)	KWh year per capita Number of travel per years (by each kind) Number, % Km per capita; T_R = road surf/total sur Km per capita; KM
	9. Level of mobility interchange: a) Kinds of public transport available (boat, train, tram, metro, buses, microbuses, …) b) Number of "Park-and-ride" parking places c) Interchange node/hub	Number of vehicles for each category Numbers Number Interchange performance evaluation
Management	**10. Food:** a) Food needed daily. b) Amount of urban farm production per person. c) Extent of municipally organised plots for cultivation	Kg. capita Kg. capita Ha
	11. Waste management: a) Amount of solid waste produced. b) Rate of waste reused* c) Rate of materials coming from re-cycling	Cubic meter per capita T = $M_{volume of reused}$ / $M_{Total Volume of waste}$ T = (%)
	12. Water management : a) Water use per person. b) Water productivity* c) Wastewater purified in a wastewater treatment plant*	Cubic meter per capita $T_{water prod.}$ = ($V_{water needed}$ / $V_{total water}$) % $T_{equivalent}$ = ($V_{water tr.}$ / $V_{water to treat}$) %

Indicators First measurement: A core set of elements based specifically on environmental themes but interlinked with other themes social, i.e., economy. They are tools used in the IMM process for objective measurement for comparing the performance of different systems (External comparison), or the System performance prior to and after the transformation design process (Internal comparison).

In other words, after the Investigation Phase, the IMM process comes out with an idea (Assumption) about a possible Local modification of the chosen subsystem (layer) and Key Category that make possible to act transforming globally the entire system (CAS). The choice of one subsystem (layer) and a ligands (Key Category) as a first driver of the transformation is the main goal of this phase, assigning respectively to the selected Subsystem (Layer) and to the selected Ligand (key Category) the Catalyst role and to the others the Reactants function. The principal outcomes of this Assumption/Interpretation phase are:

- The choice of a the Horizontal and Vertical catalysts as a supposition based on the knowledge obtained by the previous phase and dedicated to explain the CAS

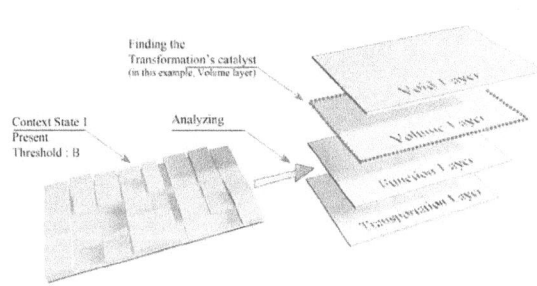

Evaluating the existing system performance, by analyzing the system components, thanks to the Key Categories. In this phase the malfunctioning layer would be the transformation catalyst. In this example, the void layer has indicated as the catalyst; however, it is solely an example it could be any of the layers.

configuration as well as its behavior and performance.

- The assignation to each Subsystem the role of Catalyst or Reactants respectively.

- The assignation to each Key Category the role of Catalyst or Reactants respectively.

- A preliminary control of the Local consequences of the choice.

	DOP Design Ordering Principles.	Key categories	Determinants
Morphology	1. Balance the ground use. 2. Fostering the local energy production; Building as components of Community Energy System.	Porosity Porosity	Compactness
	3. Promote Walkability.	Proximity	
Typology	4. Fostering Mixed used spaces. 5. Makes Biodiversity an important part of urban life.	Diversity	Complexity
	6. Create connected open spaces system, activate urban metabolism.	Interface	
Technology	7. Balancing the public transportation potential	Effectiveness	Connectivity
	8. Promote Cycling and Reinforce the public transportation		
	9. Change from multimodality to inter-modality concept.	Accessibility	
Management	10. Convert the City in a food producer. 11. Prevent the negative impact of waste. 12. Implement water management.		Governance

DOP are tools/instruments used to arrange the structure of the CAS. The application of these principles are applied affects its structure and performance. DOPs are associated with Indicators.

Choosing the Catalyst

The CAS is composed by a hierarchy of multiple levels of organization. Considering that at any particular scale, the system is actually a sub-system, the cross-scale effects have a great significance in the dynamics of CAS. Meanwhile the chosen Catalysts plays a tremendous role in the IMM.

From the selection of one layer as Horizontal Catalyzer and one Key Category, as Vertical Catalyzer the reaction of the system starts, driving the local modification and activating the system's transformation. It is clear that the choice of the catalysts depends on the Investigation Phase. The Catalyzers choice as a first driver of the transformation is the main goal of this phase, assigning respectively to the selected Subsystem and Key Category the Catalyst role and to the others the Reactants one.

The Role of the DOP (Design Ordering Principle)

In this second phase the DOP plays a great role, these tools are used to adjust the structure of the cas and its performance. The application of the DOPs is a fundamental stage of the IMM Phasing process and a way for directing the modification process of the CAS towards a more sustainable and efficient form. It is really important to recall that the 12 DOPs are associated to the 12 Indicators previously used for the estimation of the actual energy performance of the CAS (Data Collection-Step 1c) as well as for the CAS retrofitting process (Step 4a-Second measurement). The DOPs role into the design process is significant for addressing the consequence of the Investigation/Analysis Phase. As main players of the Formulation Phase, they work like active prescriptions, and, if combined, they produce an integrated action towards the final result. The DOPs are respectively:

- Balance the ground use.

- Fostering the local energy production; Building as components of Community Energy System.

- Promote Walkability.

- Fostering Mixed used spaces.

- Make Biodiversity part of urban life.

- Create connected open spaces system and activate urban metabolism.

- Balancing the public transportation potential.

- Promote Cycling and Reinforce the public transportation.

- Change from multimodality to intermodality concept.

- Convert the City in a food producer.

- Prevent the negative impact of waste.

- Implement water management.

As main players of the Formulation Phase they work as active prescriptions, which combined produce an integrated and combined action towards a final result.

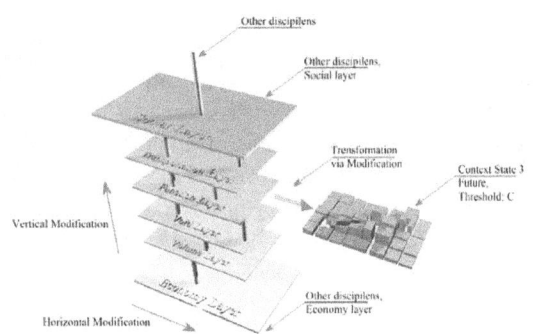

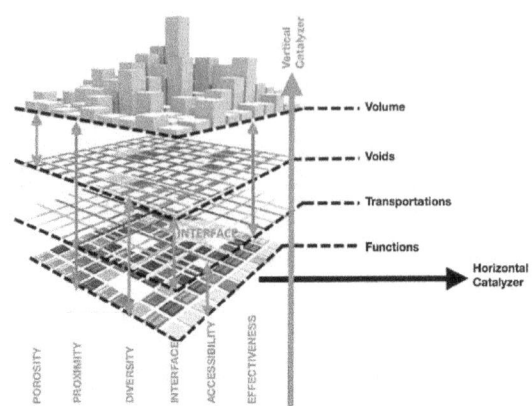

Transforming the system to the context state 3, indicated in purple color, Threshold C, via modification and integration of the four main environmental layers, the volume, the void, the function and the transportation layers, together with the other discipline layers is illustrated.

26.3.3 Phase 3 Modification/Transformation, (Intervention & Design).

The third step of the IMM is a specific design phase that involves the FLS. and applies to a multi-layer and multi-disciplinary approach. Thanks to a driver (Catalysts) a Local modification (Horizontal modification) marks the starting point of a chain reaction (Horizontal and Vertical modification) towards the global transformation of the CAS. Actually due to the fact that CAS is composed of four subsystems, we consider its state as a superposition of products of the subsystems' states. The main outcomes of this phase are:

- The design/project of the chosen catalysts layer, and Key Category in order to achieve a local modification that will be transmitted to process to the reactants layers.

- The local transformation towards a structural transformation of the CAS.

- Preliminary evaluation of the transformation.

It is composed by two inner phases, respectively:

1. Horizontal modification, the Horizontal Catalysts and Horizontal Reactants Phase (Step 3a)

2. Vertical Modification, the Vertical Catalyst and the Vertical Reactants (Step 3b)

Thanks to a driver assumed as Catalyst the Horizontal modification starts a Local modification with the goal of makings the starting point of a chain reaction (Vertical modification) towards the global transformation of the CAS. Once

Catalyst. The choice of one subsystem (layer) as Horizontal Catalyzer and one Key Category as Vertical Catalyzer makes possible to act transforming globally the entire system (CAS). The Catalyzers choice as a first driver of the transformation is the main goal of this phase, assigning respectively to the selected Subsystem and Key Category the Catalyst role and to the others the Reactants role.

the subsystems interact, their states are no longer independent. In urban term this phase is oriented to the local modification (neighborhoods/local nodes) with the aim of global transformation achievement. In this phase, the project work horizontally (modifying the local subsystems individually) and vertically (modifying the other subsystems and the architecture of their connections). Folding and superimposing the selected layers collaboratively, in a way in which the transformation of each layer changes the other one's structure/performance and characteristic, is the key factor of the main system transformation.

Horizontal Modification; the Horizontal Catalyst Phase (Step 3a)

Horizontal modification is the first step of the Design phase and its main goal is to modify the selected layer, elected as Catalyst of the transformation and the response of the others layer seen as reactants. So the design process starts with Local modifications of the Catalyst's layer structure. The local modification as designed perturbation of a system causes a series of effects that lead to macroscopic consequences starting up a chain reaction, which can transform the CAS structurally.

Vertical Modification; the Vertical Catalyst (Step 3b)

The Vertical modification is a chain reaction of the system propelled by the project. The aim of this step is to make possible the propagation of local changes towards the distant parts of the system as a consequence of connectivity, and making this propagation the cause a global change. The

	Second measurement (New CAS Energy Performances)	
Compactness	**1. Ground Use:** a) Urban Built density* b) Compactness factor* c) Number of building per hectare. d) Weighted Urban Proliferation WUP* (index of urban sprawl)	BD = $\sum$ Floor of every storey of Building / Ground level net C = Surface (Volume) / (2-3) N = Ha WUP = UP x w_1(DIS) - w_2(UD)
	2. Population and energy: a) Consumption per capita. b) Rate of energy coming from renewable sources* c) Renewable energy percentage in transport	KWh year per capita $T_{ENR} = T_{renewable en. prod.}$ / $T_{tot.en. prod.}$ (%) $T_{ENR} = T_{renewable tr.}$ / $T_{tot.tr.}$ (%)
	3. Walkability: a) Number of key function in a walking distance from residential buildings. b) Car free or minimal car traffic streets. c) High quality Street paths.	Living within 300 m from key services (in-number of inhabitants or ha) Length km $T_{length} = T_{HQ Street paths}$ / $T_{total street path}$ (%)
Complexity	**4. Uses of the space:** a) Ratio between numbers of residents and activities* b) Housing diversity* c) Ratio of place dedicated to Innovation and Knowledge*	C = $\sum$ Resident / $\sum$ Activities LBL = $\sum$ Low income housing / $\sum$ total housing (%) A-d = $\sum$ Number of th. individual / $\sum$ total waterline (%)
	5. Open Spaces: a) Ratio of green open spaces. b) Extent and number of parks. Number of trees per ha. c) Surface of and number of Public space, paved (sealed) surfaces	T = $T_{Ha Green urban area}$ / $T_{Total ground area}$ (%) N° ha Ha & N
	6. Urban biodiversity: a) Proportion of your municipality is comprised of natural or semi-natural areas. b) Number of native species (plants, animals). d) Number of Park area available to citizens. e) % vegetation cover in the urban core in the overall area	% Number Number %
Connectivity	**7. Cycling:** a) Length of biking roads (km)	Km per capita
	8. Transportation and Mobility: a) Private passenger transport energy use per person. b) Passenger number in public transport. c) Inhabitants leaving within 300m from public transport d) Length of roads per capita. Road Ratio* e) Vehicles distance travelled (VDT)	KWh year per capita Number of travel per years (for each kind) Number, % Km per capita; T_R = road surf/total sur Km per capita; KM
	9. Level of mobility interchange: a) Kinds of public transport available (boat, train, tram, metro, buses, microbuses, ...) b) Number of "Park-and-ride" parking places c) Interchange node/hub	Number of vehicles for each category Numbers Number Interchange performance evaluation
Management	**10. Food:** a) Food needed daily. b) Amount of urban farm production per person. c) Extent of municipally organised plots for cultivation.	Kg./capita Kg./capita Ha
	11. Waste management: a) Amount of solid waste produced. b) Rate of waste reused* c) Rate of materials coming from re-cycling.	Cubic meter per capita T = $M_{Volume of reused}$ / $M_{total Volume of waste}$ T = (%)
	12. Water management : a) Water use per person. b) Water productivity* c) Wastewater purified in a wastewater treatment plant*	Cubic meter per capita $T_{water prod.}$ = ($V_{water needed}$ / $V_{final water}$) % $T_{wastewater}$ = ($V_{wastewater}$ / $V_{water treated}$) %

Second measurement: A core set of elements based specifically on environmental themes but interlinked with other themes social, i.e., economy. They are tools used in the IMM process for objective measurement for comparing the performance of different systems (External comparison), or the System performance prior to and after the transformation design process (Internal comparison).

Vertical modification is driven by the Vertical Catalyst (the chosen Key Category elected as Catalyst) and the response of the others Key Categories as reactants. The action modifies the architecture of the ligands thereby activating the reaction that transform the structure of the System.

26.3.4 Phase 4. Retrofitting and Optimization Phase.

The last step is oriented towards the evaluation of the performance of the new CAS as a new energy-using complex system, which is composed of modified subsystems, in its own new formal configuration; thus, this new configuration will become the new context (formal structure) available for a new transformations, since the transformation is an endless process. The new provisional CAS will be evaluated and compared with the old one using the Indicators applied in the previous steps. The new provisional CAS will be evaluated and compared with the old one using the 12 Indicators applied in the Step 1b. After the retrofitting process the last phase Local modification/optimization is driven by the Key categories for achieving the conclusive optimization of the CAS. Morphological, typological and technological features, like the follows, express the new superimposition, or symbiotic integration. The main outcomes of this phase are:

1. Testing the new structure of the new CAS' state

2. Final valuation and comparison of the CAS performance

3. Optimizing the new CAS using the KC

New CAS measurement (Retrofitting Step 4a)

Once the transformation has occurred, a new CAS measurement as part of the retrofitting process starts. The process is based on the comparison between the new CAS performances and characteristics, and the previous one. This second measurement and comparison evaluates the transformed system's performances. Thanks to these 12 Indicators, it is possible to compare the characteristic performances of the system, before and after the transformation process. Moreover, the Indicators help to lead the complex system transformation in a correct way, as well as the result of transformation process.

New CAS Optimization (Step 4b)

The last phase is driven by the Key categories for achieving the conclusive optimization of the CAS. Of course this minor and local change[7] affects again the architecture of the CAS' performance, modifying it structurally another time but operating with a better control of the previous transformation's reaction. The final result of this optimization process is a concluding but still provisional CAS, that configures itself as the new threshold of an endless transformation process.

Universal Indicators (Comparison Step 4c)

Unlike the prior measurement processes, which evaluate the system's performances before and after the design process, Universal Indicators are tools to make a comparison between the city's performances and other cities.

26.4   See also

26.5   References

[1] S. Vahabzadeh Manesh, M. Tadi, (2013) Integrated Modification Methodology (I.M.M). A phasing process for sustainable Urban Design Issue 73 of World Academy of Science, Engineering and Technology, ISSN 2010-376X, e ISSN 2010-3778, pp 1207–1213

[2] Brownlee, J., (2007). Complex Adaptive Systems. CIS Technical Report 070302A,: p. 1–6

[3] Salat, S. and CSTB, (2011). Cities and Forms on Sustainable Urbanism: Hermann Editeurs des Sciences et des Arts

[4] S. Vahabzadeh Manesh, M. Tadi, (2013). A phasing process for sustainable Urban DesignEditRe-order section Issue 73 of World Academy of Science, Engineering and Technology, ISSN 2010-376X, eISSN 2010-3778, pp 1207–1213

[5] M. Tadi, S. Vahabzadeh Manesh, (2014) Transformation of an urban complex system into a more sustainable form via integrated modification methodology (I.M.M). The International Journal of Sustainable Development and Planning. Volume 9, Number 4. (WIT press Southampton, UK) ISSN: 1743-761x (online) and ISSN: 1743-7601 (paper format)

[6] H. Mohammad Zadeh, A. Naraghi, S. Vahabzadeh Manesh, M. Tadi, (2014). Environmental and energy performance optimization of a neighborhood in Tehran, via IMM methodology. International Journal of Engineering Science and Innovative Technology (IJESIT), ISSN No: 2319-5967. Volume 3,Issue 1: pp409-428.

[7] G. Lobaccaro, D.Palazzo, M. Tadi, A. Wyckmans (2014). Green design strategies for urban heat island mitigation in a solar optimized access Eixample via IMM methodology". WSB (World Sustainable Buildings) ISBN 978-84-697-1815-5

26.6   Further reading

• Ahern, J. (2006). "Green Infrastructure for Cities: The spatial Dimension". In Cities of the Future Towards Integrated Sustainable Water and Landscape Management, edited by Vladimir Novotny and Paul Brown, 267-283. London: WA publishing.

• Anderson, P. (1999). Complexity Theory and Organization Science Organization Science. 10(3): 216–232.

• Bennett, S., (2009), A Case of Complex Adaptive Systems Theory- Sustainable Global Governance: The Singular Challenge of the Twenty-first Century. RISC-Research Paper No.5: p. 38

• Brownlee, J., (2007), Complex Adaptive Systems. CIS Technical Report: p. 1-6.

• Backlund, A. (2000), "The definition of system". In: Kybernetes Vol. 29 nr. 4, pp. 444–451.

• Clarke, C. and P. Anzalone, Architectural Applications of Complex Adaptive Systems, XO (eXtended Office). p. 19.

• Crotti, S., (1991), Metafora, Morfogenesi e Progetto, E.D'alfonso and E.Franzini, Editors. 1991: Milano.

• Hildebrand, F. (1999), Designing the city towards a more sustainable urban form. Routledge.

• Hough, Micheal. (2004). Cities and Natural Processes: a Basis for Sustainability. London: Routledge.

• Jenks, M., E. Burton, and K. Williams, (1996), The compact city, a sustainable form?: F a FN Spon, an imprint of Chapman & Hall. 288

• Salat, S. and CSTB, (2011), Cities and Forms on Sustainable Urbanism: Hermann Editeurs des Sciences et des Arts.

• Steel, C. (2009), Hungry City: How Food Shapes Our Lives, Random House UK.

• Tadi, M. Vahabzadeh Manesh, S. A.Daysh, G. Kahraman, I. Ursu (2013) The case study of Timisoara (Romania). IMM design for a more sustainable, livable and responsible city. AST Management Pty Ltd, Nerang, QLD, Australia.

• Thom, R., (1975), Stabilite Structurelle et Morphogenese. Massachusetts: W.A.Benjamin, Inc. 348.

• Vahabzadeh Manesh, S. Tadi, M. (2013) Integrated Modification Methodology (I.M.M). A phasing process for sustainable Urban Design Issue 73 of World Academy of Science, Engineering and Technology.

• Vahabzadeh Manesh, S. M. Tadi, (2013) Neighborhood Design and Urban Morphological Transformation through Integrated Modification Methodology (IMM) part 1. The Designer Architectural Magazine Vol.8. IRAN.

Chapter 27

Landscape planning

Landscape planning is a branch of landscape architecture. According to Erv Zube (1931–2002) landscape planning is defined as an activity concerned with reconciling competing land uses while protecting natural processes and significant cultural and natural resources.

Urban park systems and greenways of the type planned by Frederick Law Olmsted are key examples of urban landscape planning. Landscape designers tend to work for clients who wish to commission construction work. Landscape planners can look beyond the 'closely drawn technical limits' and 'narrowly drawn territorial boundaries' which constrain design projects.

Landscape planners tend to work on projects which:

- are of broad geographical scope

- concern many land uses or many clients

- are implemented over a long period of time

In rural areas, the damage caused by unplanned mineral extraction was one of the early reasons for a public demand for landscape planning.

Mineral working in the Sierra Nevada, outside Granada, Spain. This is part of a Landscape, *and it can be* planned

27.1 In Asia

In India, the history of landscape planning can be traced to the Vedas and to the Vaastu Shastras. These ancient texts set forth principles for planning settlements, temples and other structures in relation to the natural landscape. Relationships with mountains (the home of the gods) and with rivers (regarded as goddesses) were of particular importance. A square form represented the earth and a circular form represented heaven. A mandala explained the relationship between heaven and earth. Square plans, for both secular and religious structures, were set out with their sides facing north, south, east and west. The earliest surviving stone temple set out in this way is Sanchi.

In China, landscape planning originated with Feng Shui, which is translated into English as 'wind and water' and is used to describe a set of general principles for the planning of development in relation to the natural landscape. The aim was to find the most auspicious environment possible, one sited in harmony with natural phenomena and the physical and psychological needs of man' (*Chinese Architecture* by Nancy Steinhardt et al. Yale University Press and New World Press 2002, p. 255)

27.2 In Europe

In Europe, the history of landscape planning can be traced to the work of Vitruvius. In discussing the planning of towns, he wrote about site planning with regard to microclimate, about the planning of streets and about the role of metaphor in design. Vitruvius' theories were revived during the renaissance and came to influence the planning of towns throughout Europe and the Americas. Alberti wrote on the need for town squares for markets. In North Europe this developed into the idea that residential squares should planned around green spaces. The first space of this type was the Place des Vosges. Residential squares were also made in Britain and their planning developed into the idea of incorporating public open space (public parks within towns). Frederick Law Olmsted gave momentum to this idea with his proposal for a park systems in Boston - the famous Emerald Necklace. Patrick Abercrombie took up this idea and incorporated it in his great 1943-4 Open Space Plan for the County of London.

27.3   In the US

Landscape architects in the United States of America are active in landscape planning. But, unlike Canada and Europe, the US does not have a national land use planning system. Frederick Law Olmsted and Ian McHarg are the most famous American landscape planners. McHarg's work on overlay landscape planning contributed to the development of GIS and to the foundation of ESRI by Jack Dangermond.

27.4   Legislation

The principles of landscape planning are now incorporated in various types of legislation and policy documents. In America, the National Environmental Policy Act was influenced by the work of Ian McHarg on Environmental impact assessment. In Germany, the Federal Nature Conservation Act requires the preparation of landscape plans. For the Europe Union as a whole, the European Landscape Convention has wide-ranging implications for the design and planning of relationships between development and the landscape. In Asia, major development projects are taking place and illustrating the need for good landscape planning. The Three Gorges Dam, for example, will have extensive impacts on the landscape. They have been planned to a degree but future monitoring of the project is likely to show that better landscape planning and design would have been possible.

27.5   Theory

Landscape planners are concerned with the 'health' of the landscape, just as doctors are concerned with bodily health. This analogy can be taken further. Medical doctors advise both on the health of individuals and on matters of public health. When individuals take actions injurious to their own health this is regarded as a private matter. But if they take actions injurious to public health, these actions are properly regulated by law. The collective landscape is a public good which should be protected and enhanced by legislation and public administration. If, for example, mineral extraction has a damaging impact on the landscape, this

is a proper field for intervention. Negative impacts on the landscape could include visual impacts, ecological impacts, hydrological impacts and recreational impacts. As well as protecting existing public goods, societies are responsible for the creation of new public goods. This can be done by positive landscape planning. There are, for example, many former mineral workings (e.g. the Norfolk Broads) which have become important public goods. Medical doctors are trained in anatomy, physiology, biochemistry etc. before becoming practitioners. Landscape doctors are trained in geomorphology, hydrology, ecology etc. before becoming practitioners in design and planning. When qualified, they can specialize in areas of landscape planning:

- Landscape of roads - The landscape treatment of roads is concerned with the planning and design of roads and highways with regard to their environmental impact on the surrounding landscape. Sylvia Crowe wrote a pioneering book on the design of roads with regard to their impact on the landscape and Ian McHarg proposed an overlay system for highway route selection in his book on *Design with nature*.

- Landscape of forestry - The **landscape treatment of forests** is concerned with the non-timber objectives which can be obtained by conserving and developing forests: scenic quality, water quality, recreation, wildlife conservation and other environmental goods. This work is done by foresters who also hold qualifications in landscape architecture and also by landscape architects and landscape planners with a specialization in forestry. The United States Forest Service, the UK Forestry Commission and other forest agencies are also employers of landscape architects. They have mitigated criticism of plantation forestry, monoculture, and clear-cutting.

- Landscape of energy - The generation of energy has become a major land use with consequent environmental impacts upon the landscape. Landscape planning for energy is concerned with designs and plans to mitigate the impact of power generation upon the landscape. This includes landscape planning for:

  - power stations
  - power transmission lines
  - hydroelectric power
  - tidal power
  - solar power
  - wind power

- Landscape of urbanization - Most of the world's cities are expanding into agricultural land. Landscape planning for urbanization is concerned with the conservation and development of landscape resources as urbanization takes place. Ian McHarg's approach to this problem was to prepare a series of overlay maps showing the different types of quality in the land around settlements: agricultural value, property value, hydrological value, scenic value, recreational value, ecological value etc. The maps of value were then overlaid to produce a composite map which looked like an X-Ray photograph - with the dark areas having the most value and the white areas the least value. His belief was that the 'white land' should be urbanized in preference to the 'black land'. The overlay mapping system which McHarg proposed is now carried out using a Geographic Information System. Other landscape planners have proposed greenways and green belts as a means of urban growth management.

- Landscape Urbanism is a theory of urbanism arguing that landscape, rather than architecture, is more capable of organizing the city and enhancing the urban experience.

- Landscape of recreation - Recreation has become a large-scale landuse. In the coastal regions of many countries (e.g. Spain and Denmark) it is the dominant land use and has had an enormous impact on the natural landscape. The landscape of recreation is concerned with mitigating the environmental impact of recreation on the landscape. Landscape architects and landscape planners work on the landscape of recreation to:

  - reclaim landscapes which have been damaged by recreational use

  - carry out environmental impact assessments of recreational development proposals

  - prepare landscape plans and designs for recreation projects

- Landscape of mineral extraction - Mineral extraction has a long history of damaging the landscape. Harm has been caused to scenic quality, water quality, habitat quality and other environmental goods. The landscape treatment of mineral workings is concerned with:

  - the reclamation of abandoned mines and quarries

  - the planning and design of future mineral workings to minimize their harmful environmental impact and achieve as many positive impacts on the environment as possible.

- Landscape of agriculture - The primary purpose of agriculture is food production but concern for other objectives (e.g. wildlife, conservation, biodiversity, recreation and scenery) have a long history and are of increasing importance in wealthy and urbanized countries. The European Union Set-Aside Policy was designed as a means of giving money to farmers to produce non-food environmental goods from farmland. Landscape planners are involved with the preparation of agricultural landscape plans for the achievement of non-food objectives from agricultural land.

- Landscape of rivers - The construction of new buildings and new roads accelerates the discharge of surface water runoff and raises flood peaks in the lower reachers of river catchments. River banks therefore have to be raised and flood channels have to be constructed. The adoption of sustainable urban drainage systems facilitates the reclamation of rivers. When water is detained, infiltrated and transpired near to the where it falls, flood peaks are lowered and rivers can be reclaimed. The high concrete walls in which they were confined can be replaced by vegetated embankments. Water quality is improved and wildlife returns to the river. Projects to achieve these objectives are described as River Reclamation or River Restoration.

The western group of Khajuraho temples (2005) set in an archaeologically inappropriate parkland landscape

- Landscape of archaeology - The landscape of archaeology is concerned with the landscape treatment of archaeological sites. This involves analysis, discussion, the formulation of policies and the preparation of landscape designs relating a number of issues. In Britain the former Ministry of Public Works and Buildings (MPBW) (now English Heritage, Cadw, Historic Scotland and Northern Ireland Environment and Heritage Service) had a policy of treating archaeological sites like highly manicured gardens. The grass was maintained almost to the quality of a bowling green. The standard of care was admirable but there was little or no regard to the former landscape character of the archaeological site. The development of the discipline of landscape archaeology has awakened interest in this question because the wider landscape is seen, correctly, as having archaeological value. Most historic societies earned their livelihood from working the land. Their buildings were an important accessory to the use of the land. In countries where the looting of archaeological sites is a problem, there is a tendency for them to be surrounded with chain links or other security fencing. This detracts from the relationship between the site and its landscape setting. In other countries, archaeological sites have become important visitor attractions and sources of revenue. This has led to the building of tourist facilities (visitor centres, hotels etc.) which can easily have a detrimental impact on the archaeological site and its landscape setting.

In each case, the aim is to take a specialist land use and make recommendations for what can be done to enhance its impact on the stock of environmental goods.

27.6 Methodology

The conventional planning process is a linear progression of activities. The common steps are:

- Identification of problems and opportunities.

- Establishment of goals.

- Inventory and analysis of the biophysical environment.

- Human community inventory and analysis.

- Development of concepts and the selection of options.

- Adoption of a plan.

- Community involvement and education.

- Detailed design.

- Plan implementation.

- Plan administration.

Landscape planning not always means an ecological planning method, for that it must be considered that "planning is a process that uses the scientific and technical information for considering and reaching consensus on a range of choices. Ecology is the study of the relationship of all living things, including people, to their biological and physical environments. Ecological planning then may be defined as the

use of biophysical and sociocultural information to suggest opportunities and constraints for decision making about the use of landscape". (Steiner, 1991)

27.7 See also

- Landscape architecture

- Landscape management

- Landscape ecology

- Landscape Institute

- Landscape of agriculture

- Landscape urbanism

- Environmental impact assessment

- Urban design

- Green roof

- Growth management

- Principles of Intelligent Urbanism

- Sustainable city

- Sustainable landscape architecture

- European Landscape Convention

- Permanent European Conference for the Study of the Rural Landscape

27.8 Footnotes

- Landscape planning education in America: retrospect and prospect

- *Ecological design and planning* George F. Thompson and Frederick R. Steiner, (Wiley, 1997)

- *Landscape planning : an introduction to theory and practice* Hackett, Brian (Oriel, 1971)

- *Landscape planning and environmental impact design* Tom Turner (2nd ed UCL Press, 1998)

- *Design with nature* Ian L. McHarg (Wiley, 1992)

- *The living landscape: an ecological approach to landscape planning* Steiner, Frederick R. (McGraw-Hill College, 1991)

27.9 External links

- European Landscape Convention (official statement by the Council of Europe).

- The "Landscape Must Become the Law" - Or Should It? by Gert Groening.

- Landscaping Planning

Chapter 28

Life Cycle Thinking

Life Cycle Thinking is a different approach to becoming mindful of how everyday life has an impact on the environment. This approach evaluates how both consuming products and engaging in activities impacts the environment but it not only evaluates them at one single step, but takes a holistic picture of an entire product or activity system. This means when talking about a product and taking a Life Cycle Thinking approach, what is actually being evaluated is the impact of the activity of consuming that product. This is because by consuming a product, a series of associated activities are required to make it happen. For example, the raw material extraction, material processing, transportation, distribution, consumption, reuse/recycling, and disposal must all be considered when evaluating the environmental impact. This is called the life cycle of a product. The overall idea of making a holistic evaluation of a systems impact can be defined as Life Cycle Thinking.

Life Cycle Thinking therefore also can be applied to the consumption of other socio-economic activities such as watching a movie, making arts and crafts, cooking dinner, or even doing homework. For example, renting a movie, which seems to be a harmless activity, would involve burning gasoline to drive to the video store, using electricity to power the television and DVD player, and consuming power from the remote's batteries. However when trying to analyze quantitatively the impacts of life cycles, limits to evaluation are subject to what assessment approach is taken because the chain reaction can become so complex that it could require decades to figure out the life cycle of a specific process. Life Cycle Thinking overall is a way to become more mindful of the complexities of consuming products and engaging in activities and how they affect the environment.

28.1 Goals

The goal of Life Cycle Thinking is to make people and companies more aware of how their actions impact the environment in a holistic sense rather than a one time pollution that

comes as a direct result of using a product or doing an activity at one specific time. Although it is nearly impossible to undergo consumption of anything with no environmental impact, Life Cycle Thinking can help people make better alternative decisions to mitigate their environmental impact. One of Life Cycle Thinking's biggest goals is to avoid burden shifting.[1] This is to make sure that reducing the environmental impact at one stage in the life cycle does not increase the impact at other places in the cycle. For example, plug in electric cars reduce the amount of gasoline burned but they increase the amount of electricity used which is usually generated by other polluting energy sources such as coal. Life Cycle Thinking can also demonstrate the benefits to technological innovation. For example, movies can now be downloaded through television service providers and gaming devices which eliminates the need to drive to a DVD rental location. By identifying pollution costs, companies can innovate to mitigate their expenses while consumers can make better alternative choices to mitigate their impact.

- Avoid burden shifting

- Reveal the complexity of the system triggered by an action which can have several negative environmental effects.

- Connect people more directly with the impacts of their life style and demonstrate how each action has a reaction which is sometimes asymmetrically worse for the environment.

- Make companies more mindful of environmental impacts of their operations.

  Help identify cost cutting possibilities

  Help identify less harmful operation strategies

- Provide people with a framework to make choices that over a life cycle have less environmental impacts.

- Create a culture focused on sustainability rather than short term gratification.

28.2 Sectors

Life-cycle thinking has applications in many sectors, such as the following:[2]

28.2.1 Agriculture

The agriculture/food sector is a big source environmental impacts that occur throughout the lifetime of a product, from farm to table to disposal. Life-cycle thinking works to reduce these impacts at all stages of food production. Nutrition, health, well being, cultural identity and lifestyle are also factors that should be addressed when looking at the impacts of choices made in food production to ensure decreases in emissions and environmental impact do not occur at the expense of consumer wellbeing.

28.2.2 Manufacturing

A product Life Cycle Analysis involves all production and service processes involved in the manufacturing of a product throughout its life-cycle. This includes the production of materials needed to make the product. Since the manufacturing sector is a big emitter of pollutants and user of natural resources, pinpointing areas in which to decrease environmental impact throughout the manufacturing process is a big part of life-cycle thinking.

28.2.3 Energy

Drastic increases in atmospheric CO_2 caused by the burning of fossil fuels, has led to the search for alternative energy sources like biofuels and renewable energy sources. To analyze whether or not these alternative sources have overall less environmental impact then conventional energy sources, life-cycle analysis is needed. Life-cycle thinking is an intricate part of finding new energy sources that have an overall smaller impact on the environment.

28.2.4 Waste management

Life-cycle thinking and analysis can help reduce negative environmental impacts of waste generation and management. This includes looking at ways to reduce waste production, increase recycling, and dispose of waste in a more environmentally friendly way. This is complicated by differences in benefits and burdens of in different geographical regions and the fact that effects usually occur over long periods of time. Furthermore benefits and burdens of different processes can occur in many different forms and can be difficult to identify, quantify and compare.

28.2.5 Retail

Retail often accounts for a significant portion of economies and thus can have huge implications in terms of environmental impacts. The life cycle of a product in retail would include the complete supply-chain of the product, its use and disposal or end-of-life treatment.

28.2.6 Construction

There are many uses for life-cycle thinking in construction, especially in terms of construction waste and waste management. Finding better ways to recycle waste and prevent waste are important to reduce negative environmental impact of the construction industry.

28.2.7 Transport

Finding alternative fuel sources are the biggest challenges to reducing negative environmental impact in the transportation sector. Biofuels are becoming increasing popular as an alternative to fossil fuels. Life cycle analysis can provide a fuller picture of the extent alternative fuel sources reduce emissions and overall environmental impact compared to conventional fuels.

28.2.8 Services

Service industries play a big part in adding environmental burdens, especially in terms of greenhouse gas emissions generated by travel and tourist industries. The service industry is expected to play a larger part in the modern economy as "dematerialization," or the replacement of manufactured goods by services in many firms, plays a bigger role in the economy.

28.3 Approaches

There are many different approaches to life cycle thinking that all involve looking at life cycle-generated impacts and ways to minimize these impacts. An important component to life cycle approaches is avoiding burden shifting, in other words, ensuring that improvements in one stage are not achieved at the expense of another stage. Approaches of impact measurement focus on decreasing environmental impact and resource use throughout all stages of a process.[3]

Commonly used approaches:

28.3.1 Life-cycle assessment

Life-cycle assessment (LCA or life cycle analysis) is a technique used to access environmental impacts of a product at different stages of its life. This technique takes a "cradle-to-grave" approach and looks at environmental impacts that occur throughout the lifetime of a product from raw material extraction, manufacturing and processing, distribution, use, repair and maintenance, disposal and recycling.

28.3.2 Life Cycle Management (LCM)

Life cycle management is a business approach to manage the total life cycle of products and services. It follows the life cycle thinking that businesses, through the activities they must perform, have environmental, social and economic impacts. LCM is used to understand and analyze life cycle stages of products and services of a business, identify potential economic, social or environmental risks and opportunities at each stage and create ways to act upon those opportunities and reduce potential risks.[4]

28.3.3 Life Cycle Costing (LCC)

Life cycle costing (or life cycle cost analysis) is the total cost analysis of a process or system. This includes costs incurred over the life of the system and is frequently used to find most cost-effective means for providing goods and services.[5]

28.3.4 Design for the Environment

DfE Logo

Design for the Environment Program (DfE) was created in 1992 by the United States Environmental Protection Agency and works to prevent pollution and the reduce the risks pollution presents to humans and the environment. The main goals of the DfE are to promote green cleaning, recognize safer industrial and consumer products through safer product labeling, define best practices in production and manufacturing, and identify safer chemicals for these processes based on life cycle thinking. Having said this they must know that the air pollution in USA has the mixing of liquid, solid, gaseous, odour and noise pollution which is dangerous for human being, animals and plants.[6]

28.3.5 Product Service System

Product Service Systems (PSS) are sets of marketable products and services that work together to fulfill a user's needs. This new approach is a result of firms realizing that services in combination with products can provide higher profits and customer satisfaction then simply selling products alone. Firms that use PSS work to find ways to maximize the use of their product throughout its lifetime, using services to supplement its usage. PSS has been seen to have smaller environmental impact than traditional business models, as the focus on services has led to a decrease in material production and consumption. This applies to life cycle thinking because it involves looking at the life-cycle cost of a product (i.e. maintenance and storage costs) for a consumer and reducing that cost by providing services with the purchased good.

28.3.6 Integrated Product Policy (IPP)

Integrated product policy works to minimize environmental degradation caused by products by looking at all phases of a product's life-cycle to pinpoint where taking action is most effective. This also uses a cradle-to-grave approach when looking at a product's life. In addition, it is important that policies avoid burden shifting and do not decrease environment emissions at one stage of development at the expense of another. Policy measures used to action upon recommendations include economic instruments, substance bans, voluntary agreements, environmental labeling and product design guidelines.[7]

28.4 Applications

There are multiple situations to which Life Cycle Thinking can be applied including every day life of consumers, business and government policy. By applying life cycle thinking to multiple aspects of the community, consumers, businesses and governments can have a largely positive aspect on the environment. This is true even if the steps taken to apply life cycle thinking are small.

28.4.1 Consumers

Each day consumers make choices as to which products they would like to use based on their needs and the different brands available. However, most consumers do not take into consideration the environmental impacts of the product when they make their choice. For example, consumers do not consider the product's energy usage, illegal labor conditions that produced it, hazardous waste from production, impacts on the ecosystem, or pollution of air or water.

Consumers can apply life cycle thinking in multiple different ways with regards to their product choices in order to reduce their impact on the environment. Firstly, consumers can chose to use products from companies who take strides towards sustainability. Many companies provide sustainability reports that consumers can read to educate themselves about the companies they buy from. By using life cycle thinking, consumers can chose a company with smaller production impacts.

Primarily, consumer usage has the largest impact on the environment throughout a products life. By using life cycle thinking this impact can be reduced. This would require educating consumers to make better choices about product usage. This can come from the companies who provide the service or product or from government agencies. For example, consumers can ask themselves what impacts they have while using the product. Ask do I really need to use this or is there a more sustainable option, such as hang drying laundry on a nice day rather than using a dryer. Consumers can educate themselves on how to become more sustainable themselves through life cycle thinking rather than relying on companies and the government to be sustainable for them.

28.4.2 Businesses

Businesses are responsible for many choices about their services and products each day. By applying life cycle thinking, businesses can recognize the potential impacts of their choices. They consider how each design and manufacturing decision has an effect on the environment and how they can make it more sustainable. Businesses not only take into consideration how the product is made, but also how the product will be used and disposed of by the user. Companies try to have a more sustainable product by making products recyclable or reusable. They challenging part is balancing cost and sustainable choices. Life cycle thinking allows them to see the best sustainable options but is limited when it comes to pricing these choices. Life cycle thinking for businesses

entails consideration of where to obtain raw material, how to manufacture the material, transporting, distributing, using, and disposing of the product. By looking at all of these phases businesses make the best choices for themselves and the consumer for a lower impact on the environment.

28.4.3 Governments

Government plays a key role in life cycle thinking by establishing policies to regulate environmental impacts. By applying life cycle thinking policy makers can set standards that businesses and consumers need to meet. They do so by gathering information as a baseline of the environmental impact and use that to set goals based on knowledge from life cycle thinking. They can also use trends from supply chains of different businesses they regulate to determine where the biggest influence can be made to majorly reduce the impacts of the businesses. Government sectors can also use life cycle thinking to better educate consumers. Requiring labels on products describing the impacts the product has and how to use the products in order to reduce the impact can be an important role for the government. Regulating supply chains and consumers with policy is motivational as negative reinforcement. Life cycle thinking provides a methodology for creating those policies in order to have the most effective and most cost efficient means of reducing environmental impacts.

28.5 Life Cycle Thinking in Policy

Many consumers, when making decisions on what to buy and what not to buy, consider the environmental impact of the particular product. Policy makers recognize this desire, and act to create policy that not only helps consumers do this, but will do so while keeping a growing economy in mind.

28.5.1 European Policy

JRC logo

There are many aspects of Life Cycle Thinking incorporated into European policy. The Sustainable Consumption and Production Action Plan is a piece of legislation that aims to reduce environmental impact and consumption of resources associated with the complete life cycles of goods and services.[8] On July 16, 2008 the European Commission presented this legislation. This proposal suggests plans on how to not only reduce the environmental impacts of goods and services, but also encourages the use of more sustainable goods and production technologies. This action plan also encourages the European Union to seek out every opportunity to innovate in industry.[9]

The Integrated Product Policy is another legislative action that Europe has taken in order to facilitate life cycle thinking. The Integrated Product Policy seeks to minimize the environmental degradation caused from the manufacturing, use and disposal of all products. This legislation looks at all aspects of the product's life cycle and takes action where necessary to reduce.[10]

The Thematic Strategy on Sustainable Use of Natural Resources was implemented on the 21st of December in 2005 to reduce environmental impacts associated with resource use and to do this in a growing economy. The objective can be described as "ensuring that the consumption of resources and their associated impacts do not exceed the carrying capacity of the environment and breaking linkages between economic growth and resource use".[11]

28.5.2   United States Policy

While the term "life cycle thinking" is not as prominent in United States policy, there are considerations of the life cycle process throughout governmental policies and programs. There are Environmental Product Declarations that are used to incorporate life cycle thinking into companies and organizations. They communicate to the consumer the environmental performance of a product or system. These declarations are based on the Life Cycle Assessment and once the assessment is complete a product or system can be certified EPD.[12]

The Environmental Protection Agency's program, Design for the Environment works with individual industry sectors to compare and improve the performance and human health and environmental risks and costs of existing and alternative products, processes, and practices. DfE partnership projects promote integrating cleaner, cheaper, and smarter solutions into everyday business practices.[13] The Design for the Environment program is also equipped with a labeling program. They allow safer products to carry these labels and they are an indication to consumers that buying these products will be safer for the environment and their families.[14]

Also, the Energy Independence and Security Act of 2007 is a piece of legislation that incorporates life cycle thinking. While this exact phrase isn't listed. This act includes sections on advanced biofuels. In Title II of the act, it requires the creation of Biomass-based diesel which is the addition of renewable biofuels to diesel fuel and will reduce emissions by 50% as compared to petroleum biofuel. In Title III improved standards will be implemented.

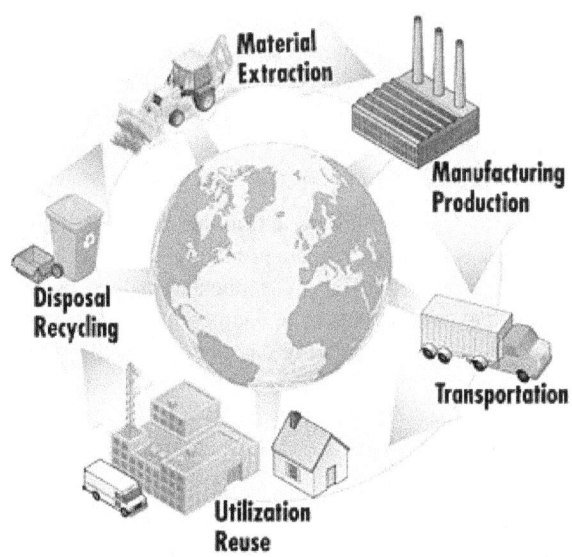

Life Cycle Thinking Product System

28.6   Importance of Life Cycle Thinking

Since Life Cycle Thinking can be involved in the choices of individual consumers, as well as policy makers and businesses, it is very important that people are well informed about the subject and its uses.[1] Increasing awareness of the Life Cycle Analysis technique would allow companies as well as individuals to consider multiple options for a new product. After consideration of all available options, Life Cycle Thinking would encourage selection of the most sustainable option. If more individuals practiced Life Cycle Thinking when looking for new materials or methods, they would be more aware of how the environmental cost of ownership of products can be influenced by the running costs in energy and consumables.[15]

Life Cycle Thinking can help people find new ways to improve environmental performance, image, and economic benefits.[1] Since the decisions of global businesses and government organizations have such a large impact on the environment, incorporating Life Cycle Thinking into their actions could greatly reduce negative environmental effects and improve sustainability. Many businesses do not always consider their supply chains or the "end-of-life" processes associated with their products; likewise, government actions frequently consider their own country or region and do not take into account the impact that they could have on other regions.[1]

Not only could Life Cycle Thinking help the environment, it can also save the company more money and improve their reputation. If a company knows where their materials come from as well as where they will end up after they have reached the end of their useful life, economic performance could be further enhanced. Also, since presently so much emphasis is placed on sustainable actions, the more a company shows its concern and respect for the environment, the better its reputation will be.

In a case study on laundry detergents, it was found that washing clothes at lower temperatures resulted in energy savings and improvements in several environmental indicators, like climate change, acidification and photochemical ozone creation. Because the company understood the importance of Life Cycle Thinking, they made the decision to conduct a Life Cycle Analysis to find the benefits of developing a different laundry detergent. Not only did the new detergents reduce environmental impact by decreasing energy consumption, it also benefitted the consumer by reducing electricity bills and helped the company by becoming a leader in the industry. (3)[16]

28.7 References

[1] European Commission. "Life Cycle Thinking and Assessment". Retrieved 18 November 2011.

[2] "Life Cycle Thinking in sectors". European Commission. Retrieved 16 November 2011.

[3] "Why Take a Life Cycle Approach?" (PDF). UNEP. Retrieved 16 November 2011.

[4] "What is Life Cycle Management (LCM)". EPA. Retrieved 16 November 2011.

[5] "Life Cycle Costing". Sandia National Laboratories. Retrieved 16 November 2011.

[6] "Pollution in USA". What is USA. 23 April 2013. Retrieved 2012-12-25.

[7] "What is Integrated Product Policy". European Commission on Environment. Retrieved 16 November 2011.

[8] "Life Cycle Thinking in European Policy". *Life Cycle Thinking and Assessment*.

[9] "European Commission - Environment - Sustainable Development". Ec.europa.eu. Retrieved 2011-12-16.

[10] "European Commission - Environment - Integrated Product Policy". Ec.europa.eu. Retrieved 2011-12-16.

[11] "European Commission - Environment - Sustainable Development". Ec.europa.eu. Retrieved 2011-12-16.

[12] "Environmental Product Declaration Home Page". Environmentalproductdeclarations.com. Retrieved 2011-12-16.

[13] "LCA Resources". *Environmental Protection Agency*.

[14] "Design for the Environment". *Environmental Protection Agency*.

[15] Tarr, Martin. [1. http://www.ami.ac.uk/courses/topics/0109_lct/ "Life cycle thinking"]. Retrieved 18 November 2011.

[16] European Commission (2010). *Making sustainable consumption and production a reality*. Luxembourg: Publications Office of the European Union. ISBN 978-92-79-14357-1.

28.8 External links

- http://lct.jrc.ec.europa.eu/pdf-directory/Making%20sustainable%20consumption%20and%20production%20a%20reality-A%20guide%20for%20business%20and%20policy%20makers%20to%20Life%20Cycle%20Thinking%20and%20Assessment.pdf

- http://www.setac.org/node/90

Chapter 29

NABERS

The NABERS logo

29.1 Introduction

NABERS, the **National Australian Built Environment Rating System**, is a government initiative to measure and compare the environmental performance of Australian buildings and tenancies. There are NABERS rating tools for commercial office buildings to measure greenhouse gas emissions, energy efficiency, water efficiency, waste efficiency and indoor environment quality. There are also energy/greenhouse and water rating tools for hotels, shopping centres and data centres.

29.2 Third Party Assessments

A key feature of the program is the use of independent 'Accredited Assessors' to conduct ratings. Assessors are required to attend training, pass an exam and complete two supervised assessments before they receive full accreditation.[1] While there are no formal pre-requisites to attend the training, most Assessors have experience in the building services, property or energy management industries. Building owners and tenants can use the online 'self-assessment' tool, however they cannot promote these results. Only ratings that have been certified by the NABERS National Administrator can be promoted using the NABERS trademark.

29.3 Calculating a rating

The vision statement of NABERS is 'To support a more sustainable built environment through a relevant, reliable and practical measure of building performance.' To this end the NABERS tools attempt to provide an accurate measurement of how efficiently building owners and tenants are providing their services *without penalising them for factors that are beyond their control.* So, for example, the primary service that an office building owner provides is safe, lit and comfortable office space. Therefore the NABERS Energy for offices tool considers how much space is being used, how much energy is being used to supply services to the space, and then statistically adjusts for factors like the climate - which will influence how much energy is used for heating and cooling.

29.3.1 Office Energy

To obtain a NABERS Energy for offices rating, consumption data for the building (such as electricity and gas bills) is collected by Accredited Assessors along with data about a number of other aspects of the building such as its size, hours of occupation, climate location and density of occupation. Data requirements are set out in a document called 'The NABERS Energy and Water for offices Rules for Collecting and Using Data v.3.0'. This data is then input into the NABERS calculator [2] which statistically adjusts for these factors so that the building can have its consumption fairly benchmarked against its peers. The result of this calculation is a star rating on a six star scale, where zero is very poor performance and six is market leading.

29.3.2 Office Water

The procedure for an office water rating is similar to conducting an office energy rating. The main differences are that it is water rather than energy bills that are used, and some data such as the hours of operation are not required. Unlike office energy ratings, which can either be for the base building, tenancies or whole building, office water ratings are only available for whole buildings.

29.3.3 Data Centre Energy

Like NABERS for offices, NABERS Energy for data centres has three distinct rating types to reflect the different interests and responsibilities from data centres owners, operators and tenants - Infrastructure (co-location owner), Whole Facility (data centre owner) and IT Equipment (data centre tenant) ratings. The tool is designed to rate the majority of data centres in Australia, provide a direct comparison with other rateable data centres, and allow an individual data centre to measure and compare performance over time.

A NABERS Data Centre - IT Equipment Rating is designed for organisations that control and manage their own IT equipment (servers, storage and networking devices). The IT Equipment rating measures feastures that are closely related to the primary functions of a data centre (processing, storage and networking) and that all data centres provide, regardless of how they provide them. NABERS uses Processing Capacity (the sum of the number of server cores multiplied by clock speed in gigahertz (GHz) and Storage Capacity (total unformatted storage capacity in terabytes) as the two IT equipment metrics. The NABERS performance benchmark model predicts the industry median greenhouse gas emissions for a given amount of data centre processing and storage capacity. This means that if a data centre consumes more energy than the benchmark model predicts, the site is less energy efficient than the industry median (set at 3 stars), while if it consumes less energy it is more efficient than the median. To obtain a NABERS Energy for data centres IT Equipment rating, energy consumption data for the IT equipment over a 28-40 day period is collected by Accredited Assessors along with data about the total unformatted storage capacity and total processing capacity. Data requirements are set out in a document called [3]

A NABERS Data Centre - Infrastructure Rating measures the energy efficiency in delivering support services to the IT equipment, using the widely accepted industry Power Usage Effectiveness (PUE) ratio that is converted into kilogram of emissions with some modification for climate and shared cooling services. To obtain a NABERS Energy for data centres Infrastructure rating, 12 months of energy consumption data for IT equipment and infrastructure services is collected by Accredited Assessors along with the climate location of the data centre. Data requirements are set out in a document called [4]

A NABERS Data Centre - Whole Facility rating measures the energy efficiency of the whole data centre by assessing the processing and storage capacity and the industry median energy efficiency for infrastructure services compared with the overall energy consumption of the data centre. It is a combination of both the IT Equipment and Infrastructure rating benchmarks To obtain a NABERS Energy for data centres Whole Facility rating, 12 months of energy consumption data for the data centre is collected by Accredited Assessors along with the processing and storage capacity and climate location of the data centre. Data requirements are set out in a document called [5]

29.4 Comparison to other building rating systems

There are a number of building environmental certification systems across the world, such as LEED, Green Star, BRE Environmental Assessment Method (BREEAM) and Display Energy Certificates (DECs). The key features of NABERS as a system are that it is based on performance rather than design, assessments are carried out by third party 'Accredited Assessors', it is based on third party verifiable data (such as utility bills), ratings undergo government quality assurance checks and it distinguishes between the environmental impact of a building's shared services and its tenancies. While other rating systems across the world share some of these features, none share all of them.

29.5 Program Success

NABERS Energy for offices is considered by many to have been successful, as over 72% of the Australian national office market has now been rated with either a base building or whole building rating. Factors behind the success of the tool are largely attributed to its ability to differentiate between the base building and tenants energy end uses and strong government support.[6][7] Far fewer tenancy energy ratings have been conducted however and there has also been far less uptake of the other tools.

29.6   Use in Australian Energy Programs & Policy

While NABERS Energy is a voluntary rating scheme for buildings, its success has been at least partly driven by its extensive use in energy initiatives by government and industry throughout Australia. Some programs include:

- **The NSW Government Resource Efficiency Policy (GREP):** The most recent iteration of a series of NSW government procurement policies that set out NABERS targets in government leasing criteria. In the GREP, government tenants require a building to have a 4.5 star NABERS rating. It also states a 4.5 star energy rating as a minimum criteria for government data centres. Similar policies are in place other states and territories, as well as the Australian government (the 'Energy Efficiency in Government Operations' policy).

- **Emissions Reduction Fund:** the centrepiece of Australia's carbon abatement strategy, to begin operating in early 2015.[8] NABERS Energy is used in the commercial buildings methodology, to calculate and ensure the carbon abatement achieved by project proponents is real [9]

- **Energy Savings Scheme (ESS):** a New South Wales state energy program where commercial buildings can obtain Energy Saving Certificates (ESC) for energy efficiency projects, which can be sold to the market. A NABERS Energy ratings are used to demonstrate the energy savings achieved by the project.[10]

- **Green Building Fund:** a former Australian Government program, where commercial buildings could obtain up to 50% of capital funding for energy efficiency projects. The program used NABERS Energy ratings to ensure the savings effectively occurred, as well as to calculate the total amount of energy and emissions saved.[11]

- **City Switch:** an initiative that supports commercial office tenants to improve energy efficiency, run by a coalition of local councils throughout Australia. City Switch uses NABERS Energy as its key indicator of energy performance, and provides assistance to its members to achieve a rating of 4 stars or higher [12]

29.7   Use in Australian Legislation

- **The Building Energy Efficiency Disclosure Act 2010:** Australian government legislation that requires owners of office buildings to disclose the energy efficiency of the building to prospect tenants or buyers. Known operationally as the Commercial Building Disclosure (CBD) program, a certified NABERS Energy rating is the main energy efficiency indicator required of building owners.[13]

29.8   Use Internationally

- **NABERSNZ:** The Energy Efficiency and Conservation Authority (EECA) in New Zealand licensed NABERS in 2013 to create NABERSNZ.

- **The Global Real Estate Sustainability Benchmark (GRESB):** The GRESB is a global standard for portfolio-level sustainability assessment in real estate.The GRESB benchmark addresses issues including corporate sustainability strategy, policies and objectives, environmental performance monitoring, and the use of high-quality voluntary rating tools such as NABERS.

- **The Climate Bonds Initiative (CBI):** The CBI creates Climate Bonds Standards, which provide a Fair Trade-like labelling system for bonds, designed to make it easier for investors to work out what sorts of investments genuinely contribute to addressing climate change. Data from NABERS Energy rating reports can be used in Climate Bond reporting under the Climate Bonds Standard for Low Carbon Commercial Buildings.

- **NABERS IE in India:** One NABERS Indoor Environment rating has been conducted in India, at the Paraharpur Business Centre. The rating was certified in May 2015.

Australian government legislation that requires owners of office buildings to disclose the energy efficiency of the building to prospect tenants or buyers. Known operationally as the Commercial Building Disclosure (CBD) program, a certified NABERS Energy rating is the main energy efficiency indicator required of building owners.[14]

29.9   References

[1] http://www.nabers.gov.au/public/WebPages/ ContentStandard.aspx?module=30&template=3&include= AccreditAssessor.htm&side=AssessorTertiary.htm

[2] The NABERS Rating calculator

[3] http://www.nabers.gov.au/public/WebPages/
DocumentHandler.ashx?docType=3&id=85&attId=0

[4] http://www.nabers.gov.au/public/WebPages/
DocumentHandler.ashx?docType=3&id=85&attId=0

[5] http://www.nabers.gov.au/public/WebPages/
DocumentHandler.ashx?docType=3&id=85&attId=0

[6] http://www.buildingrating.org/content/
what-explains-success-nabers

[7] http://icebo2012.com/cms/resources/uploads/files/papers/
43%20NABERS%2012%20years%20paper%20-%
20final.pdf

[8] http://www.abc.net.au/news/2014-11-02/
emissions-reduction-fund-direct-action-begin-allocating-money/
5860762

[9] http://www.environment.gov.au/climate-change/
emissions-reduction-fund/methods

[10] http://www.ess.nsw.gov.au/Projects_and_equipment/
Commercial_buildings_NABERS

[11] http://www.business.gov.au/grants-and-assistance/
closed-programs/gbf/Pages/default.aspx

[12] http://www.cityswitch.net.au/Nabers/Howmuchdoesitcost/
CitySwitchNABERSEnergyGrantprogram.aspx

[13] http://www.cbd.gov.au/get-and-use-a-rating/
nabers-energy-for-offices-star-ratings

[14] http://www.cbd.gov.au/get-and-use-a-rating/
nabers-energy-for-offices-star-ratings

Chapter 30

New Suburbanism

New Suburbanism is an urban design movement which intends to improve on existing suburban or exurban designs.[1] New Suburbanists seek to establish an alternative between a dichotomy of the centripedal city and centrifugal suburb,[2] by features such as rear-loading garages and walking-focused landscaping.[3]

30.1 Locations

- Kyle, Texas

30.2 References

[1]

[2] STEPHEN MARSHALL (1978). "The Emerging 'Silicon Savanna': From Old Urbanism to New Suburbanism" **32**. Alexandrine Press. pp. 267–280. Retrieved 26 February 2015.

[3] "New Suburbanism Growing In Southern California". Barternews.com. Retrieved 2013-09-01.

Chapter 31

New Urbanism

"Neotraditionalism" redirects here. For other uses, see Neotraditional.

New Urbanism is an urban design movement which pro-

Seaside, Florida

Market Street, Celebration, Florida

motes walkable neighborhoods containing a range of housing and job types.[1] It arose in the United States in the early 1980s, and has gradually influenced many aspects of real estate development, urban planning, and municipal land-use strategies.

New Urbanism is strongly influenced by urban design practices that were prominent until the rise of the automobile prior to World War II; it encompasses principles such as traditional neighborhood design (TND) and transit-oriented development (TOD).[2] It is also related to regionalism, environmentalism, and smart growth.

The organizing body for New Urbanism is the Congress for the New Urbanism, founded in 1993. Its foundational text is the *Charter of the New Urbanism*, which begins:

> We advocate the restructuring of public policy and development practices to support the following principles: neighborhoods should be diverse in use and population; communities should be designed for the pedestrian and transit as well as the car; cities and towns should be shaped

by physically defined and universally accessible public spaces and community institutions; urban places should be framed by architecture and landscape design that celebrate local history, climate, ecology, and building practice.[3]

New Urbanists support regional planning for open space, context-appropriate architecture and planning, and the balanced development of jobs and housing. They believe their strategies can reduce traffic congestion, increase the supply of affordable housing, and rein in suburban sprawl. The *Charter of the New Urbanism* also covers issues such as historic preservation, safe streets, green building, and the re-development of brownfield land. The ten Principles of Intelligent Urbanism also phrase guidelines for new urbanist approaches.

Architecturally, new urbanist developments are often accompanied by New Classical, postmodern, or vernacular styles, although that is not always the case.

New Broad Street, Baldwin Park, Florida

Beach Drive, St. Petersburg, Florida

31.1 Background

Until the mid 20th century, cities were generally organized into and developed around mixed-use walkable neighborhoods. For most of human history this meant a city that was entirely walkable, although with the development of mass transit the reach of the city extended outward along transit lines, allowing for the growth of new pedestrian communities such as streetcar suburbs. But with the advent of cheap automobiles and favorable government policies, attention began to shift away from cities and towards ways of growth more focused on the needs of the car.[4] Specifically, after World War II urban planning largely centered around the use of municipal zoning ordinances to segregate residential from commercial and industrial development, and focused on the construction of low-density single-family detached houses as the preferred housing format for the growing middle class. The physical separation of where people live from where they work, shop and frequently spend their recreational time, together with low housing density, which often drastically reduced population density relative to historical norms, made automobiles indispensable for practical transportation and contributed to the emergence of a culture of automobile dependency.

This new system of development, with its rigorous separation of uses, arose after World War II and became known as "conventional suburban development"[5] or pejoratively as urban sprawl. The majority of U.S. citizens now live in suburban communities built in the last fifty years, and automobile use per capita has soared.

Although New Urbanism as an organized movement would

Celebration, FL Post Office, designed by architect Michael Graves

only arise later, a number of activists and thinkers soon began to criticize the modernist planning techniques being put into practice. Social philosopher and historian Lewis Mumford criticized the "anti-urban" development of postwar America. *The Death and Life of Great American Cities,* written by Jane Jacobs in the early 1960s, called for planners to reconsider the single-use housing projects, large car-dependent thoroughfares, and segregated commercial centers that had become the "norm."

Rooted in these early dissenters, the ideas behind New Urbanism began to solidify in the 1970s and 80s with the urban visions and theoretical models for the reconstruction of the "European" city proposed by architect Leon Krier, and the pattern language theories of Christopher Alexander. The term "new urbanism" itself started being used in this context in the mid-1980s,[6][7] but it wasn't until the early 1990s that it was commonly written as a proper noun capitalized.[8]

In 1991, the Local Government Commission, a private nonprofit group in Sacramento, California, invited architects Peter Calthorpe, Michael Corbett, Andrés Duany, Elizabeth Moule, Elizabeth Plater-Zyberk, Stefanos Polyzoides, and Daniel Solomon to develop a set of community principles for land use planning. Named the *Ahwahnee Principles* (after Yosemite National Park's Ahwahnee Hotel), the commission presented the principles to about one hundred government officials in the fall of 1991, at its first Yosemite Conference for Local Elected Officials.[9]

Calthorpe, Duany, Moule, Plater-Zyberk, Polyzoides, and Solomon founded the Chicago-based Congress for the New Urbanism in 1993. The CNU has grown to more than 3,000 members, and is the leading international organization promoting New Urbanist design principles. It holds annual Congresses in various U.S. cities.

In 2009, co-founders Elizabeth Moule, Hank Dittmar, and Stefanos Polyzoides authored the Canons of Sustainable Architecture and Urbanism to clarify and detail the relationship between New Urbanism and sustainability. The Canons are "a set of operating principles for human settlement that reestablish the relationship between the art of building, the making of community, and the conservation of our natural world." They promote the use of passive heating and cooling solutions, the use of locally obtained materials, and in general, a "culture of permanence." [10]

New Urbanism is a broad movement that spans a number of different disciplines and geographic scales. And while the conventional approach to growth remains dominant, New Urbanist principles have become increasingly influential in the fields of planning, architecture, and public policy.[11]

31.2 Defining elements

Andrés Duany and Elizabeth Plater-Zyberk, two of the founders of the Congress for the New Urbanism, observed mixed-use streetscapes with corner shops, front porches, and a diversity of well-crafted housing while living in one of the Victorian neighborhoods of New Haven, Connecticut. They and their colleagues observed patterns including the following:

Prospect New Town in Longmont, Colorado, showing a mix of aggregate housing and traditional detached homes

A park in Celebration, Florida

- The neighborhood has a discernible center. This is often a square or a green and sometimes a busy or memorable street corner. A transit stop would be located at this center.

- Most of the dwellings are within a five-minute walk of the center, an average of roughly 0.25 miles (0.40 km).

- There are a variety of dwelling types — usually houses, rowhouses, and apartments — so that younger and older people, singles and families, the poor and the wealthy may find places to live.

- At the edge of the neighborhood, there are shops and offices of sufficiently varied types to supply the weekly needs of a household.

- A small ancillary building or garage apartment is permitted within the backyard of each house. It may be used as a rental unit or place to work (for example, an office or craft workshop).

- An elementary school is close enough so that most children can walk from their home.

- There are small playgrounds accessible to every dwelling — not more than a tenth of a mile away.

- Streets within the neighborhood form a connected network, which disperses traffic by providing a variety of pedestrian and vehicular routes to any destination.

- The streets are relatively narrow and shaded by rows of trees. This slows traffic, creating an environment suitable for pedestrians and bicycles.

- Buildings in the neighborhood center are placed close to the street, creating a well-defined outdoor room.

- Parking lots and garage doors rarely front the street. Parking is relegated to the rear of buildings, usually accessed by alleys.

- Certain prominent sites at the termination of street vistas or in the neighborhood center are reserved for civic buildings. These provide sites for community meetings, education, and religious or cultural activities.

31.3 Terminology

A Mediterranean Revival house in Celebration, Florida

Several terms are viewed either as synonymous, included in, or overlapping with the New Urbanism. The terms Neo-traditional Development[12] or Traditional Neighborhood Development are often associated with the New Urbanism. These terms generally refer to complete New Towns or new neighborhoods, often built in traditional architectural styles, as opposed to smaller infill and redevelopment projects. The term Traditional Urbanism has also been used to describe the New Urbanism by those who object to the "new" moniker. The term "Walkable Urbanism" was proposed as an alternative term by developer and professor Christopher Leinberger.[13] Many debate whether Smart Growth and the New Urbanism are the same or whether substantive differences exist between the two; overlap exists in membership and content between the two movements. Placemaking is another term that is often used to signify New Urbanist efforts or those of like-minded groups. The term Transit-Oriented Development is sometimes cited as being coined by prominent New Urbanist Peter Calthorpe[14] and is heavily promoted by New Urbanists. The term Sustainable development is sometimes associated with the New Urbanism as there has been an increasing focus on the environmental benefits of New Urbanism associated with the rise of the term sustainability in the 2000s, however, this has caused some confusion as the term is also used by the United Nations and Agenda 21 to include human development issues (e.g., developing country) that exceed the scope of land development intended to be addressed by the New Urbanism or Sustainable Urbanism. The term livability or livable communities has also become popular under U.S. President Barack Obama's administration[15] though it dates back at least to the mid-1990s when the term was used by the Local Government Commission.[16]

Planning magazine discussed the proliferation of "urbanisms" in an article in 2011 titled "A Short Guide to 60 of the Newest Urbanisms." [17] Several New Urbanists have popularized terminology under the umbrella of the New Urbanism including Sustainable Urbanism and Tactical Urbanism[18] (of which Guerrilla Urbanism can be viewed as a subset). The term Tactical Urbanism was coined by Frenchman Michel de Certau in 1968 and revived in 2011 by New Urbanist Mike Lydon and the co-authors of the Tactical Urbanism Guide.[19] In 2011 Andres Duany authored a book that used the term Agrarian Urbanism to describe an agriculturally-focused subset of New Urbanist town design.[20] In 2013 a group of New Urbanists led by CNU co-founder Andres Duany began a research project under the banner of Lean Urbanism[21] which purported to provide a bridge between Tactical Urbanism and the New Urbanism.

Other terms have surfaced in reaction to the New Urbanism intended to provide a contrast, alternative to, or a refine-

A Key West style house in Baldwin Park, Florida

New urbanist Sankt Eriksområdet quarter in Stockholm, Sweden, built in the 1990s. (More photos)

Mixed use pedestrian-friendly street in Bitola, Republic of Macedonia.

ment of the New Urbanism. Some of these terms include Everyday Urbanism by Harvard Professor Margaret Crawford, John Chase, and John Kaliski,[22] Ecological Urbanism, and True Urbanism by architect Bernard Zyscovich. Landscape urbanism was popularized by Charles Waldheim who explicitly defined it as in opposition to the New Urbanism in his lectures at Harvard University.[23] A book called Landscape Urbanism and its Discontents, edited by Andres Duany and Emily Talen, specifically addressed the tension between these two views of urbanism.[24] Michael E. Arth promotes what he describes as a variant of the New Urbanism called the New Pedestrianism, which is intended to be a more pedestrian-oriented and traces its origins to a 1929 planned community in Radburn, New Jersey.[25]

31.4 Organizations

The primary organization promoting the New Urbanism in the United States is the Congress for the New Urbanism (CNU). The Congress for the New Urbanism (CNU) is the leading organization promoting walkable, mixed-use neighborhood development, sustainable communities and healthier living conditions. CNU members promote the principles of CNU's Charter and the hallmarks of New Urbanism, including:

- Livable streets arranged in compact, walkable blocks.

- A range of housing choices to serve people of diverse ages and income levels.

- Schools, stores and other nearby destinations reachable by walking, bicycling or transit service.

- An affirming, human-scaled public realm where appropriately designed buildings define and enliven streets and other public spaces.

The CNU has met annually since 1993 when they held their first general meeting in Alexandria, Virginia, with approximately 100 attendees. By 2008 the Congress was drawing 2,000 to 3,000 attendees to the annual meetings.

The CNU began forming local and regional chapters circa 2004 with the founding of the New England and Florida Chapters.[26] By 2011 there were 16 official chapters and interest groups for 7 more. As of 2013, Canada hosts two full CNU Chapters, one in Ontario (CNU Ontario), and one in British Columbia (Cascadia) which also includes a portion of the north-west US states.

While the CNU has international participation in Canada, sister organizations have been formed in other areas of the world including the Council for European Urbanism

(CEU),[27] the Movement for Israeli Urbanism (MIU) and the Australian Council for the New Urbanism.

By 2002 chapters of Students for the New Urbanism began appearing at universities including the Savannah College of Art and Design, University of Georgia, University of Notre Dame, and the University of Miami. In 2003, a group of younger professionals and students met at the 11th Congress in Washington, D.C. and began developing a "Manifesto of the Next Generation of New Urbanists". The Next Generation of New Urbanists held their first major session the following year at the 12th meeting of the CNU in Chicago in 2004. The group has continued meeting annually as of 2014 with a focus on young professionals, students, new member issues, and ensuring the flow of fresh ideas and diverse viewpoints within the New Urbanism and the CNU. Spinoff projects of the Next Generation of the New Urbanists include the Living Urbanism publication first published in 2008 and the first Tactical Urbanism Guide.[28]

The CNU has spawned publications and research groups. Publications include the *New Urban News* and the *New Town Paper*. Research groups have formed independent nonprofits to research individual topics such as the Form-Based Codes Institute, The National Charrette Institute and the Center for Applied Transect Studies.

In the United Kingdom New Urbanist and European urbanism principles are practised and taught by the The Prince's Foundation for the Built Environment. Other organisations promote New Urbanism as part of their remit, such as INTBAU, A Vision of Europe, and others.

The CNU and other national organizations have also formed partnerships with like-minded groups. Organizations under the banner of Smart Growth also often work with the Congress for the New Urbanism. In addition the CNU has formed partnerships on specific projects such as working with the United States Green Building Council and the Natural Resources Defense Council to develop the LEED for Neighborhood Development standards,[1] and with the Institute of Transportation Engineers to develop a Context Sensitive Solutions (CSS) Design manual.

31.5   Film

The New Urbanism Film Festival[29] was held in 2013 and 2014 in Los Angeles to highlight films and short films about the New Urbanism and related topics. The 2011 film Urbanized by Gary Hustwit featured then CNU Board Chair Ellen Dunham-Jones[30] and other urban thinkers on the international story of urbanization including the New Urbanist efforts in the United States.

The 2004 documentary *The End of Suburbia: Oil De-*

pletion and the Collapse of the American Dream argues that the depletion of oil will result in the demise of the sprawl-type development.[31] *New Urban Cowboy: Toward a New Pedestrianism*, a feature length 2008 documentary about urban designer Michael E. Arth, explains the principles of his New Pedestrianism, a more ecological and pedestrian-oriented version of New Urbanism.[25][32] The film also gives a brief history of New Urbanism, and chronicles the rebuilding of an inner city slum into a model of New Pedestrianism and New Urbanism called The Garden District.[33][34]

31.6   Criticism

New Urbanism has drawn both praise and criticism from all parts of the political spectrum.[35] It has been criticized both for being a social engineering scheme and for failing to address social equity and for both restricting private enterprise and for being a deregulatory force in support of private sector developers.

In an interview in *Reason*, a right-libertarian magazine, professor Peter Gordon, a professor of Urban Planning from University of Southern California, spoke in favor of suburbanization, stating that the New Urbanism ignores consumer preference the free market and that cities have moved towards car-oriented development because that is what people want.[36]

On the other hand, journalist Alex Marshall has decried New Urbanism as essentially a marketing scheme that repackages conventional suburban sprawl behind a façade of nostalgic imagery and empty, aspirational slogans.[37] In a 1996 article in Metropolis Magazine, Marshall denounced New Urbanism as "a grand fraud".[38] The attack continued in numerous articles, including an opinion column in the Washington *Post* in September of the same year,[39] and in Marshall's first book, *How Cities Work: Suburbs, Sprawl, and the Roads Not Taken*[40]

Critics have asserted that the effectiveness claimed for the New Urbanist solution of mixed income developments lacks statistical evidence.[41] Independent studies have supported the idea of addressing poverty through mixed-income developments,[42][43] but the argument that New Urbanism produces such diversity has been challenged from findings from one community in Canada.[44]

Some parties have criticized the New Urbanism for being too accommodating of motor vehicles and not going far enough to promote walking, cycling, and public transport. The Charter of the New Urbanism states that "communities should be designed for the pedestrian and transit as well as the car".[45] Some critics suggest that communities should exclude the car altogether in favor of car-free developments. Michael E Arth proposes new pedestrianism as a way to further elevate the status of pedestrians by focusing on pedestrian-only paths. Steve Melia proposes the idea of "filtered permeability" (see Permeability (spatial and transport planning)) which increases the connectivity of the pedestrian and cycling network resulting in a time and convenience advantage over drivers while still limited the connectivity of the vehicular network and thus maintaining the safety benefits of cul de sacs and horseshoe loops in resistance to property crime.[46]

In response to critiques of a lack of evidence for the New Urbanism's claimed environmental benefits, a rating system for neighborhood environmental design, LEED-ND, was developed by the U.S. Green Building Council, Natural Resources Defense Council, and the Congress for the New Urbanism,[47] to quantify the sustainability of New Urbanist neighborhood design.[1][48] New Urbanist and board member of CNU, Doug Farr has taken a step further and coined Sustainable Urbanism, which combines New Urbanism and LEED-ND to create walkable, transit-served urbanism with high performance buildings and infrastructure.

New Urbanism has been criticized for being a form of centrally planned, large-scale development, "instead of allowing the initiative for construction to be taken by the final users themselves".[49] It has been criticized for asserting universal principles of design instead of attending to local conditions.[1][50]

31.7 Examples

Main article: Examples of New Urbanism

31.7.1 United States

New Urbanism is having a growing influence on how and where metropolitan regions choose to grow. At least fourteen large-scale planning initiatives are based on the principles of linking transportation and land-use policies, and using the neighborhood as the fundamental building block of a region. Miami, Florida, has adopted the most ambitious New Urbanist-based zoning code reform yet undertaken by a major U.S. city.[51]

More than six hundred new towns, villages, and neighborhoods in the U.S. following New Urbanist principles are planned or under construction. Hundreds of new, small-scale, urban and suburban infill projects are under way to reestablish walkable streets and blocks. In Maryland and several other states, New Urbanist principles are an integral part of *smart growth* legislation.

In the mid-1990s, the U.S. Department of Housing and Urban Development (HUD) adopted the principles of the New Urbanism in its multibillion-dollar program to rebuild public housing projects nationwide. New Urbanists have planned and developed hundreds of projects in infill locations. Most were driven by the private sector, but many, including HUD projects, used public money.

University Place in Memphis

In 2010 University Place in Memphis became the second only U.S. Green Building Council (USGBC) LEED certified neighborhood. LEED ND (neighborhood development) standards integrates principles of smart growth, urbanism and green building and were developed through a collaboration between USGBC, Congress for the New Urbanism, and the Natural Resources Defense Council. University Place, developed by McCormack Baron Salazar, is a 405-unit, 30-acre, mixed-income, mixed use, multigenerational, HOPE VI grant community that revitalized the severely distressed Lamar Terrace public housing site.[52]

The Cotton District

The Cotton District in Starkville, Mississippi, was the first New Urbanist development, begun in 1968 long before the New Urbanism movement was organized.[53] The District borders Mississippi State University, and consists mostly of residential rental units for college students along with restaurants, bars and retail. The Cotton District got its name because it is built in the vicinity of an old cotton mill.

Seaside

Seaside, Florida, the first fully New Urbanist town, began development in 1981 on eighty acres (324,000 m^2) of Florida Panhandle coastline. It was featured on the cover of the Atlantic Monthly in 1988, when only a few streets were completed, and has become internationally famous for its architecture, and the quality of its streets and public spaces.

Seaside is now a tourist destination and appeared in the 1998 movie *The Truman Show*. Lots sold for $15,000 in the early 1980s, and slightly over a decade later, the price had escalated to about $200,000. Today, most lots sell for more than a million dollars, and some houses top $5 million.

Stapleton

The site of the former Stapleton International Airport in Denver, Colorado, closed in 1995, is now being redeveloped by Forest City Enterprises. Stapleton is expected to be home to at least 30,000 residents, six schools and 2 million square feet (180,000 m^2) of retail. Construction began in 2001.[54][55] Northfield Stapleton, one of the development's major retail centers, recently opened.

San Antonio

In 1997 San Antonio, Texas, as part of a new master plan, created new regulations called the Unified Development Code (UDC), largely influenced by New Urbanism. One feature of the UDC is six unique land development patterns that can be applied to certain districts: Conservation Development, Commercial Center Development, Office or Institutional Campus Development, Commercial Retrofit Development, Tradition Neighborhood Development, Transit Oriented Development. Each district has specific standards and design regulation. The six development patterns were created to reflect existing development patterns.[56]

Mountain House

Mountain House, one of the latest New Urbanist projects in the United States, is a new town located near Tracy, California. Construction started in 2001. Mountain House will consist of 12 villages, each with its own elementary school, park, and commercial area. In addition, a future train station, transit center and bus system are planned for Mountain House.

Mesa del Sol

Mesa del Sol, New Mexico—the largest New Urbanist project in the United States—was designed by architect Peter Calthorpe, and is being developed by Forest City Enterprises. Mesa del Sol may take five decades to reach full build-out, at which time it should have 38,000 residential units, housing a population of 100,000; a 1,400-acre (5.7 km^2) industrial office park; four town centers; an urban center; and a downtown that would provide a twin city within Albuquerque.

I'On

Located in Mount Pleasant, South Carolina, I'On is a traditional neighborhood development, mixed with a new urbanism styled architecture, reflecting on the building designs of the nearby downtown areas of Charleston, South Carolina. Founded on April 30, 1995, I'On was designed by the town planning firms of Dover, Kohl & Partners and Duany Plater-Zyberk & Company, and currently holds over 750 single family homes. Features of the community include extensive sidewalks, shared public greens and parks,

trails and a grid of narrow, traffic calming streets. Most homes are required to have a front porch of not less than eight feet (2.46 m) in depth. Floor heights of 10 feet (3.1 m), raised foundations and smaller lot sizes give the community a dense, vertical feel.

Haile Plantation

Haile Plantation, Florida, is a 2,600 household (1,700 acres (6.9 km^2)) development of regional impact southwest of the city of Gainesville, within Alachua County. Haile Village Center is a traditional neighborhood center within the development. It was originally started in 1978 and completed in 2007. In addition to the 2,600 homes the neighborhood consists of two merchant centers (one a New England narrow street village and the other a chain grocery strip mall). There are also two public elementary schools and an 18-hole golf course.

Celebration, Florida

In June 1996, the Walt Disney Company unveiled its 5,000 acre (20 km^2) town of Celebration, near Orlando, Florida. Celebration opened its downtown in October 1996, relying heavily on the experiences of Seaside, whose downtown was nearly complete. Disney shuns the label New Urbanism, calling Celebration simply a "town."

Celebration's Downtown has become one of the area's most popular tourist destinations making the community a showcase for New Urbanism as a prime example of the creation of a "sense of place".[57]

Jersey City

The construction of the Hudson Bergen Light Rail in Hudson County, New Jersey has spurred transit-oriented development. In Jersey City, two projects are planned to transform brownfield sites, both of which have required remediation of toxic waste by previous owners. Bayfront, once site of a Honeywell plant is a 100 acres (0.40 km^2) site on the Hackensack River, and is nearby the planned West Campus of New Jersey City University. Canal Crossing, named for the former Morris Canal, was once partially owned by PPG Industries, and is a 117 acres (0.47 km^2) site west of Liberty State Park.

Old York Village, Chesterfield Township, New Jersey

The sparsely developed agricultural Township of Chesterfield in New Jersey covers approximately 21.61 square miles (56.0 km^2) and has made farmland preservation a priority since the 1970s. Chesterfield has permanently preserved more than 7,000 acres (28 km^2) of farmland through state and county programs and a township-wide transfer of development credits program that directs future growth to a designated "receiving area" known as Old York Village. Old York Village is a neo-traditional, new urbanism town on 560 acres (2.3 km^2) incorporating a variety of housing types, neighborhood commercial facilities, a new elementary school, civic uses, and active and passive open space areas with preserved agricultural land surrounding the planned village. Construction began in the early 2000s and a significant percentage of the community is now complete. Old York Village was the winner of the American Planning Association National Outstanding Planning Award in 2004.[58][59][60]

Civita

Civita is a sustainable, transit-oriented 230-acre masterplanned village under development in the Mission Valley area of San Diego, California, United States. Located on a former quarry site, the urban-style village is organized around a 19-acre community park that cascades down the terraced property.[61]

Civita development plans call for 60 to 70 acres of parks and open space, 4,780 residences (including approximately 478 affordable units), an approximately 480,000-square-foot retail center, and 420,000 square feet for an office/business campus.[62][63]

Del Mar Station

Del Mar (Los Angeles Metro station), which won a Congress for the New Urbanism Charter Award in 2003,[64] is a transit-oriented development surrounding a prominent Metro Rail stop on the Gold Line, which connects Los Angeles and Pasadena. Located at the southern edge of downtown Pasadena, it serves as a gateway to the city with 347 apartments, out of which 15% are affordable units. Approximately 20,000 square feet of retail is linked with a network of public plazas, paseos and private courtyards. The 3.4-acre, $77 million project sits above a 1,200-car multilevel subterranean parking garage, with 600 spaces dedicated to transit. The light rail right of way, detailed as a public street, bisects the site. It was designed by Moule & Polyzoides.[65][66]

31.7.2 Other countries

New Urbanism is closely related to the Urban village movement in Europe. They both occurred at similar times

and share many of the same principles although urban villages has an emphasis on traditional city planning. In Europe many brown-field sites have been redeveloped since the 1980s following the models of the traditional city neighbourhoods rather than Modernist models. One well-publicized example is Poundbury in England, a suburban extension to the town of Dorchester, which was built on land owned by the Duchy of Cornwall under the overview of Prince Charles. The original masterplan was designed by Leon Krier. A report carried out after the first phase of construction found a high degree of satisfaction by residents, although the aspirations to reduce car dependency had not been successful. Rising house prices and a perceived premium have made the open market housing unaffordable for many local people.[67]

The Council for European Urbanism (CEU), formed in 2003, shares many of the same aims as the U.S.'s New Urbanists. CEU's Charter is a development of the Congress for the New Urbanism Charter revised and reorganised to relate better to European conditions. An Australian organisation, Australian Council for New Urbanism has since 2001 run conferences and events to promote New Urbanism in that country. A New Zealand Urban Design Protocol was created by the Ministry for the Environment in 2005.

There are many developments around the world that follow New Urbanist principles to a greater or lesser extent:

Europe

Example of Neo-Traditionalism at Le Plessis Robinson

- Le Plessis-Robinson, a 21st-century example of neo-traditionalism,[68] in the south-west of Paris. This city is in the process of transforming itself, destroying old modern blocklike buildings and replacing them with traditional buildings and houses in one of the biggest worldwide projects with Val d'Europe. In 2008 the city was nominated best architectural project of the European Union.[69]

The new marketplace of Le Plessis-Robinson

Jakriborg, started in the late 1990s near Malmö

- Val d'Europe, east of Paris, France. Developed by Disneyland Resort Paris, this town is a kind of European counterpart to Walt Disney World Celebration City.

- Jakriborg, in Southern Sweden, is a recent example of the New Urbanist movement.

- *Sankt Eriksområdet* quarter in Stockholm, Sweden, built in the 1990s.

- Other developments can be found in the Netherlands, at Heulebrug, part of Knokke-Heist, in Belgium, and Fonti di Matilde in San Bartolomeo (outside of Reggio Emilia),[70] Italy.

Americas

- Orchid Bay, Belize, is one of the largest New Urbanist projects in Central America and the Caribbean.

- Las Catalinas, Costa Rica, is a coastal town in the Guanacaste Province of Northwest Costa Rica. Envisioned as a compact, walkable beach town, Las Catalinas was founded in 2006 by Charles Brewer (businessman) and incorporates many of the principles of New Urbanism.

- McKenzie Towne is a New Urbanist development which commenced in 1995 by Carma Developers LP in Calgary.

- Cornell, within the city of Markham, Ontario, was designed with walkable neighborhoods, density to support public transit, a variety of housing types and retail.[71]

- New Amherst is a new urbanist development in the town of Cobourg, Ontario.

- UniverCity, beside the Simon Fraser University campus on Burnaby Mountain in Burnaby, British Columbia, is an award-winning sustainable community that is designed to be walkable, dense, and well connected to public transit networks.

Asia

- The structure plan for Thimphu, Bhutan, follows Principles of Intelligent Urbanism, which share underlying axioms with the New Urbanism.

- Tullimbar Village, NSW Australia, is a new development which follows the principles of New Urbanism.[72]

Africa

There are several such developments in South Africa. The most notable is Melrose Arch in Johannesburg. Triple Point is a comparable mixed-use development in East London, in Eastern Cape province. The development, announced in 2007, comprises 30 hectares. It is made up of three apartment complexes together with over 30 residential sites as well as 20,000 sq m of residential and office space. The development is valued at over R2 billion ($250 million).[73]

31.8   See also

31.9   References

[1] Boeing; et al. (2014). "LEED-ND and Livability Revisited". *Berkeley Planning Journal* **27**: 31–55. Retrieved 2015-04-15.

[2] Kelbaugh, Douglas S. 2002. Repairing the American Metropolis: Common Place Revisited. Seattle: University of Washington Press. 161.

[3] Charter of the New Urbanism

[4] Kunstler, James Howard. 1998. *Home from nowhere: remaking our everyday world for the twenty-first Century*. A Touchstone book. New York, NY: Simon & Schuster. p.28.

[5] David Gordon and Shayne Vipond: "Gross Density and New Urbanism: Comparing Conventional and New Urbanist Suburbs in Markham, Ontario". Journal of the American Planning Association, 1939-0130, Volume 71, Issue 1, 2005, pages 41–54

[6] Reid, Barton (1985). *The New Urbanism as a Way of Life: The Relationship Between Inner City Revitalization in Canada and the Rise of the New Middle Class*. Retrieved 2014-09-06.

[7] Meinig, Donald (1986). *The Shaping of America: A Geographical Perspective on 500 Years of History, Volume 2*. Yale University Press. p. 255. ISBN 9780300173949. Retrieved 2014-09-06.

[8] *Urban Design Update: Newsletter of the Institute for Urban Design, Volumes 7-15*. Institute for Urban Design. 1991. Retrieved 2014-09-06.

[9] http://www.lgc.org/about/ahwahnee/principles

[10] The Canons of Sustainable Architecture and Urbanism |date=October 8, 2014

[11] Cozens, Paul Michael. 2008. New Urbanism, Crime and the Suburbs: A Review of the Evidence. Urban Policy and Research. 26(4):429-444.

[12] http://facweb.arch.ohio-state.edu/jnasar/crpinfo/research/NeoTradJPER2003.pdf

[13] Leinberger, Christopher (2009). *The Option of Urbanism*. District of Columbia: Island Press. ISBN 1597261378.

[14] http://www.archdaily.com/409612/does-china-s-urbanization-spell-doom-or-salvation-peter-calthorpe-weighs-in/

[15] http://usa.streetsblog.org/2011/05/04/president-obamas-transportation-bill-prioritizes-livability-high-speed-rail/

[16] http://www.newpartners.org/about/about-the-event

[17] Barnett, Jonathan (April 2014). "A Short Guide to 60 of the Newest Urbanisms" **77** (4). pp. 19–21. Retrieved 23 October 2014.

[18] http://www.citylab.com/design/2012/03/guide-tactical-urbanism/1387/

[19] http://www.cbsnews.com/news/tactical-urbanism-citizen-projects-go-mainstream/

[20] Duany, Andres (2011). *Garden Cities: Theory & Practice of Agrarian Urbanism*. The Prince's Foundation for the Built Environment. ISBN 1906384045.

[21] http://www.citylab.com/design/2014/03/why-andres-duany-so-focused-making-lean-urbanism-thing/8635/

[22] Chase, John (1999). *Everyday Urbanism*. The Monacelli Press. ISBN 1885254814.

[23] Mehaffy, Micheal. "The Landscape Urbanism: Sprawl in a Pretty Green Dress?". *Planetizen*.

[24] http://www.newsociety.com/Books/L/Landscape-Urbanism-and-its-Discontents

[25] Arth, Michael E. (2010). *Democracy and the Common Wealth: Breaking the Stranglehold of the Special Interests* Golden Apples Media, ISBN 978-0-912467-12-2. pp. 120-139, 363-386

[26] http://cnuflorida.org/

[27] http://www.ceunet.org/

[28] http://tacticalurbanismguide.com/

[29] http://newurbanismfilmfestival.com/

[30] http://www.hustwit.com/about-urbanized/

[31] http://www.endofsuburbia.com link to official website

[32] Website about *Democracy and the Common Wealth*

[33] Teri Pruden, "The New Urban Cowboy: Michael E. Arth transforms Cracktown into Historic Garden District in DeLand" DeLand Magazine, Jan-Feb, 2008. Pages 8, 9.

[34] New Urban Cowboy review in Carbusters Magazine, issue #32, Winter 2007/2008, page 26.

[35] Sharifi, Ayyoob (September 2015). "From Garden City to Eco-urbanism: The quest for sustainable neighborhood development". *Sustainable Cities and Society*. doi:10.1016/j.scs.2015.09.002.

[36] "Plan Obsolescence", *Reason*, June 1998

[37] See, *e.g.*, Alex Marshall, "Building New Urbanism: Less Filling, But Not So Tasty", *Builder Magazine*, 30 November 1999, p. ___. Print; archived on Marshall's web site, http://www.alexmarshall.org/2006/08/02/building-new-urbanism-less-filling-but-not-so-tasty/. Retrieved 1 November 2013.

[38] Alex Marshall, "Suburbs in Disguise", *Metropolis Magazine*, July 1996, p. 70, republished as "New Urbanism" in Busch, Akiko, ed., *Design is ... Words, Things, People, Buildings and Places* (New York:Metropolis Books/Princeton Architectural Press, 2002), p. 272; and as "Suburbs in Disguise" on Marshall's web site, http://www.alexmarshall.org/2007/08/31/suburbs-in-disguise/, retrieved 2 October 2013.

[39] Alex Marshall, "Putting Some 'City' Back In the Sub-urbs", Washington (D.C.) *Post*, 1 September 1996, p. C1, print, http://www.washingtonpost.com/wp-srv/local/longterm/library/growth/solutions/nokent.htm; retrieved 2 October 2013.

[40] U. of Texas Press 2000.

[41] Popkin, S. et al. (2004) A Decade of HOPE VI. The Urban Institute

[42] Goetz, Edward G. (2003) Clearing the Way: Deconcentrating the Poor in Urban America, The Urban Institute Press: Washington, DC

[43] Chaskin, R.J., Joseph, M.L., Webber, H.S. (2007) The Theoretical Basis for Addressing Poverty Through Mixed-Income Development. Urban Affairs Review 42 (3): 369-409.

[44] Grant, J. and K. Perrott (2009) Producing diversity in a new urbanism community. Town Planning Review 80 (3): 267-289.

[45] http://www.cnu.org/charter

[46] "Neighbourhoods Should be Made Permeable for Walking and Cycling but not for Cars", Steve Melia, Local Transport Today, January 23, 2008

[47] http://www.cnu.org/leednd

[48] http://www.usgbc.org/articles/getting-know-leed-neighborhood-development

[49] "A brief history of Peer-to-peer Urbanism", Nikos Salingaros and Federico Mena-Quintero, October 2010

[50] Grant, J. (2006) Planning the Good Community: New Urbanism in Theory and Practice. London: Routledge

[51] Miami Reforms

[52] Architecture Inc. Celebrates LEED-ND Certification of University Place in Memphis, Multi Housing News, May 18, 2011.

[53] The Town Paper, Vol. 4, No. 1 — December 2001/ January 2002

[54] DSST Web site

[55] http://www.usatoday.com/news/nation/2006-10-26-100-million_x.htm USA Today

[56] Greenburg, Ellen, 2004. Codifying New Urbanism: How to Reform Municipal Land Development Regulations. American Planning Association PAS Report Number 526

[57] Celebration Business Alliance, Sept 2010

[58] "Old York Village, Chesterfield Wins an American Planning Association Award for an Outstanding Project/ Program/ Tool"

[59] "Old York Village Implementing Smart Growth"

[60] "Master Plan Amendment: Township of Chesterfield"

[61] "Sand and gravel quarry becoming a sustainable, urban village". U-T San Diego. Retrieved 16 April 2014.

[62] Kirk, Patricia. "Civita: San Diego's New City within the City". Urban Land Magazine. Retrieved 16 April 2014.

[63] Leung, Lily. "Mission Valley's 230-acre Civita to debut". San Diego Union-Tribune. Retrieved 16 April 2014.

[64] "Charter Award Recipients", Congress for the New Urbanism (accessed 8 April 2015).

[65] "Del Mar Station Transit Village". Moule & Polyzoides. Retrieved 8 October 2014.

[66] Del Mar Station, Congress for the New Urbanism, September 14, 2007.

[67] WATSON, G., BENTLEY, I., ROAF, S. and SMITH, P., 2004. Learning from Poundbury, Research for the West Dorset District Council and the Duchy of Cornwall. Oxford Brookes University.

[68] http://www.planetizen.com/node/57600

[69] http://www.jeunesarchi.com

[70] A TASTE FOR THE PAST. Architect Pier Carlo Bontempi

[71] "Is new urbanism the answer to suburbia's dying communities?". Canadian Geographic. Retrieved 2011-01-31.

[72] Art, Screen. "Tullimbar Village :: Wollongong NSW :: Contemporary Lifestyle Community". *www.tullimbarvillage.com.au*. Retrieved 2015-05-29.

[73] "EAST LONDON GETS OWN MELROSE ARCH", *eProp.co.za*, 12 December 2007

31.10   Further reading

- Arth, Michael E., *The Labors of Hercules: Modern Solutions to 12 Herculean Problems*. 2007 Online edition. Labor IX: Urbanism Link to book

- Arth, Michael E. (2010). *Democracy and the Common Wealth: Breaking the Stranglehold of the Special Interests* Golden Apples Media, ISBN 978-0-912467-12-2. pp. 120–139, 363-386 Website about, and excerpts from *Democracy and the Common Wealth*

- Bohl, Charles C. "New Urbanism in the City: Potential Applications and Implications for Distressed Inner-City Neighborhoods." *Housing Policy Debate* 11.4 (2000): 761-801. (http://www.botsfor.no/publikasjoner/litteratur/new%20urbanism/new%20urbanism%20and%20the%20city%20by%20charles%20bohl.pdf)

- Brooke, Steven (1995). *Seaside*. Gretna, La.: Pelican Publishing Company. ISBN 0-88289-997-X

- Calthorpe, Peter (1993). *The Next American Metropolis: Ecology, Community, and the American Dream*. New York: Princeton Architectural Press. ISBN 1-878271-68-7

- Calthorpe, Peter and William Fulton (2001). *The Regional City: Planning for the End of Sprawl*. Washington, DC: Island Press. ISBN 1-55963-784-6

- Congress for the New Urbanism (1999). Leccese, Michael; and McCormick, Kathleen (Eds.), ed. *Charter of the New Urbanism*. McGraw-Hill Professional. ISBN 0-07-135553-7.

- Duany, Andres; Plater-Zyberk, Elizabeth; & Alminana, Robert (2003). *The New Civic Art: Elements of Town Planning*. New York: Rizzoli Publications. ISBN 0-8478-2186-2.

- Duany, Andres; Plater-Zyberk, Elizabeth; & Speck, Jeff (2000). *Suburban Nation: The Rise of Sprawl and the Decline of the American Dream*. North Point Press. ISBN 0-86547-557-1.

- Dutton, John A. (2001). *New American Urbanism: Re-forming the Suburban Metropolis*. Milano: Skira editore. ISBN 88-8118-741-8

- El Nasser, Haya (November 14, 2005). "Miss. Wal-Marts may apply 'new urbanism' in rebuilding". USA Today.

- Gallini, Jared. 2010. "Demographics and Their Relationship to the Characteristics of New Urbanism: A Preliminary Study" . Applied Research Projects, Texas State University-San Marcos. Paper 340.http://ecommons.txstate.edu/arp/340

- Jacobs, Jane (1992). *The Death and Life of Great American Cities*. New York: Vintage Books. ISBN 0-679-74195-X. Originally published: New York: Random House, (1961).

- Katz, Peter (1994). *The New Urbanism: Toward an Architecture of Community*. New York: McGraw-Hill. ISBN 0-07-033889-2

- Kunstler, James Howard (1994). *Geography Of Nowhere: The Rise And Decline of America's Man-Made Landscape*. New York: Simon & Schuster. ISBN 0-671-88825-0

- The New American Landscape: A New Urbanist's Perspective on Sildeshare

- Talen, Emily (2005). *New Urbanism & American Planning: The Conflict of Cultures*. New York: Routledge. ISBN 0-415-70133-3.

- Tagliaventi, Gabriele (2002). *New Urbanism*. Florence: Alinea. ISBN 88-8125-602-9.

- Steuteville, Robert, ed. (2009). *New Urbanism Best Practices Guide*. Ithaca: New Urban News. ISBN 09-7450-216-2.

- Waugh, David. 2004 Buying New Urbanism: A Study of New Urban Characteristics that Residents Value. Applied Research Project. Texas State University. http://ecommons.txstate.edu/arp/22/

31.11 External links

- Congress for the New Urbanism

- Australian Council for New Urbanism

- Council for European Urbanism

- NewUrbanism.org

- The Next Generation of New Urbanists

- A Vision of Europe

- Sustainable Urban Development Resource Guide

- Polis: Building Really Compact

Chapter 32

Permaculture

Permaculture is a system of agricultural and social design principles centered around simulating or directly utilizing the patterns and features observed in natural ecosystems. The term *permaculture* (as a systematic method) was first coined by Australians David Holmgren, then a graduate student, and his professor, Bill Mollison, in 1978. The word *permaculture* originally referred to "permanent agriculture",[1] but was expanded to stand also for "permanent culture", as it was seen that social aspects were integral to a truly sustainable system as inspired by Masanobu Fukuoka's natural farming philosophy.

It has many branches that include but are not limited to ecological design, ecological engineering, environmental design, construction and integrated water resources management that develops sustainable architecture, regenerative and self-maintained habitat and agricultural systems modeled from natural ecosystems.[2][3]

Mollison has said: "Permaculture is a philosophy of working with, rather than against nature; of protracted and thoughtful observation rather than protracted and thoughtless labor; and of looking at plants and animals in all their functions, rather than treating any area as a single product system."[4]

32.1 History

In 1929, Joseph Russell Smith took up an antecedent term as the subtitle for *Tree Crops: A Permanent Agriculture*, a book in which he summed up his long experience experimenting with fruits and nuts as crops for human food and animal feed.[5] Smith saw the world as an inter-related whole and suggested mixed systems of trees and crops underneath. This book inspired many individuals intent on making agriculture more sustainable, such as Toyohiko Kagawa who pioneered forest farming in Japan in the 1930s.[6]

The definition of permanent agriculture as that which can be sustained indefinitely was supported by Australian P. A. Yeomans in his 1964 book *Water for Every Farm*. Yeomans introduced an observation-based approach to land use in Australia in the 1940s, and the keyline design as a way of managing the supply and distribution of water in the 1950s.

Stewart Brand's works were an early influence noted by Holmgren.[7] Other early influences include Ruth Stout and Esther Deans, who pioneered no-dig gardening, and Masanobu Fukuoka who, in the late 1930s in Japan, began advocating no-till orchards, gardens and natural farming.[8]

32.2 Core tenets and principles of design

The three core tenets of permaculture are:[9][10][11]

- *Care for the earth*: Provision for all life systems to continue and multiply. This is the first principle, because without a healthy earth, humans cannot flourish.

- *Care for the people*: Provision for people to access those resources necessary for their existence.

- *Return of surplus*: Reinvesting surpluses back into the system to provide for the first two ethics. This includes returning waste back into the system to recycle into usefulness.[12] The third ethic is sometimes referred to as Fair Share to reflect that each of us should take no more than what we need before we reinvest the surplus.

Permaculture design emphasizes patterns of landscape, function, and species assemblies. It determines where these elements should be placed so they can provide maximum benefit to the local environment. The central concept of permaculture is maximizing useful connections between components and synergy of the final design. The focus of permaculture, therefore, is not on each separate element, but rather on the relationships created among elements by the way they are placed together; the whole becoming greater than the sum of its parts. Permaculture design therefore seeks to minimize waste, human labor, and energy input by building systems with maximal benefits between design elements to achieve a high level of synergy. Permaculture designs evolve over time by taking into account these relationships and elements and can become extremely complex systems that produce a high density of food and materials with minimal input.[13]

The design principles which are the conceptual foundation of permaculture were derived from the science of systems ecology and study of pre-industrial examples of sustainable land use. Permaculture draws from several disciplines including organic farming, agroforestry, integrated farming, sustainable development, and applied ecology.[14] Permaculture has been applied most commonly to the design of housing and landscaping, integrating techniques such as agroforestry, natural building, and rainwater harvesting within the context of permaculture design principles and theory.

32.3 Theory

32.3.1 Twelve design principles

Twelve Permaculture design principles articulated by David Holmgren in his *Permaculture: Principles and Pathways Beyond Sustainability*:[15]

1. *Observe and interact*: By taking time to engage with nature we can design solutions that suit our particular situation.

2. *Catch and store energy*: By developing systems that collect resources at peak abundance, we can use them in times of need.

3. *Obtain a yield*: Ensure that you are getting truly useful rewards as part of the work that you are doing.

4. *Apply self-regulation and accept feedback*: We need to discourage inappropriate activity to ensure that systems can continue to function well.

5. *Use and value renewable resources and services*: Make the best use of nature's abundance to reduce our consumptive behavior and dependence on non-renewable resources.

6. *Produce no waste*: By valuing and making use of all the resources that are available to us, nothing goes to waste.

7. *Design from patterns to details*: By stepping back, we can observe patterns in nature and society. These can form the backbone of our designs, with the details filled in as we go.

8. *Integrate rather than segregate*: By putting the right things in the right place, relationships develop between those things and they work together to support each other.

9. *Use small and slow solutions*: Small and slow systems are easier to maintain than big ones, making better use of local resources and producing more sustainable outcomes.

10. *Use and value diversity*: Diversity reduces vulnerability to a variety of threats and takes advantage of the unique nature of the environment in which it resides.

11. *Use edges and value the marginal*: The interface between things is where the most interesting events take place. These are often the most valuable, diverse and productive elements in the system.

12. *Creatively use and respond to change*: We can have a positive impact on inevitable change by carefully observing, and then intervening at the right time.

32.3.2 Layers

Suburban permaculture garden in Sheffield, UK with different layers of vegetation

Layers are one of the tools used to design functional ecosystems that are both sustainable and of direct benefit to humans. A mature ecosystem has a huge number of relationships between its component parts: trees, understory, ground cover, soil, fungi, insects, and animals. Because plants grow to different heights, a diverse community of life is able to grow in a relatively small space, as each layer is stacked one on top of another. There are generally seven recognized layers in a food forest, although some practitioners also include fungi as an eighth layer.[16]

1. The canopy: the tallest trees in the system. Large trees dominate but typically do not saturate the area, i.e. there exist patches barren of trees.

2. Understory layer: trees that revel in the dappled light under the canopy.

3. Shrub layer: a diverse layer of woody perennials of limited height. includes most berry bushes.

4. Herbaceous layer: Plants in this layer die back to the ground every winter (if winters are cold enough, that is). They do not produce woody stems as the Shrub layer does. Many culinary and medicinal herbs are in this layer. A large variety of beneficial plants fall into this layer. May be annuals, biennials or perennials

5. Soil surface/Groundcover: There is some overlap with the Herbaceous layer and the Groundcover layer; however plants in this layer grow much closer to the ground, grow densely to fill bare patches of soil, and often can tolerate some foot traffic. Cover crops retain soil and lessen erosion, along with green manures that add nutrients and organic matter to the soil, especially nitrogen

6. Rhizosphere: Root layers within the soil. The major components of this layer are the soil and the organisms that live within it such as plant roots (including root crops such as potatoes and other edible tubers), fungi, insects, nematodes, worms, etc.

7. Vertical layer: climbers or vines, such as runner beans and lima beans (vine varieties)[16][17]

32.3.3 Guilds

There are many forms of guilds, including guilds of plants with similar functions (that could interchange within an ecosystem), but the most common perception is that of a mutual support guild. Such a guild is a group of species where each provides a unique set of diverse functions that work in conjunction, or harmony. Mutual support guilds are groups of plants, animals, insects, etc. that work well together. Some plants may be grown for food production, some have tap roots that draw nutrients up from deep in the soil, some are nitrogen-fixing legumes, some attract beneficial insects, and others repel harmful insects. When grouped together in a mutually beneficial arrangement, these plants form a guild. See Dave Jacke's work on edible forest gardens for more information on other guilds, specifically resource-partitioning and community-function guilds.[18][19][20]

32.3.4 Edge effect

The edge effect in ecology is the effect of the juxtaposition or placing side by side of contrasting environments on an ecosystem. Permaculturists argue that, where vastly differing systems meet, there is an intense area of productivity and useful connections. An example of this is the coast; where the land and the sea meet there is a particularly rich area that meets a disproportionate percentage of human and animal needs. So this idea is played out in permacultural designs by using spirals in the herb garden or creating ponds that have wavy undulating shorelines rather than a simple circle or oval (thereby increasing the amount of edge for a given area).

32.3.5 Zones

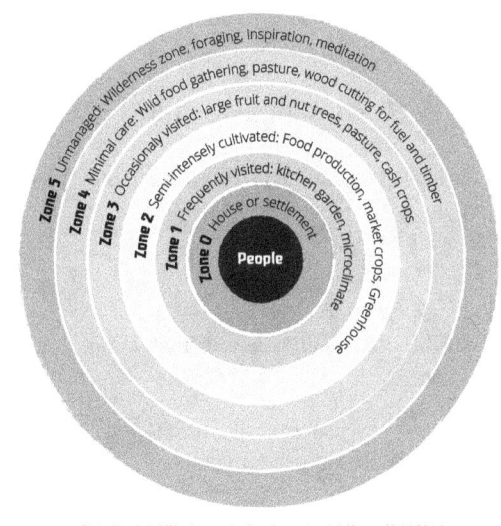

Permaculture Zones 0-5.

Zones are a way of intelligently organizing design elements in a human environment on the basis of the frequency of human use and plant or animal needs. Frequently manipulated or harvested elements of the design are located close to the house in zones 1 and 2. Less frequently used or manipulated elements, and elements that benefit from isolation (such as wild species) are farther away. Zones are about positioning things appropriately, and are numbered from 0 to 5.[21]

Zone 0 The house, or home center. Here permaculture principles would be applied in terms of aiming to reduce energy and water needs, harnessing natural resources such as sunlight, and generally creating a harmonious, sustainable environment in which to live and work. Zone 0 is an informal designation, which is not specifically defined in Bill Mollison's book.

Zone 1 The zone nearest to the house, the location for those elements in the system that require frequent attention, or that need to be visited often, such as salad crops, herb plants, soft fruit like strawberries or raspberries, greenhouse and cold frames, propagation area, worm compost bin for kitchen waste, etc. Raised beds are often used in zone 1 in urban areas.

Zone 2 This area is used for siting perennial plants that require less frequent maintenance, such as occasional weed control or pruning, including currant bushes and orchards, pumpkins, sweet potato, etc. This would also be a good place for beehives, larger scale composting bins, and so on.

Zone 3 The area where main-crops are grown, both for domestic use and for trade purposes. After establishment, care and maintenance required are fairly minimal (provided mulches and similar things are used), such as watering or weed control maybe once a week.

Zone 4 A semi-wild area. This zone is mainly used for forage and collecting wild food as well as production of timber for construction or firewood.

Zone 5 A wilderness area. There is no human intervention in zone 5 apart from the observation of natural ecosystems and cycles. Through this zone we build up a natural reserve of bacteria, moulds and insects that can aid the zones above it.[22]

32.3.6 People and permaculture

Permaculture uses observation of nature to create regenerative systems, and the place where this has been most visible has been on the landscape. There has been a growing awareness though that firstly, there is the need to pay more attention to the peoplecare ethic, as it is often the dynamics of people that can interfere with projects, and secondly that the principles of permaculture can be used as effectively to create vibrant, healthy and productive people and communities as they have been in landscapes.

32.3.7 Domesticated animals

Domesticated animals are often incorporated into site design.[23]

32.4 Common practices

32.4.1 Agroforestry

Agroforestry is an integrated approach of using the interactive benefits from combining trees and shrubs with crops and/or livestock. It combines agricultural and forestry technologies to create more diverse, productive, profitable, healthy and sustainable land-use systems.[24] In agroforestry systems, trees or shrubs are intentionally used within agricultural systems, or non-timber forest products are cultured in forest settings.

Forest gardening is a term permaculturalists use to describe systems designed to mimic natural forests. Forest gardens, like other permaculture designs, incorporate processes and relationships that the designers understand to be valuable in natural ecosystems. The terms forest garden and food forest are used interchangeably in the permaculture literature. Numerous permaculturists are proponents of forest gardens, such as Graham Bell, Patrick Whitefield, Dave Jacke, Eric Toensmeier and Geoff Lawton. Bell started building his forest garden in 1991 and wrote the book *The Permaculture Garden* in 1995, Whitefield wrote the book *How to Make a Forest Garden* in 2002, Jacke and Toensmeier co-authored the two volume book set *Edible Forest Gardening* in 2005, and Lawton presented the film *Establishing a Food Forest* in 2008.[13][25][26]

Tree Gardens, such as Kandyan tree gardens, in South and Southeast Asia, are often hundreds of years old. Whether they derived initially from experiences of cultivation and forestry, as is the case in agroforestry, or whether they derived from an understanding of forest ecosystems, as is the case for permaculture systems, is not self-evident. Many studies of these systems, especially those that predate the term permaculture, consider these systems to be forms of agroforestry. Permaculturalists who include existing and ancient systems of polycropping with woody species as examples of food forests may obscure the distinction between permaculture and agroforestry.

Food forests and agroforestry are parallel approaches that sometimes lead to similar designs.

32.4.2 Hügelkultur

Hügelkultur is the practice of burying large volumes of wood to increase soil water retention. The porous structure of wood acts as a sponge when decomposing underground. During the rainy season, masses of buried wood can absorb enough water to sustain crops through the dry season.[27] This technique has been used by permaculturalists Sepp Holzer, Toby Hemenway, Paul Wheaton and Masanobu Fukuoka.[28][29]

32.4.3   Natural building

A natural building involves a range of building systems and materials that place major emphasis on sustainability. Ways of achieving sustainability through natural building focus on durability and the use of minimally processed, plentiful or renewable resources, as well as those that, while recycled or salvaged, produce healthy living environments and maintain indoor air quality.

The basis of natural building is the need to lessen the environmental impact of buildings and other supporting systems, without sacrificing comfort, health or aesthetics. To be more sustainable, natural building uses primarily abundantly available, renewable, reused or recycled materials. In addition to relying on natural building materials, the emphasis on the architectural design is heightened. The orientation of a building, the utilization of local climate and site conditions, the emphasis on natural ventilation through design, fundamentally lessen operational costs and positively impact the environment. Building compactly and minimizing the ecological footprint is common, as are on-site handling of energy acquisition, on-site water capture, alternate sewage treatment and water reuse.

32.4.4   Rainwater harvesting

Rainwater harvesting is the accumulating and storing of rainwater for reuse before it reaches the aquifer.[30] It has been used to provide drinking water, water for livestock, water for irrigation, as well as other typical uses. Rainwater collected from the roofs of houses and local institutions can make an important contribution to the availability of drinking water. It can supplement the subsoil water level and increase urban greenery. Water collected from the ground, sometimes from areas which are especially prepared for this purpose, is called stormwater harvesting.

Greywater is wastewater generated from domestic activities such as laundry, dishwashing, and bathing, which can be recycled on-site for uses such as landscape irrigation and constructed wetlands. Greywater is largely sterile, but not potable (drinkable). Greywater differs from water from the toilets which is designated sewage or blackwater, to indicate it contains human waste. Blackwater is septic or otherwise toxic and cannot easily be reused. There are, however, continuing efforts to make use of blackwater or human waste. The most notable is for composting through a process known as humanure; a combination of the words human and manure. Additionally, the methane in humanure can be collected and used similar to natural gas as a fuel, such as for heating or cooking, and is commonly referred to as biogas. Biogas can be harvested from the human waste and the remainder still used as humanure. Some of the sim-

plest forms of humanure use include a composting toilet or an outhouse or dry bog surrounded by trees that are heavy feeders which can be coppiced for wood fuel. This process eliminates the use of a standard toilet with plumbing.

32.4.5   Sheet mulching

In agriculture and gardening, mulch is a protective cover placed over the soil. Any material or combination can be used as mulch, such as stones, leaves, cardboard, wood chips, gravel, etc., though in permaculture mulches of organic material are the most common because they perform more functions. These include: absorbing rainfall, reducing evaporation, providing nutrients, increasing organic matter in the soil, feeding and creating habitat for soil organisms, suppressing weed growth and seed germination, moderating diurnal temperature swings, protecting against frost, and reducing erosion. Sheet mulching is an agricultural no-dig gardening technique that attempts to mimic natural processes occurring within forests. Sheet mulching mimics the leaf cover that is found on forest floors. When deployed properly and in combination with other Permacultural principles, it can generate healthy, productive and low maintenance ecosystems.[31][32]

Sheet mulch serves as a "nutrient bank," storing the nutrients contained in organic matter and slowly making these nutrients available to plants as the organic matter slowly and naturally breaks down. It also improves the soil by attracting and feeding earthworms, slaters and many other soil microorganisms, as well as adding humus. Earthworms "till" the soil, and their worm castings are among the best fertilizers and soil conditioners. Sheet mulching can be used to reduce or eliminate undesirable plants by starving them of light, and can be more advantageous than using herbicide or other methods of control.

32.4.6   Intensive rotational grazing

Grazing has long been blamed for much of the destruction we see in the environment. However, it has been shown that when grazing is modeled after nature, the opposite effect can be seen.[33][34] Also known as cell grazing, managed intensive rotational grazing (MIRG) is a system of grazing in which ruminant and non-ruminant herds and/or flocks are regularly and systematically moved to fresh pasture, range, or forest with the intent to maximize the quality and quantity of forage growth. This disturbance is then followed by a period of rest which allows new growth. MIRG can be used with cattle, sheep, goats, pigs, chickens, rabbits, geese, turkeys, ducks and other animals depending on the natural ecological community that is being mimicked. Sepp Holzer and Joel Salatin have shown how the disturbance caused by the animals can be the spark needed to start ecological succession or prepare ground for planting. Allan Savory's holistic management technique has been likened to "a permaculture approach to rangeland management".[35][36] One variation on MIRG that is gaining rapid popularity is called eco-grazing. Often used to either control invasives or re-establish native species, in eco-grazing the primary purpose of the animals is to benefit the environment and the animals can be, but are not necessarily, used for meat, milk or fiber.[37][38][39][40][41][42][43]

32.4.7 Keyline design

Keyline design is a technique for maximizing beneficial use of water resources of a piece of land developed in Australia by farmer and engineer P. A. Yeomans. The *Keyline* refers to a specific topographic feature linked to water flow which is used in designing the drainage system of the site.[44]

32.4.8 Fruit tree management

> The no-pruning option is usually ignored by fruit experts, though often practised by default in people's back gardens! But it has its advantages. Obviously it reduces work, and more surprisingly it can lead to higher overall yields.
>
> — Whitefield, Patrick, *How to make a forest garden*, p. 16

Masanobu Fukuoka, as part of early experiments on his family farm in Japan, experimented with no-pruning methods, noting that he ended up killing many fruit trees by simply letting them go, which made them become convoluted and tangled, and thus unhealthy.[45][46] Then he realised this is the difference between natural-form fruit trees and the process of change of tree form that results from abandoning previously-pruned unnatural fruit trees.[45][47] He concluded that the trees should be raised all their lives without pruning, so they form healthy and efficient branch patterns that follow their natural inclination. This is part of his implementation of the Tao-philosophy of Wú wéi translated in part as no-action (against nature), and he described it as no unnecessary pruning, nature farming or "do-nothing" farming, of fruit trees, distinct from non-intervention or literal no-pruning. He ultimately achieved yields comparable to or exceeding standard/intensive practices of using pruning and chemical fertilisation.[45][47][48]

Another proponent of the no, or limited, pruning method is Sepp Holzer who used the method in connection with Hügelkultur berms. He has successfully grown several varieties of fruiting trees at altitudes (approximately 9,000 feet (2,700 m)) far above their normal altitude, temperature, and snow load ranges. He notes that the Hügelkultur berms kept and/or generated enough heat to allow the roots to survive during alpine winter conditions. The point of having unpruned branches, he notes, was that the longer (more naturally formed) branches bend over under the snow load until they touched the ground, thus forming a natural arch against snow loads that would break a shorter, pruned, branch.

32.4.9 Mollison and Holmgren

Bill Mollison in January 2008.

In the mid-1970s, Bill Mollison and David Holmgren started developing ideas about stable agricultural systems on the southern Australian island state of Tasmania. This was a result of the danger of the rapidly growing use of industrial-agricultural methods. In their view, highly dependent on non renewable resources, these methods were additionally poisoning land and water, reducing biodiversity, and removing billions of tons of topsoil from previously fertile landscapes. A design approach called *permaculture* was their response and was first made public with the publication of their book *Permaculture One* in 1978.

By the early 1980s, the concept had broadened from agricultural systems design towards sustainable human habitats. After *Permaculture One*, Mollison further refined and developed the ideas by designing hundreds of permaculture sites and writing more detailed books, notably *Permaculture: A Designers Manual*. Mollison lectured in over 80 countries and taught his two-week Permaculture Design Course (PDC) to many hundreds of students. Mollison "encouraged graduates to become teachers themselves and set up their own institutes and demonstration sites. This multiplier effect was critical to permaculture's rapid expansion."[49]

In 1991, a four-part television documentary by ABC productions called "The Global Gardener" showed permaculture applied to a range of worldwide situations, bringing the concept to a much broader public. In 2012, the UMass Permaculture Initiative won the White House "Champions of Change" sustainability contest, which declared that "they demonstrate how permaculture can feed a growing population in an environmentally sustainable and socially responsible manner".[50]

In 1997, Holmgren explained that the primary agenda of the permaculture movement is to assist people to become more self-reliant through the design and development of productive and sustainable gardens and farms.[14]

In 2014, Holmgren endorsed and helped launch a new Australian permaculture magazine, Pip Magazine.[51]

32.5   Notable permaculturists

Joseph Russell Smith took up an antecedent term as the subtitle for *Tree Crops: A Permanent Agriculture*, a book in which he summed up his long experience experimenting with fruits and nuts as crops for human food and animal feed. By that year (1929), Smith saw the world as an interrelated whole and suggested mixed systems of trees and crops underneath. This book inspired many individuals intent on making permaculture a valid means of sustainable food production. Bill Mollison and David Holmgren developed it further, and permaculturists were trained under the umbrella of Bill Mollison's train the trainer system.

Geoff Lawton, Toby Hemenway and P. A. Yeomans - creator of the keyline design each have more than 20 years experience teaching and promoting permaculture as a sustainable way of growing food.

The permaculture movement also spread throughout Central America, with Rony Lec leading the foundation of the Mesoamerican Permaculture Institute (IMAP) in Guatemala and Juan Rojas co-founding the Permaculture Institute of El Salvador.

32.6   Trademark and copyright issues

There has been contention over who, if anyone, controls legal rights to the word *permaculture*: is it trademarked or copyrighted? and if so, who holds the legal rights to the use of the word? For a long time Bill Mollison claimed to have copyrighted the word, and his books said on the copyright page, "The contents of this book and the word PERMACULTURE are copyright." These statements were largely accepted at face-value within the permaculture community. However, copyright law does not protect names, ideas, concepts, systems, or methods of doing something; it only pro-

tects the expression or the description of an idea, not the idea itself. Eventually Mollison acknowledged that he was mistaken and that no copyright protection existed for the word *permaculture*.[52]

In 2000, Mollison's US based Permaculture Institute sought a service mark (a form of trademark) for the word, permaculture, when used in educational services such as conducting classes, seminars, or workshops.[53] The service mark would have allowed Mollison and his two Permaculture Institutes (one in the US and one in Australia) to set enforceable guidelines regarding how permaculture could be taught and who could teach it, particularly with relation to the PDC, despite the fact that he had instituted a system of certification of teachers to teach the PDC in 1993. The service mark failed and was abandoned in 2001. Also in 2001 Mollison applied for trademarks in Australia for the terms "Permaculture Design Course"[54] and "Permaculture Design".[54] These applications were both withdrawn in 2003. In 2009 he sought a trademark for "Permaculture: A Designers' Manual"[54] and "Introduction to Permaculture",[54] the names of two of his books. These applications were withdrawn in 2011. There has never been a trademark for the word *permaculture* in Australia.[54]

32.7   Criticisms

32.7.1   General criticisms

In 2011, Owen Hablutzel argued that "permaculture has yet to gain a large amount of specific mainstream scientific acceptance," and that "the sensitiveness to being perceived and accepted on scientific terms is motivated in part by a desire for permaculture to expand and become increasingly relevant." Bec-Hellouin permaculture farm engaged in a research program in partnership with INRA and AgroParisTech to collect scientific data.[55][56]

In his books *Sustainable Freshwater Aquaculture* and *Farming in Ponds and Dams*, Nick Romanowski expresses the view that the presentation of aquaculture in Bill Mollison's books is unrealistic and misleading.

32.7.2   Agroforestry

Greg Williams argues that forests cannot be more productive than farmland because the net productivity of forests decline as they mature due to ecological succession.[57] Proponents of permaculture respond that this is true only if one compares data from between woodland forest and climax vegetation, but not when comparing farmland vegetation with woodland forest.[58] For example, ecological succession generally results in a forest's productivity rising after its establishment only until it reaches the *woodland state* (67% tree cover), before declining until *full maturity*.[13]

32.8   See also

- Agrarianism
- Aquaponics
- Bill Mollison
- Biomimicry
- Climate-friendly gardening
- David Holmgren
- Ecoagriculture
- Forest gardening
- Geoff Lawton
- Holzer Permaculture
- Hügelkultur
- List of permaculture projects
- Microponics
- Paul Wheaton
- Permaforestry
- Seed saving
- Sepp Holzer
- Biointensive agriculture
- Zaï
- Agroecology
- Agroforestry
- Biodynamics

32.9   References

[1] King 1911.

[2] Hemenway 2009, p. 5.

[3] Mars, Ross (2005). *The Basics of Permaculture Design.* Chelsea Green. p. 1. ISBN 978-1-85623-023-0.

[4] Mollison, B. (1991). *Introduction to permaculture.* Tasmania, Australia: Tagari.

[5] Smith, Joseph Russell; Smith, John (1987). *Tree Crops: A permanent agriculture.* Island Press. ISBN 978-1-59726873-8.

[6] Hart 1996, p. 41.

[7] Holmgren, David (2006). "The Essence of Permaculture". Holmgren Design Services. Retrieved 10 September 2011.

[8] Mollison, Bill (September 15–21, 1978). "The One-Straw Revolution by Masanobu Fukuoka". *Nation Review.* p. 18.

[9] Greenblott, Kara; Nordin, Kristof (2012), *Permaculture Design for Orphans and Vulnerable Children Programming: Low-Cost, Sustainable Solutions for Food and Nutrition Insecure Communities,* AIDS Support and Technical Assistance Resources, AIDSTAR-One (Task Order 1), Arlington, VA: USAID.

[10] Mollison 1988, p. 2.

[11] Holmgren, David (2002). *Permaculture: Principles & Pathways Beyond Sustainability.* Holmgren Design Services. p. 1. ISBN 0-646-41844-0.

[12] Mollison, Bill. "Permaculture: A Quiet Revolution". *Scott London* (interview). Retrieved 17 May 2013.

[13] "Edible Forest Gardening".

[14] Holmgren, David (1997). "Weeds or Wild Nature" (PDF). Permaculture International Journal. Retrieved 10 September 2011.

[15] "Permaculture: Principles and Pathways Beyond Sustainability". Holmgren Design. Retrieved 2013-10-21.

[16] *Nine layers of the edible forest garden,* TC permaculture, May 27, 2013.

[17] "Seven layers of a forest", *Food forests,* CA: Permaculture school.

[18] Simberloff, D; Dayan, T (1991). "The Guild Concept and the Structure of Ecological Communities". *Annual Review of Ecology and Systematics* **22**: 115. doi:10.1146/annurev.es.22.110191.000555.

[19] "Guilds". *Encyclopaedia Britannica.* Retrieved 2011-10-21.

[20] Williams, SE; Hero, JM (1998). "Rainforest frogs of the Australian Wet Tropics: guild classification and the ecological similarity of declining species". *Proceedings. Biological sciences* (The Royal Society) **265** (1396): 597–602. doi:10.1098/rspb.1998.0336. PMC 1689015. PMID 9881468.

[21] Burnett 2001.

[22] *Permacultuur course,* NL: WUR.

[23] Mollison 1988, p. 5: 'Deer, rabbits, sheep, and herbivorous fish are very useful to us, in that they convert unusable herbage to acceptable human food. Animals represent a valid method of storing inedible vegetation as food.'

[24] "USDA National Agroforestry Center (NAC)". UNL. 2011-08-01. Retrieved 2011-10-21.

[25] "Graham Bell's Forest Garden". *Permaculture*. Media mice.

[26] "Establishing a Food Forest" (film review). Transition culture. Feb 11, 2009.

[27] Wheaton, Paul. "Raised garden beds: hugelkultur instead of irrigation" Richsoil. Retrieved 2012-07-15.

[28] Hemenway 2009, pp. 84–85.

[29] Feineigle, Mark. "Hugelkultur: Composting Whole Trees With Ease". Permaculture Research Institute of Australia. Retrieved 2012-07-15.

[30] "Rainwater harvesting". DE: Aramo. 2012. Retrieved 2012.

[31] "Sheet Mulching: Greater Plant and Soil Health for Less Work". Agroforestry. 2011-09-03. Retrieved 2011-10-21.

[32] Mason, J (2003), *Sustainable Agriculture*, Landlinks.

[33] "Prince Charles sends a message to IUCN's World Conservation Congress". *International Union for Conservation of Nature*. Retrieved 6 April 2013.

[34] Undersander, Dan; et al. "Grassland birds: Fostering habitat using rotational grazing" (PDF). University of Wisconsin-Extension. Retrieved 5 April 2013.

[35] Fairlie, Simon (2010). *Meat: A Benign Extravagance*. Chelsea Green. pp. 191–93. ISBN 978-1-60358325-1.

[36] Bradley, Kirsten. "Holistic Management: Herbivores, Hats, and Hope". Milkwood. Retrieved 25 March 2014.

[37] "Munching sheep replace lawn mowers in Paris". *The Sunday Times*. Apr 4, 2013. Retrieved 7 April 2013.

[38] Ash, Andrew, *The Ecograze Project – developing guidelines to better manage grazing country* (PDF), et al., CSIRO, ISBN 0-9579842-0-0, retrieved 7 April 2013

[39] McCarthy, Caroline. "Things to make you happy: Google employs goats". *CNET*. Retrieved 7 April 2013.

[40] Gordon, Ian. "A systems approach to livestock/resource interactions in tropical pasture systems" (PDF). *The James Hutton Institute*. Retrieved 7 April 2013.

[41] Littman, Margaret. "Getting your goat: Eco-friendly mowers". *Chicago Tribune News*. Retrieved 7 April 2013.

[42] Stevens, Alexis. "Kudzu-eating sheep take a bite out of weeds". *The Atlanta Journal-Constitution*. Retrieved 7 April 2013.

[43] Klynstra, Elizabeth. "Hungry sheep invade Candler Park". *CBS Atlanta*. Retrieved 7 April 2013.

[44] Tipping, Don (4 January 2013). "Creating Permaculture Keyline Water Systems" (video). UK: Beaver State Permaculture.

[45] Masanobu, Fukuoka (1987) [1985], *The Natural Way of Farming – The Theory and Practice of Green Philosophy* (rev ed.), Tokyo: Japan Publications, p. 204

[46] Fukuoka 1978, pp. 13, 15–18, 46, 58–60.

[47] Fukuoka 1978.

[48] "Masanobu Fukuoka", *Public Service* (biography), PH: The Ramon Magsaysay Award Foundation, 1988, retrieved 2011-03-02.

[49] Lillington, Ian; Holmgren, David; Francis, Robyn; Rosenfeldt, Robyn. "The Permaculture Story: From 'Rugged Individuals' to a Million Member Movement" (PDF). *Pip Magazine*. Retrieved 9 July 2015.

[50] "UMass Amherst permaculture project wins White House award". *Boston*. Mar 13, 2012.

[51] "Pip Launches to Great Success". *Pip Magazine*. Retrieved 1 July 2015.

[52] Grayson, Russ (2011). "The Permaculture Papers 5: time of change and challenge — 2000-2004". Pacific edge. Retrieved 8 September 2011.

[53] United States Patent and Trademark Office (2011). "Trademark Electronic Search System (TESS)". US Department of Commerce. Retrieved 8 September 2011.

[54] "Result". Commonwealth of Australia. 2011. Retrieved 8 September 2011.

[55] Paul, Willi (2011). "Symbols & Patterns. Interview with Owen Hablutzel, Director, Permaculture Research Institute, USA". Retrieved 2012-06-21.

[56] "Why permaculture needs accurate data and measurement to persuade the mainstream". 2012-05-02.

[57] Williams, Greg (2001). "Gaia's Garden: A Guide to Home-Scale Permaculture". *Whole Earth*.

[58] "A toolbox, not a tool".

32.9.1 Bibliography

- Bell, Graham (2004) [1992, Thorsons, ISBN 0-7225-2568-0], *The Permaculture Way* (2nd ed.), UK: Permanent Publications, ISBN 1-85623-028-7.

- ——— (2004), *The Permaculture Garden*, UK: Permanent, ISBN 1-85623-027-9.

- Burnett, G (2001), *Permaculture: a Beginner's Guide*, UK: Spiralseed, ISBN 978-0-95534921-8.

- Fern, Ken (1997), *Plants For A Future*, UK: Permanent, ISBN 1-85623-011-2.

- Fukuoka, Masanobu (1978), *The One–Straw Revolution*, Holistic Agriculture Library, US: Rodale Books.

- Hart, Robert (1996), *Forest Gardening*, UK: Green Books, p. 41, ISBN 978-1-60358050-2; ISBN 1-900322-02-1.

- Hemenway, Toby (2009) [2001, ISBN 1-890132-52-7], *Gaia's Garden: A Guide to Home-Scale Permaculture*, US: Chelsea Green, ISBN 978-1-60358-029-8

- Holmgren, David, *Melliodora (Hepburn Permaculture Gardens): A Case Study in Cool Climate Permaculture 1985–2005*, AU: Holmgren Design Services.

- ——, *Collected Writings & Presentations 1978–2006*, AU: Holmgren Design Services.

- —— (2009), *Future Scenarios*, White River Junction: Chelsea Green.

- ——, *Permaculture: Principles and Pathways Beyond Sustainability*, AU: Holmgren Design Services.

- ——, *Update 49: Retrofitting the suburbs for sustainability*, AU: CSIRO Sustainability Network.

- Jacke, Dave with Eric Toensmeier. *Edible Forest Gardens. Volume I: Ecological Vision and Theory for Temperate-Climate Permaculture, Volume II: Ecological Design and Practice for Temperate-Climate Permaculture*. Edible Forest Gardens (US) 2005

- King, Franklin Hiram (1911), *Farmers of Forty Centuries: Or Permanent Agriculture in China, Korea and Japan*.

- Law, Ben (2005), *The Woodland House*, UK: Permanent, ISBN 1-85623-031-7.

- ——, *The Woodland Way*, UK: Permanent Publications, ISBN 1-85623-009-0.

- Loofs, Mona. *Permaculture, Ecology and Agriculture: An investigation into Permaculture theory and practice using two case studies in northern New South Wales* Honours thesis, Human Ecology Program, Department of Geography, Australian National University 1993

- Macnamara, Looby. *People and Permaculture: caring and designing for ourselves, each other and the planet.* [Permanent Publications] (UK) (2012) ISBN 1-85623-087-2.

- Mollison, Bill (1979), *Permaculture Two*, Australia: Tagari Press, ISBN 0-908228-00-7.

- —— (1988), *Permaculture: A Designer's Manual*, AU: Tagari Press, ISBN 0-908228-01-5.

- ——; Holmgren, David (1978), *Permaculture One*, AU: Transworld Publishers, ISBN 0-552-98060-9.

- Odum, H.T., Jorgensen, S.E. and Brown, M.T. 'Energy hierarchy and transformity in the universe', in *Ecological Modelling*, 178, pp. 17–28 (2004).

- Paull, J. "Permanent Agriculture: Precursor to Organic Farming", Journal of Bio-Dynamics Tasmania, no.83, pp. 19–21, 2006. Organic eprints.

- Rosemary, Morrow, *Earth User's Guide to Permaculture*, ISBN 0-86417-514-0.

- Shepard, Mark: *Restoration Agriculture – Redesigning Agriculture in Nature's Image* Acres US, 2013, ISBN 1-60173035-7

- Whitefield, Patrick (1993), *Permaculture In A Nutshell*, UK: Permanent, ISBN 1-85623-003-1.

- —— (2004), *The Earth Care Manual*, UK: Permanent Publications, ISBN 1-85623-021-X.

- Woodrow, Linda. *The Permaculture Home Garden*. Penguin Books (Australia).

- Yeomans, P.A. *Water for Every Farm: A practical irrigation plan for every Australian property*, KG Murray, Sydney, NSW, Australia (1973).

- *The Same Planet a different World* (free ebook), FR.

32.10 External links

- *Permaculture for agroecology: design, movement, practice, and worldview* (review), Springer – The first systematic review of the permaculture literature, from the perspective of agroecology.

- The Permaculture Research Institute – Permaculture Forums, Courses, Information, News and Worldwide Reports.

- The Worldwide Permaculture Network – Database of permaculture people and projects worldwide.

- The Permaculture Association, UK.

- The 15 pamphlets based on the 1981 Permaculture Design Course given by Bill Mollison (co-founder of permaculture) all in 1 PDF file.

- David Holmgren's web site (co-founder of permaculture)

- Ethics and principles of permaculture (Holmgren's)

- Permaculture a Beginners Guide – a 'pictorial walk-through'

- Permaculture – Sustainability and sustainable development

- Urban Permaculture Design – a city lot with over a hundred perennial edible varieties. Permaculture land acquisition discussion.

- A quarter acre suburban property in Eugene, Oregon – grass to garden, reclaim automobile space, elevated/edible landscape, rain water catchment, passive solar design, education

- The Permaculture Activist is a co-evolving quarterly produced by a dedicated handful of entirely part-time folks

Chapter 33

Principles of intelligent urbanism

Principles of intelligent urbanism (**PIU**) is a theory of urban planning composed of a set of ten axioms intended to guide the formulation of city plans and urban designs. They are intended to reconcile and integrate diverse urban planning and management concerns. These axioms include environmental sustainability, heritage conservation, appropriate technology, infrastructure-efficiency, placemaking, "Social Access," transit oriented development, regional integration, human scale, and institutional-integrity. The term was coined by Prof. Christopher Charles Benninger.

The **PIU** evolved from the city planning guidelines formulated by the *International Congress of Modern Architecture* (CIAM), the urban design approaches developed at Harvard's pioneering Urban Design Department under the leadership of Josep Lluis Sert, and the concerns enunciated by Team Ten. It is most prominently seen in plans prepared by Christopher Charles Benninger and his numerous colleagues in the Asian context (Benninger 2001). They form the elements of the planning curriculum at the School of Planning, Ahmedabad, which Benninger founded in 1971. They were the basis for the new capital plan for Thimphu, Bhutan.[1]

33.1 Axioms

33.1.1 Principle one: a balance with nature

According to proponents of intelligent urbanism, balance with nature emphasizes the distinction between utilizing resources and exploiting them. It focuses on the thresholds beyond which deforestation, soil erosion, aquifer depletion, siltation and flooding reinforce one another in urban development, saving or destroying life support systems. The principle promotes environmental assessments to identify fragile zones, threatened ecosystems and habitats that can be enhanced through conservation, density control, land use planning and open space design (McCarg: 1975). This principle promotes life cycle building energy consumption and pollutant emission analysis.[2]

This principle states there is a level of human habitation intensity wherein the resources that are consumed will be replaced through the replenishing natural cycles of the seasons, creating environmental equilibrium. Embedded in the principle is contention that so long as nature can resurge each year; so long as the biomass can survive within its own eco-system; so long as the breeding grounds of fauna and avifauna are safe; so long as there is no erosion and the biomass is maintained, nature is only being utilized.

Underlying this principle is the supposition that there is a fragile line that is crossed when the fauna, which cross-fertilizes the flora, which sustains the soil, which supports the hillsides, is no longer there. Erosion, siltation of drainage networks and flooding result. After a point of no return, utilization of natural resources will outpace the natural ability of the eco-system to replenish itself. From there on degradation accelerates and amplifies. Deforestation, desertification, erosion, floods, fires and landslides all increase.

The principle states that blatant "acts against nature" include cutting of hillside trees, quarrying on slopes, dumping sewage and industrial waste into the natural drainage system, paving and plinthing excessively, and construction on steep slopes. This urban theory proposes that the urban ecological balance can be maintained when fragile areas are reserved, conservation of eco-systems is pursued, and low intensity habitation precincts are thoughtfully identified. Thus, the principles operate within the balance of nature, with a goal of protecting and conserving those elements of the ecology that nurture the environment. Therefore, the first principle of intelligent urbanism is that urbanization be in balance with nature.

33.1.2 Principle two: a balance with tradition

Balance with tradition is intended to integrate plan interventions with existing cultural assets, respecting traditional practices and precedents of style (Spreiregen: 1965). This urban planning principle demands respect for the cultural heritage of a place. It seeks out traditional wisdom in the layout of human settlements, in the order of building plans, in the precedents of style, in the symbols and signs that transfer meanings through decoration and motifs. This principle respects the order engendered into building systems through years of adaptation to climate, to social circumstances, to available materials and to technology. It promotes architectural styles and motifs designed to communicate cultural values.

This principle calls for orienting attention toward historic monuments and heritage structures, leaving space at the ends of visual axis to "frame" existing views and vistas. Natural views and vistas demand respect, assuring that buildings do not block major sight lines toward visual assets.

Embedded in the principle is the concern for unique cultural and societal iconography of regions, their signs and symbols. Their incorporation into the spatial order of urban settings is promoted. Adherents promote the orientation and structuring of urban plans using local knowledge and meaning systems, expressed through art, urban space and architecture.

Planning decisions must operate within the balance of tradition, aggressively protecting, promoting and conserving generic components and elements of the urban pattern.

33.1.3 Principle three: appropriate technology

Appropriate technology emphasizes the employment of building materials, construction techniques, infrastructural systems and project management which are consistent with local contexts (situation, setting or circumstances). People's capacities, geo-climatic conditions, locally available resources, and suitable capital investments all temper technology. Where there are abundant craftspeople, labour-intensive methods are appropriate. Where there is surplus savings, capital intensive methods are appropriate. For every problem there is a range of potential technologies, which can be applied, and an appropriate fit between technology and other resources must be established. Proponents argue that accountability and transparency are enhanced by overlaying the physical spread of urban utilities and services upon electoral constituencies, such that people's representatives are interlinked with the urban technical systems needed for a civil society. This principle is in sync with "small is beautiful" concepts and with the use of local resources.

33.1.4 Principle four: conviviality

The fourth principle sponsors social interaction through public domains, in a hierarchy of places, devised for personal solace, companionship, romance, domesticity, "neighborliness," community and civic life (Jacobs:1993). According to proponents of intelligent urbanism, vibrant societies are interactive, socially engaging and offer their members numerous opportunities for gathering and meeting one another.[3] The PIU maintain that this can be achieved through design and that society operates within hierarchies of social relations which are space specific. The hierarchies can be conceptualized as a system of social tiers, with each tier having a corresponding physical place in the settlement structure.

A place for the individual

A goal of intelligent urbanism is to create places of solitude. These may be in urban forests, along urban hills, beside quiet streams, in public gardens and in parks where one can escape to meditate and contemplate. According to proponents, these are the quiet places wherein the individual consciousness dialogues with the rational mind. Idle and random thought sorts out complexities of modern life and allows the obvious to emerge. It is in these natural settings that the wandering mind finds its measure and its balance. Using ceremonial gates, directional walls and other "silent devices" these spaces are denoted and divined. Places of the individual cultivate introspection. These spaces may also be the forecourts and interior courtyards of public buildings, or even the thoughtful reading rooms of libraries. Meditation focuses one's thought. Intelligent urbanism creates a domain for the individual to mature through self-analysis and self-realization.

A place for friendship

The axiom insists that in city plans there must be spaces for "beautiful, intimate friendship" where unfettered dialogue can happen. This principle insists that such places will not exist naturally in a modern urban fabric. They must be a part of the conscientious design of the urban core, of the urban hubs, of urban villages and of neighborhoods, where people can meet with friends and talk out life's issues, sorrows, joys and dilemmas. This second tier is important for the emotional life of the populace. It sponsors strong mental health within the people, creating places where friendship can unfold and grow.

A place for householders

There must be spaces for householders, which may be in the form of dwellings for families, or homes for intimate companions, and where young workmates can form a common kitchen. Whatever their compositions, there must be a unique domain for social groups, familiar or biological, which have organized themselves into households. These domestic precincts are where families live and carry out their day-to-day functions of life. This third tier of conviviality is where the individual socializes into a personality.

Housing clusters planned according to this axiom create a variety of household possibilities, which respond to a range of household structures and situations. It recognizes that households transform through the years, requiring a variety of dwellings types that respond to a complex matrix of needs and abilities, which are provided for in city plans.

A place for the neighborhood

Smaller household domains must cluster into a higher social domain, the neighborhood social group. Good city planning practice sponsors, through design, such units of social space. It is in this fourth tier of social life that public conduct takes on new dimensions and groups learn to live peacefully among one another. It is through neighborhoods that the "social contract" amongst diverse households and individuals is sponsored. This social contract is the rational basis for social relations and negotiations within larger social groups. Within neighborhoods basic amenities like creches, early learning centers, preventive health care and rudimentary infrastructure are maintained by the community.

A place for communities

The next social tier, or hierarchy, is the community. Historically, communities were tribes who shared social mores and cultural behavioral patterns. In contemporary urban settings communities are formed of diverse people. But these are people who share the common need to negotiate and manage their spatial settings. In plans created through the principles of intelligent urbanism these are called urban villages. Like a rural village, social bonds are found in the community management of security, common resources and social space. Urban villages will have defined social spaces, services and amenities that need to be managed by the community. According to proponents of intelligent urbanism these urban villages optimally become the administrative wards, and therefore the constituencies, of the elected members of municipal bodies. Though there are no physical barriers to these communities, they have their

unique spatial social domain. Intelligent urbanism calls for the creation of dense, walkable zones in which the inhabitants recognize each other's faces, share common facilities and resources, and often see each other at the village centre. This fifth tier of social space is where one needs initiative to join into various activities. It is intended to promote initiative and constructive community participation. There are opportunities for one to be involved in the management of services, and amenities and to meet new people. They accommodate primary education and recreation areas. Good planning practice promotes the creation of community places, where community-based organizations can manage common resources and resolve common problems.

A place for the city domain

The principles of intelligent urbanism call for city level domains. These can be plazas, parks, stadia, transport hubs, promenades, "passages" or gallerias. These are social spaces where everyone can go. In many cities one has to pay an entrance fee to access "public spaces" like malls and museums. Unlike the lower tiers of the social hierarchy, this tier is not defined by any biological, familiar, face-to-face or exclusive characteristic. One may find people from all continents, from nearby districts and provinces and from all parts of the city in such places. By nature these are accessible and open spaces, with no physical, social or economic barriers. According to this principle it is the rules of human conduct that order this domain's behavior. It is civility, or civilization, which protects and energizes such spaces. At the lower tiers, one meets people through introductions, through family ties, and through neighborhood circumstances.

These domains would include all freely accessible large spaces. These are places where outdoor exhibits are held, sports matches take place, vegetables are sold and goods are on display. These are places where visitors to the city meander amongst the locals. Such places may stay the same, but the people are always changing. Most significant, these city scale public domains foster public interaction; they sponsor unspoken ground rules for unknown people to meet and to interact. They nurture civic understanding of the strength of diversity, variety, a range of cultural groups and ethnic mixes. It is this higher tier of social space which defines truly urbane environments.

Every social system has its own hierarchy of social relations and interactions. Intelligent urbanism sees cyberspace as a macro tier of conviviality, but does not discount physical places in forging relationships due to the Internet. These are reflected through a system of 'places' that respond to them. Good urban planning practice promotes the planning and design of such 'places' as elemental components of the urban structure.

33.1.5   Principle five: efficiency

The principle of efficiency promotes a balance between the consumption of resources such as energy, time and fiscal resources, with planned achievements in comfort, safety, security, access, tenure, productivity and hygiene. It encourages optimum sharing of public land, roads, facilities, services and infrastructural networks, reducing per household costs, while increasing affordability, productivity, access and civic viability.

Intelligent urbanism promotes a balance between perfor mance and consumption. Intelligent urbanism promotes efficiency in carrying out functions in a cost effective manner. It assesses the performance of various systems required by the public and the consumption of energy, funds, administrative time and the maintenance efforts required to perform these functions.

A major concern of this principle is transport. While recognizing the convenience of personal vehicles, it attempts to place costs (such as energy consumption, large paved areas, parking, accidents, negative balance of trade, pollution and related morbidity) on the users of private vehicles.

Good city planning practice promotes alternative modes of transport, as opposed to a dependence on personal vehicles. It promotes affordable public transport. It promotes medium to high-density residential development along with complementary social amenities, convenience shopping, recreation and public services in compact, walkable mixed-use settlements. These compact communities have shorter pipe lengths, wire lengths, cable lengths and road lengths per capita. More people share gardens, shops and transit stops.[4]

These compact urban nodes are spaced along regional urban transport corridors that integrate the region's urban nodes, through public transport, into a rational system of growth. Good planning practice promotes clean, comfortable, safe and speedy, public transport, which operates at dependable intervals along major origin and destination paths. Such a system is cheaper, safer, less polluting and consumes less energy.

The same principle applies to public infrastructure, social facilities and public services. Compact, high-density communities result in more efficient urban systems, delivering services at less cost per unit to each citizen.

There is an appropriate balance to be found somewhere on the line between wasteful low-density individual systems and over-capitalized mega systems. Individual septic tanks and water bores servicing individual households in low-density fragmented layouts, allow the use of filtered grey

water for free irrigation of gardens, but, if not maintained, can cause a local pollution of subterranean aquifer systems. The bores can dramatically lower ground water levels especially during droughts. The vantage of septic tanks an bores is to be managed by the very users, at no cost for the community. Alternatively, large-scale, citywide sewerage systems and regional water supply systems are capital intensive and prone to management and maintenance dysfunction, if not corruption or extortion by private companies. Operating costs, user fees and cost recovery expenses are high. There is a balance wherein medium-scale systems, covering compact communities, utilize modern technology, without the pitfalls of large-scale infrastructure systems. This principle of urbanism promotes the middle path with regard to public infrastructure, facilities, services and amenities.

When these appropriate facilities and service systems overlap electoral constituencies, the "imagery" between user performance in the form of payments for services, systems dependability through managed delivery, and official response through effective representation, should all become obvious and transparent.

Good city planning practices promote compact settlements along dense urban corridors, and within populated networks, such that the numbers of users who share costs are adequate to support effective and efficient infrastructure systems. Intelligent urbanism is intended to foster movement on foot, linking pedestrian movement with public transport systems at strategic nodes and hubs. Medium-scale infrastructural systems, whose catchment areas overlap political constituencies and administrative jurisdictions, result in transparent governance and accountable urban management.

33.1.6   Principle six: human scale

Intelligent urbanism encourages ground level, pedestrian oriented urban patterns, based on anthropometric dimensions. Walkable, mixed use urban villages are encouraged over single-function blocks, linked by motor ways, and surrounded by parking lots.

An abiding axiom of urban planning, urban design and city planning has been the promotion of people friendly places, pedestrian walkways and public domains where people can meet freely. These can be parks, gardens, glass-covered galleries, arcades, courtyards, street side cafes, river- and hill-side stroll ways, and a variety of semi-covered spaces.

Intelligent urbanism promotes the scale of the pedestrian moving on the pathway, as opposed to the scale of the automobile on the expressway. Intelligent urbanism promotes the ground plan of imaginable precincts, as opposed to the imagery of façades and the monumentality of the section.

It promotes the personal visibility of places moving on foot at eye level.

Intelligent urbanism advocates removing artificial barrier and promotes face-to-face contact. Proponents argue that the automobile, single use zoning and the construction of public structures in isolated compounds, all deteriorate the human condition and the human scale of the city.

According to PIU proponents, the trend towards urban sprawl can be overcome by developing pedestrian circulation networks along streets and open spaces that link local destinations. Shops, amenities, day care, vegetable markets and basic social services should be clustered around public transport stops, and at a walkable distance from work places, public institutions, high and medium density residential areas. Public spaces should be integrated into residential, work, entertainment and commercial areas. Social activities and public buildings should orient onto public open spaces. These should be the interchange sites for people on the move, where they can also revert into the realm of "slowness," of community life and of human interaction.

Human scale can be achieved through building masses that "step down" to human scale open spaces; by using arcades and pavilions as buffers to large masses; by intermixing open spaces and built masses sensitively; by using anthropometric proportions and natural materials. Traditional building precedents often carry within them a human scale language, from which a contemporary fabric of build may evolve.

The focus of intelligent urbanism is the ground plane, pedestrian movement and interaction along movement channels, stems, at crossing nodes, at interactive hubs and within vibrant urban cores. The PIU holds many values in common with Transit Oriented Development, but the PIU goal is not merely to replace the automobile, nor to balance it. These are mundane requirements of planning, which the PIU assumes are found in every design and urban configuration. The PIU goal is to enrich the human condition and to enhance the realm of human possibilities.

Intelligent urbanism conceives of urbanity as a process of facilitating human behavior toward more tolerant, more peaceful, more accommodating and more sensitive modalities of interaction and conflict resolution. Intelligent urbanism recognizes that 'urbanity' emerges where people mix and interact on a face-to-face basis, on the ground, at high densities and amongst diverse social and economic groups. Intelligent urbanism nurtures 'urbanity' through designs and plans that foster human scale interaction.

33.1.7 Principle seven: opportunity matrix

The PIU envisions the city as a vehicle for personal, social, and [economic development], through access to a range of organizations, services, facilities and information providing a variety of opportunities for enhanced employment, economic engagement, education, and recreation. This principle aims to increase access to shelter, health care and human resources development. It aims to increase safety and hygienic conditions. The city is an engine of economic growth. This is generally said with regard to urban annual net product, enriched urban economic base, sustained employment generation and urban balance of trade. More significantly this is true for the individuals who settle in cities. Moreover, cities are places where individuals can increase their knowledge, skills and sensitivities. Cities provide access to health care and preventive medicine. They provide a great umbrella of services under which the individual can leave aside the struggle for survival, and get on with the finer things of life.[5]

The PIU sees cities as catalysts for personal definition and self-discovery. In cities people get inspired, build a drive to achieve, discover aspects of their personalities, skills and intellectual curiosity which they use to craft their identity.

The city provides a range of services and facilities, whose realization in villages are the all-consuming struggle of rural inhabitants. Potable water; sewerage management; energy for cooking, heat and lighting are all piped and wired in; solid waste disposal and storm water drainage are taken for granted. The city offers access through roads, public transit, telephones and the Internet. The peace and security provided by effective policing systems, and the courts of law, are just assumed to be there in the city. Then there are the schools, the recreation facilities, the health services and a myriad of professional services offered in the city market place.

Intelligent urbanism views the city as an opportunity system. Yet these opportunities are not equally distributed. Security, health care, education, shelter, hygiene, and most of all employment, are not equally accessible. Proponents of intelligent urbanism see the city as playing an equalizing role allowing citizens to grow according to their own essential capabilities and efforts. If the city is an institution, which generates opportunities, intelligent urbanism promotes the concept of equal access to opportunities within the urban system.

Intelligent urbanism promotes a guaranteed access to education, health care, police protection, and justice before the law, potable water, and a range of basic services. Perhaps this principle, more than any other, distinguishes intelligent urbanism from other elitist, efficiency oriented urban charters and regimes.

Intelligent urbanism does not say every household will stay in an equivalent house, or travel in the same vehicle, or consume the same amount of electricity.

Intelligent urbanism recognizes the existence of poverty, of ignorance, of ill health, of malnutrition, of low skills, of gender bias and ignorance of the urban system itself. Intelligent urbanism is courageous in confronting these forms of inequality, and backlogs in social and economic development. Intelligent urbanism sees an urban plan, not only as a physical plan, but also as a social plan and as an economic plan.

The ramifications of this understanding are that the people living in intelligent cities should not experience urban development in "standard doses". In short, people may be born equal or unequal, but they grow inequitably. An important role of the city is to provide a variety of paths and channels for each individual to set right their own future, against the inequity of their past, or the special challenges they face. According to proponents of this principle this is the most salient aspect of a free society; than even voting rights access to opportunity is the essence of self-liberation and human development (Sen:2000).

According to proponents of intelligent urbanism, there will be a variety of problems faced by urbanites and they need a variety of opportunity channels for resolution. If there are ten problem areas where people are facing stresses, like economic engagement, health, shelter, food, education, recreation, transport, etc., there must be a variety of opportunities through which individuals and households can resolve each of these stresses. There must be ten channels to resolve each of ten stresses! If this opportunity matrix is understood and responded to, the city is truly functioning as an opportunity matrix. For example, opportunities for shelter could be through the channels of lodges, rented rooms, studio apartments, bedroom apartments and houses. It could be through the channels of ownership, through a variety of tendencies. It could be through opportunities for self-help, or incremental housing. It could be through the upgradation of slums. Intelligent urbanism promotes a wide range of solutions, where any stress is felt. It therefore promotes a range of problem statements, options, and variable solutions to urban stresses.

Intelligent urbanism sees cities as processes. Proponents argue that good urban plans facilitate those processes and do not place barriers before them. For example, it does not judge a "slum" as a blight on society; it sees the possibility that such a settlement may be an opportunity channel for entry into the city. Such a settlement may be the only affordable shelter, within easy access to employment and education, for a new immigrant household in the city. According to intelligent urbanism, if the plan ignores, or destroys such settlements, it is creating a city of barriers and

despair wherein a poor family, offering a good service to the city, is denied a modicum of basic needs for survival. Alternatively, if the urban plan recognizes that the "slum" is a mechanism for self development, a spring-board from which children have access to education, a place which can be up-graded with potable water, basic sanitary facilities, street lights and paving...then it is a plan for opportunity. Intelligent urbanism believes that there are slums of hope and slums of despair. It promotes slums of hope, which contribute, not only to individual opportunities, but also to nation building.

The opportunity matrix must also respond to young professionals, to skilled, well-paid day laborers, to the upper middle class and to affluent entrepreneurs. If a range of needs, of abilities to pay, of locational requirements, and of levels of development of shelter, is addressed, then opportunities are being created.

Intelligent urbanism believes that private enterprise is the logical provider of opportunities, but that alone it will not be just or effective. The regime of land, left to market forces alone, will create an exclusive, dysfunctional society. Intelligent urbanism believes that there is an essential role for the civil society to intervene in the opportunity matrix of the city.

Intelligent urbanism promotes opportunities through access to:

- Basic and primary education, skill development and knowledge about the urban world;

- Basic health care, potable water, solid waste disposal and hygiene;

- Urban facilities like storm drainage, street lights, roads and footpaths;

- Recreation and entertainment;

- Transport, energy, communications;

- Public participation and debate;

- Finance and investment mechanisms;

- Land and/or built-up space where goods and services can be produced;

- Rudimentary economic infrastructure;

- Intelligent urbanism provides a wide range of zones, districts and precincts where activities and functions can occur without detracting from one another.

Intelligent urbanism proposes that enterprise can only flourish where a public framework provides opportunities for enterprise. This system of opportunities operates through public investments in economic and social infrastructure; through incentives in the form of appropriate finance, tax inducements, subsidized skill development for workers, and: regulations which protect the environment, safety, hygiene and health. To ensure a stable playing field where one can make an investment with predictable returns, a modicum of regulation is necessary. Proponents argue that it is through government regulations that private investment can be protected from fraud. It is through government regulation that the under-pinning conditions for free enterprise can be protected.

33.1.8 Principle eight: regional integration

Intelligent urbanism envisions the city as an organic part of a larger environmental, socio-economic and cultural-geographic system, essential for its sustainability. This zone of influence is the region. Likewise, it sees the region as integrally connected to the city. Intelligent urbanism sees the planning of the city and its hinterland as a single holistic process. Proponents argue if one does not recognize growth as a regional phenomenon, then development will play a hopscotch game of moving just a bit further along an arterial roads, further up valleys above the municipal jurisdiction, staying beyond the path of the city boundary, development regulations and of the urban tax regime.

The region may be defined as the catchment area from which employees and students commute into the city on a daily basis. It is the catchment area from which people choose to visit one city, as opposed to another, for retail shopping and entertainment. Economically the city region may include the hinterland that depends on its wholesale markets, banking facilities, transport hubs and information exchanges. The region needing integration may be seen as the zone from which perishable foods, firewood and building materials supply the city. The economic region can also be defined as the area managed by exchanges in the city. Telephone calls to the region go through the city's telecom exchange; post goes through the city's general post office; money transfers go through the city's financial institutions and internet data passes electronically through the city's servers. The area over which "city exchanges" disperse matter can well be called the city's economic hinterland or region. Usually the region includes dormitory communities, airports, water reservoirs, perishable food farms, hydro facilities, out-of-doors recreation and other infrastructure that serves the city. Intelligent urbanism sees the integrated planning of these services and facilities as part of the city planning process.

Intelligent urbanism understands that the social and economic region linked to a city also has a physical form, or a geographic character. A hierarchy of watersheds, creat-

ing valleys and defining edges of neighborhoods, may define the geographic character. Forest ranges, fauna and avi-fauna habitats are set within such regions and are connected by natural corridors for movement and cross-fertilization. Within this larger, environmental scenario, one must conceptualize urbanism in terms of watersheds, subterranean aquifer systems, and other natural systems that operate across the entire region. Economic infrastructure, such as roads, hydro basins, irrigation channels, water reservoirs and related distribution networks usually follow the terrain of the regional geography. The region's geographic portals, and lines of control, may also define defense and security systems deployment.

Intelligent urbanism recognizes that there is always a spillover of population from the city into the region, and that population in the region moves into the city for work, shopping, entertainment, health care and education. With thoughtful planning the region can take pressure off of the city. Traditional and new settlements within the urban region can be enhanced and densified to accommodate additional urban households. There are many activities within the city, which are growing and are incompatible with urban habitat. Large, noisy and polluting workshops and manufacturing units are amongst these. Large wholesale markets, storage sheds, vehicular maintenance garages, and waste management facilities need to be housed outside of the city's limits in their own satellite enclaves. In larger urban agglomerations a number of towns and cities are clustered around a major urban center forming a metropolitan region.

Intelligent urbanism is not just planning for the present; it is also planning for the distant future. Intelligent urbanism is not Utopian, but futuristic in its need to forecast the scenarios to come, within its own boundaries, and within the boundaries of the distant future.

33.1.9 Principle nine: balanced movement

Intelligent urbanism advocates integrated transport systems comprising walkways, cycle paths, bus lanes, light rail corridors, under-ground metros and automobile channels. A balance between appropriate modes of movement is proposed. More capital intensive transport systems should move between high density nodes and hubs, which interchange with lower technology movement options. These modal split nodes become the public domains around which cluster high density, pedestrian, mixed-use urban villages (Taniguchi:2001).

The PIU accepts that the automobile is here to stay, but that it should not be made essential by design. A well planned metropolis would densify along mass transit corridors and around major urban hubs. Smaller, yet dense, urban nodes are seen as micro-zones of medium level density, public amenities and pedestrian access. At these points lower level nodal split will occur, such as between bus loops and cycle tracts. The PIU views nodal split points as places of urban conviviality and access to services and facilities. Modal split can be between walking, cycling, driving, and mass transit. Bus loops may feed larger rail-based rapid-movement corridors. Social and economic infrastructure becomes more intensive as movement corridors become more intense.

33.1.10   Principle ten: institutional integrity

Intelligent urbanism holds that good practices inherent in considered principles can only be realized through accountable, transparent, competent and participatory local governance, founded on appropriate data bases, due entitlements, civic responsibilities and duties. The PIU promotes a range of facilitative and promotive urban development management tools to achieve appropriate urban practices, systems and forms(Islam:2000). None of the principles or practices the PIU promotes can be implemented unless there is a strong and rational institutional framework to define, channel and legalize urban development, in all of its aspects. Intelligent urbanism envisions the institutional framework as being very clear about the rules and regulations it sponsors and that those using discretion in implementing these measures must do so in a totally open, recorded and transparent manner.

Intelligent urbanism facilitates the public in carrying out their honest objectives. It does not regulate and control the public. It attempts to reduce the requirements, steps and documentation required for citizens to process their proposals.

Intelligent urbanism is also promotive in furthering the interests of the public in their genuine utilization of opportunities. It promotes site and services schemes for households who can construct their own houses. It promotes up-gradation of settlements with inadequate basic services. It promotes innovative financing to a range of actors who can contribute to the city's development. Intelligent urbanism promotes a limited role for government, for example in "packaging" large-scale urban development schemes, so that the private sector is promoted to actually build and market urban projects, which were previously built by the government.

Intelligent urbanism does not consider itself naïve. It recognizes that there are developers and promoters who have no long term commitment to their own constructions, and their only concern is to hand over a dwelling, gain their profit and move on. For these players it is essential to have Development Control Regulations, which assure the public that the products they invest in are safe, hygienic, orderly,

durable and efficient. For the discerning citizen, such rules also lay out the civil understanding by which a complex society agrees to live together.

The PIU contends that there must be a cadastral System wherein all of the land in the jurisdiction of cities is demarcated, surveyed, characterized and archived, registering its legal owner, its legal uses, and the tax defaults against it.

The institutional framework can only operate where there is a Structure Plan, or other document that defines how the land will be used, serviced, and accessed. The Structure Plan tells landowners and promoters what the parameters of development are, which assures that their immediate investments are secure, and that the returns and use of such efforts are predictable. A Structure Plan is intended to provide owners and investors with predictable future scenarios. Cities require efficient patterns for their main infrastructure systems and utilities. According to PIU proponents, land needs to be used in a judicious manner, organizing complementary functions and activities into compact, mixed use precincts and separating out non-compatible uses into their own precincts. In a similar manner, proponents argue it is only through a plan that heritage sites and the environment can be legally protected. Public assets in the form of nature, religious places, heritage sites and open space systems must be designated in a legal plan.

Intelligent urbanism proposes that the city and its surrounding region be regulated by a Structure Plan, or equivalent mechanism, which acts as a legal instrument to guide the growth, development and enhancement of the city.

According to proponents, there must be a system of participation by the "Stake Holders" in the preparation of plans. Public meetings, hearings of objections and transparent processes of addressing objections, must be institutionalized. Intelligent urbanism promotes Public Participation. Local Area Plans must be prepared which address local issues and take into account local views and sentiments regarding plan objectives, configurations, standards and patterns. Such plans lay out the sites of plots showing the roads, public open spaces, amenities areas and conservation sites. Land Pooling assures the beneficiaries from provision of public infrastructure and amenities proportionally contribute and that a few individuals do not suffer from reservations in the plan.

According to proponents, there must be a system of Floor Area Ratios to assure that the land and the services are not over pressured. No single plot owner should have more than the determined "fair share" of utilization of the access roads, amenities and utilities that service all of the sites. Floor Area Ratios temper this relationship as regulated the manner in which public services are consumed. According to PIU proponents, Transfer of Development Rights benefits land owners whose properties have been reserved under the plan. It also benefits the local authorities that lack the financial resources to purchase lands to implement the Structure Plans. It benefits concentrated, city center project promoters who have to amortize expensive land purchases, by allowing them to purchase the development rights from the owners of reserved lands and to hand over those properties to the plan implementing authority. This allows the local authority to widen roads and to implement the Structure Plan. The local authority then transfers the needed development right to city center promoters.

Intelligent urbanism supports the use of Architectural Guidelines where there is a tradition to preserve and where precedents can be used to specify architectural elements, motifs and language in a manner, which intended to reinforce a cultural tradition. Building designs must respect traditional elements, even though the components may vary greatly to integrate contemporary functions. Even in a greenfield setting Architectural Guidelines are required to assure harmony and continuity of building proportions, scale, color, patterns, motifs, materials and facades.

Intelligent urbanism insists on safety, hygiene, durability and utility in the design and construction of buildings. Where large numbers of people gather in schools, hospitals, and other public facilities that may become emergency shelters in disasters, special care must be exercised. A suitable Building Code is the proposed instrument to achieve these aims.

PIU proponents state that those who design buildings must be professionally qualified architects; those who design the structures (especially of more than ground plus two levels) must be professionally qualified structural engineers; those who build buildings must be qualified civil engineers; and, those who supervise and control construction must be qualified construction managers. Intelligent urbanism promotes the professionalisation of the city making process. While promoting professionalism, intelligent urbanism proposes that this not become a barrier in the development process. Small structures, low-rise structures, and humble structures that do not house many people can be self designed and constructed by the inhabitants themselves. Proponents maintain that there must be recognized Professional Accrediting Boards, or Professional Bodies, to see that urban development employs adequate technical competence.

Finally, there must be legislation creating statutory local authorities, and empowering them to act, manage, invest, service, protect, promote and facilitate urban development and all of the opportunities that a modern city must sponsor.

Intelligent urbanism insists that cities, local authorities, regional development commissions and planning agencies be professionally managed. City Managers can be hired to manage the delivery of services, the planning and management of planned development, the maintenance of utilities

and the creation of amenities.

Intelligent urbanism views plans and urban designs and housing configurations as expressions of the people for whom they are planned. The processes of planning must therefore be a participatory involving a range of stakeholders. The process must be a transparent one, which makes those privileged to act as guardians of the people's will accountable for their decisions and choices. Intelligent urbanism sees urban planning and city governance as the most salient expressions of civility. Intelligent urbanism fosters the evolution of institutional systems that enhance transparency, accountability and rational public decision making.

33.2 Movements implementing the ten principles

Though not necessarily related to the principles of intelligent urbanism, there are examples representing all or some of them in urban design theory and practice. Concurrently, the recent movements of New Urbanism and New Classical Architecture promote a sustainable approach towards construction, that appreciates and develops smart growth, architectural tradition and classical design.[6][7] This in contrast to modernist and globally uniform architecture, as well as leaning against solitary housing estates and suburban sprawl.[8] Both trends started in the 1980s. The Driehaus Architecture Prize is an award that recognizes efforts in New Urbanism and New Classical Architecture, and is endowed with a prize money twice as high as that of the modernist Pritzker Prize.[9]

33.3 See also

- Michael E. Arth
- Christopher Charles Benninger
- Broadacre City
- Garden city movement
- New Classical Architecture
- New Pedestrianism
- New Urbanism
- Josep Lluis Sert
- Smart Growth
- Social Access

- Structure Plan

- Sustainable city

- Team Ten

- Theories of Urban Planning

- Transit Oriented Development

- Urban design

- Urban studies

- Urban village

- Vitruvius

33.4 Notes

[1] Thimphu Structural Plan 2002-2027

[2] McHarg I. (1975): Design with Nature, Wiley, John and Sons, New York.

[3] Jacobs Jane (1993): The Death and Life of Great American Cities, Random House, New York.

[4] Urban Land Institute (1998): Smart Growth, Urban Land Institute, Washington D.C.

[5] Sen A. (2000): Development as Freedom: Knopf, New York.

[6] Charter of the New Urbanism

[7] "Beauty, Humanism, Continuity between Past and Future". Traditional Architecture Group. Retrieved 23 March 2014.

[8] Issue Brief: Smart-Growth: Building Livable Communities. American Institute of Architects. Retrieved on 2014-03-23.

[9] "Driehaus Prize". Together, the $200,000 Driehaus Prize and the $50,000 Reed Award represent the most significant recognition for classicism in the contemporary built environment.. Notre Dame School of Architecture. Retrieved 23 March 2014.

33.5 References

- Principles of Intelligent Urbanism

- Caves Roger, Ed. (2004):"Principles of Intelligent Urbanism," "Encyclopedia of the City", London: Routledge.

- Thimphu Structure Plan Interest in the concept of Intelligent Urbanism has spread to other contexts (Williams, 2003) and its application is being widely discussed (Graz Biennial, 2001).

- Benninger C. (2001): "Principles of Intelligent Urbanism," in Ekistics, Volume 69, Number 412, pp. 39–65, Athens.

- Benninger C. (2002): "Principles of Intelligent Urbanism," Thimphu Structure Plan, Royal Government of Bhutan, Thimphu.

- Graz Biennal Committee (2001): "Imagineering and Urban Design," C. Benninger, in Proceedings of the Graz Biennial, Graz.

- Islam Nazrul (2000): Urban Governance in Asia, Pathak Samabesh, Dhaka.

- Jacobs Jane (1993): The Death and Life of Great American Cities, Random House, New York.

- Leccese M. Ed. (1999): Charter of the New Urbanism, McGraw Hill Professional, New York.

- Lewis P. (1996): Tomorrow by Design, Wiley, John and Sons, New York.

- Kingsley Dennis, John Urry (2009): After the Car, Polity Press, Cambridge, UK

- Marshall A. (2000): How Cities Work: University of Texas Press, Austin, Texas.

- McHarg I. (1975): Design with Nature, Wiley, John and Sons, New York.

- Sen A. (2000): Development as Freedom: Knopf, New York.

- Spreiregen P. (1965): Urban Design: the Architecture of Towns and Cities, McGraw-Hill, New York.

- Taniguchi E. (2001): City Logistics: Network Modeling and Intelligent Transport Systems, Elsevier Science and Technology Books, Hoboken.

- Urban Land Institute (1998): Smart Growth, Urban Land Institute, Washington D.C.

- Williams T. (2003): "Smart Advice for Urban Growth," Regeneration and Renewal, 6 June 2003, London.

33.6 External links

- Thimphu Structure Plan

- Keynote Address at the Conference on Urban Policy at Bangkok on 6 January 2007 delivered by Lyonpo Jigme Thinley, Minister of Culture, Royal Government of Bhutan

- Article on PIU and the Middle Path

- Paper Published in the Proceedings of the Centre for Bhutan Studies, on "From principles to Action: Creating Happy places to Live In"

- Article in the HINDU by Harsh Kabra, titled "Living Architecture"

- The Labors of Hercules: Modern Solutions to 12 Herculean Problems. Labor IX: Urbanism

- Nature in the City

- CNU Congress of new urbanism

- http://www.lancs.ac.uk/fass/sociology/papers/ word%20docs/dennis-cityvisionsmobilityfutures. doc.

- http://www.dudh.gov.bt/Thimphustructural/ intelegenturbanism/1.2/1.2.html

- http://www.benproject.org/repository/WP3/ recommendations_spatial.pdf

- http://www.klimentovska.as/fotky1/fotos/_s_ 122Microsoft%20Word%20-%20Model% 20udryziteln-oho%20rozvoje%20ve%20morstech% 20a%20obc-uch_final_AJ.pdf

Chapter 34

Public interest design

Public interest design is a human-centered[1] and participatory design practice[2] that places emphasis on the "triple bottom line" of sustainable design that includes ecological, economic, and social issues and on designing products, structures, and systems that address issues such as economic development and the preservation of the environment.

Starting in the late 1990s, several books, convenings, and exhibitions have generated new momentum and investment in public interest design. Since then, public interest design—frequently described as a movement or field—has gained public recognition.[3]

34.1 History

Public interest design grew out of the community design movement, which got its start in 1968 after American civil rights leader Whitney Young issued a challenge to attendees of the American Institute of Architects (AIA) national convention:[4]

"*. . . you are not a profession that has distinguished itself by your social and civic contributions to the cause of civil rights, and I am sure this does not come to you as any shock. You are most distinguished by your thunderous silence and your complete irrelevance.*[5]"

The response to Young's challenge was the establishment of community design centers (CDCs) across the United States.[6] CDCs, which were often established with the support of area universities,[7] provided a variety of design services – such as affordable housing - within their own neighborhoods.

In architecture schools, "design/build programs" provided outreach to meet local design needs, particularly in low-income and underserved areas.[7] One of the earliest design/build programs was Yale University's Vlock Building Project. The project, which was initiated by students at Yale University School of Architecture in 1967, requires graduate students to design and build low-income housing.[8]

One of the most publicized programs is the Auburn University Rural Studio design/build program, which was founded in 1993.[2][9][10] The Rural Studio's first project, Bryant House, was completed in 1994 for $16,500.[11]

34.2 Public Interest Design from the 1990s – Present

Interest in public interest design – particularly socially responsible architecture – began to grow during the 1990s and continued into the first decade of the new millennium. Conferences, books, and exhibitions began to showcase the design work being done beyond the community design centers,[2] which had greatly decreased in numbers since their peak in the seventies.[7]

Non-profit organizations – including Architecture for Humanity, BaSiC Initiative, Design Corps, Public Architecture, Project H, and MASS Design Group – began to provide design services that served a larger segment of the population than had been served by traditional design professions.[2][12]

Many public interest design organizations also provide training and service-learning programs for architecture students and graduates. In 1999, the Enterprise Rose Architectural Fellowship was established,[2] giving young architects the opportunity to work on three-year-long design and community development projects in low-income communities.[13]

Other organizations providing professional training through fellowships include Code for America,[14] Design Corps,[7] Design Impact,[15] Gulf Coast Community Design Studio,[16] bcWORKSHOP,[17] IDEO.org.[18]

Two of the earliest formal public interest design programs include the Gulf Coast Community Design Studio at Mississippi State University[2][19] and the Public Interest Design Summer Program at the University of Texas[2] .[20]In February 2015, Portland State University launched the first graduate certificate program in Public Interest Design in the United States. [21]

The first professional-level training was conducted in July 2011 by the Public Interest Design Institute (PIDI) and held at the Harvard Graduate School of Design.[22]

Also in 2011, a survey of American Institute of Architects (AIA), 77% of AIA members agreed that the mission of the professional practice of public interest design could be defined as the belief that every person should be able to live in a socially, economically, and environmentally healthy community.[23][24]

34.3 Publications, Conferences, and Exhibits

Several books have been published that showcase a variety of public interest design projects and practitioners:

- *Good Neighbors, Affordable Family Housing." Tom Jones, William Pettus, and Michael Pyatok, 1997 (ISBN 978-0070329133)[25]*

- *Learning by Building: Design and Construction in Architectural Education." William J. Carpenter, 1997 (ISBN 978-0471287933)[26]*

- *Good Deeds, Good Design: Community Service through Architecture." Bryan Bell, 2003 (ISBN 978-1568983912)[27]*

- *Design Like You Give a Damn: Architectural Responses to Humanitarian Crises.* Kate Stohr and Cameron Sinclair, ed., 2006 (ISBN 978-1933045252)[28]

- *Expanding Architecture, Design as Activism." Bryan Bell and Katie Wakeford, ed., 2008 (ISBN 978-1933045788)[29]*

- *Design Revolution: 100 Products that Empower People.* Emily Pilloton, 2009 (ISBN 978-1933045955)[30]

- *The Power of Pro Bono: 40 Stories about Design for the Public Good by Architects and Their Clients.* John Cary, 2010 (ISBN 978-1935202189)[31]

- *Design Like You Give a Damn 2: Building Change from the Ground Up." Kate Stohr and Cameron Sinclair, ed., 2012 (ISBN 978-0810997028)[32]*

The annual Structures for Inclusion conference showcases public interest design projects from around the world. The first conference, which was held in 2000, was called "Design for the 98% Without Architects."[2] Speaking at the conference, Rural Studio co-founder Samuel Mockbee challenged

attendees to serve a greater segment of the population: "I believe most of us would agree that American architecture today exists primarily within a thin band of elite social and economic conditions[33]...in creating architecture, and ultimately community, it should make no difference which economic or social type is served, as long as the status quo of the actual world is transformed by an imagination that creates a proper harmony for both the affluent and the disadvantaged.[33]" The Structures for Inclusion conference is now entering its 14th year, to take place March 23–25, 2013, hosted by the University of Minnesota College of Design. The event is part of a larger, first-of-its-kind Public Interest Design Week. Other recurring events, chief among them the Design Like You Give a Damn: LIVE,[34] coordinated by Architecture for Humanity, and the Design Access Summit,[35] coordinated by Public Architecture, have also taken shape in recent years.

In 2007, the Cooper Hewitt National Design Museum held an exhibition, titled "Design for the Other 90%," curated by Cynthia Smith. Following the success of this exhibit, Smith developed the "Design Other 90"[36] initiative into an ongoing series, the second of which was titled "Design for the Other 90%: CITIES"[37] (2011), held at the United Nations headquarters. In 2010, Andres Lipek of the Museum of Modern Art in New York curated an exhibit, called "Small Scale, Big Change: New Architectures of Social Engagement.[38]"[2] In 2012, the U.S. Pavilion of the 13th International Venice Architecture Biennale, curated by Cathy Lang Ho, focused on "Spontaneous Interventions: Design Actions for the Common Good.[39]" Also in 2012, "Public Interest Design: Products, Places, & Processes,[40]" was curated by John Cary and Courtney E. Martin, at the Autodesk Gallery[41] in San Francisco.

34.4 Professional Networks

One of the oldest professional networks related to public interest design is the professional organization is the Association for Community Design (ACD), which was founded in 1977.[2][42]

In 2005, adopting a term coined by architect Kimberly Dowdell, the Social Economic Environmental Design (SEED) Network[43] was co-founded by a group of community design leaders,[2] during a meeting hosted by the Loeb Fellowship at the Harvard Graduate School of Design. The SEED Network established a common set of five principles[44] and criteria for practitioners of public interest design. An evaluation tool called the SEED Evaluator is available to assist designers and practitioners in developing projects that align with SEED Network goals and criteria.[45]

In 2006, the Open Architecture Network[46] was launched by Architecture for Humanity in conjunction with co-founder Cameron Sinclair's TED Wish.[47] Taking on the name Worldchanging[48] in 2011, the network is an open-source community dedicated to improving living conditions through innovative and sustainable design. Designers of all persuasions can share ideas, designs and plans as well as collaborate and manage projects. while protecting their intellectual property rights using the Creative Commons "some rights reserved" licensing system.

In 2007, DESIGN 21: Social Design Network,[49] an on-line platform built in partnership with UNESCO,[50] was launched.

In 2011, the Design Other 90 Network[51] was launched by the Cooper-Hewitt, National Design Museum, in conjunction with its Design with the Other 90%: CITIES exhibition.

In 2012, IDEO.org,[52] with the support of The Bill & Melinda Gates Foundation,[53] launched HCD Connect,[54] a network for social sector leaders committed to human-centered design. In this context, human-centered design begins with the end-user of a product, place, or system — taking into account their needs, behaviors and desires. The fast-growing professional network of 15,000 builds on "The Human-Centered Design Toolkit,"[55] which was designed specifically for people, nonprofits, and social enterprises that work with low-income communities throughout the world. People using the HCD Toolkit or human-centered design in the social sector now have a place to share their experiences, ask questions, and connect with others working in similar areas or on similar challenges.

34.5   See also

- Design/Build

- Healthy community design

- Leadership in Energy and Environmental Design

- Sustainable architecture

- Sustainable Design

34.6   References

[1] HCD Connect Methods

[2] Cary, John. "Infographic: Public Interest Design". PublicInterestDesign.org. Retrieved 23 October 2012.

[3] Cary, John; Courtney E. Martin (October 6, 2012). "Dignifying Design". The New York Times. Retrieved 23 October 2012.

[4] Leavitt, Jacqueline; Kara Hoffernan (2006). Mary C. Hardin, ed. From the Studio to the Streets: Service Learning in Planning and Architecture. Sterling, VA: Stylus Publishing. p. 103. ISBN 978-1563771002.

[5] Whitney Young 1968 Speech to the AIA

[6] Design Coalition, "Our Roots"

[7] Pearson, Jason (2002). Mark Robbins, ed. University-Community Design Partnerships: Innovations in Practice (PDF). New York, New York: Princeton Architectural Press. pp. 6–7. ISBN 1-56898-379-4. Retrieved 23 October 2012.

[8] Hill, David (March 2012). "The New Frontier in Education". Architectural Record. Retrieved 23 October 2012.

[9] Clemence, Sara (April 2012). "Avant-Garde in Alabama". The Wall Street Journal. Retrieved 23 October 2012.

[10] Bostwick, William. ""Citizen Architect": The Humble Origins of Socially-Responsible Design". Fast Company. Retrieved 23 October 2012.

[11] "Samuel Mockbee and the Rural Studio: Community Architecture". National Building Museum. Retrieved 23 October 2012.

[12] Hughes, C.J. (March 2012). "Does "Doing Good" Pay the Bills?". Architectural Record. Retrieved 23 October 2012.

[13] Enterprise Community website, "About the Fellowship"

[14] Code for America Fellowship Program webpage

[15] Design Impact website, "Fellowship"

[16] Gulf Coast Community Design Center, "Public Interest Design Program"

[17] bcWORKSHOP website, "Summer 2012 bcFELLOWs"

[18] IDEO.org, "Innovators in Residence"

[19] Coast Community Design Center home page

[20] Overview of Public Interest Design program, University of Texas School of Architecture

[21] http://www.pdx.edu/the-arts/recent-and-upcoming/cpid-launches-graduate-certificate-in-public-interest-design

[22] Harvard School of Design course description

[23] Bell, Bryan. "Public Interest Design Takes Shape". Metropolismag.com. Retrieved 23 October 2012.

[24] Bell, Bryan; Roberta Feldman; Sergio Palleroni; Davide Perkes (31 December 2011). "2011 Latrobe Prize Progress Report: Public Interest Practices in Architecture" (PDF). Retrieved 23 October 2012.

[25] *Good Neighbors, Affordable Family Housing* book description on Amazon.com

[26] *Learning by Building: Design and Construction in Architectural Education* book description on Amazon.com

[27] *Good Deeds, Good Design: Community Service through Architecture* book description on Amazon.com

[28] *Design Like You Give a Damn: Architectural Responses to Humanitarian Crises* book description on Amazon.com

[29] *Expanding Architecture, Design as Activism* book description on Amazon.com

[30] *Design Revolution: 100 Products that Empower People* book description on Amazon.com

[31] *The Power of Pro Bono* book description on Amazon.com

[32] *Design Like You Give a Damn 2: Building Change from the Ground Up* book description on Amazon.com

[33] Bell, Bryan (2003). *Good Deeds, Good Design: Community Service Through Architecture.* New York, New York: Princeton Architectural Press. p. 156. ISBN 978-1568983912.

[34] Design Like You Give a Damn: LIVE webpage

[35] Design Access Summit webpage

[36] Design Other 90 Network

[37] Design for the Other 90%: CITIES

[38] MoMA Exhibition: "Small Scale, Big Change: New Architectures of Social Engagement"

[39] 2012 U.S. Pavilion Exhibition: "Spontaneous Interventions: Design Actions for the Common Good"

[40] Exhibition: "Public Interest Design: Products, Places, & Processes

[41] Autodesk Gallery

[42] Association for Community Design website, "About" page

[43] SEED Network website

[44] SEED Network principles

[45] SEED Evaluator

[46] Open Architecture Network website

[47] Cameron Sinclair's TED Prize talk and wish for the Open Architecture Network

[48] Worldchanging website

[49] Design 21 Social Design Network website

[50] UNESCO website

[51] Design Other 90 Network website

[52] IDEO.org website

[53] The Bill & Melinda Gates Foundation website

[54] HCD Connect website

[55] The Human-Centered Design Toolkit from IDEO

34.7 External links

- Infographic: "From Idealism to Realism: The History of Public Interest Design"

- PublicInterestDesign.org daily blog and website

- Public Interest Design Institute

- The SEED Network

Chapter 35

Rain garden

Business parking lot that drains to a rain garden. The curb retains the asphalt pavement, yet lets water flow off the edges.

A **rain garden** is a planted depression or a hole that allows rainwater runoff from impervious urban areas, like roofs, driveways, walkways, parking lots, and compacted lawn areas, the opportunity to be absorbed. This reduces rain runoff by allowing stormwater to soak into the ground (as opposed to flowing into storm drains and surface waters which causes erosion, water pollution, flooding, and diminished groundwater).[1] They should be designed for specific soils and climates.[2] The purpose of a rain garden is to improve water quality in nearby bodies of water and to ensure that rainwater becomes available for plants as groundwater rather than being sent through stormwater drains straight out to sea. Rain gardens can cut down on the amount of pollution reaching creeks and streams by up to 30%.[3]

Native and adapted plants are recommended for rain gardens because they are more tolerant of one's local climate, soil, and water conditions; have deep and variable root systems for enhanced water infiltration and drought tolerance; habitat value and diversity for local ecological communities; and overall sustainability once established. There can be trade-offs associated with using native plants, including lack of availability for some species, late spring emergence, short blooming season, and relatively slow establishment.

The plants — a selection of wetland edge vegetation, such as wildflowers, sedges, rushes, ferns, shrubs and small trees — take up excess water flowing into the rain garden. Water filters through soil layers before entering the groundwater system. Root systems enhance infiltration, maintain or even augment soil permeability, provide moisture redistribution, and sustain diverse microbial populations involved in biofiltration.[4] Also, through the process of transpiration, rain garden plants return water vapor to the atmosphere.[5] A more wide-ranging definition covers all the possible elements that can be used to capture, channel, divert, and make the most of the natural rain and snow that falls on a property. The whole garden can become a rain garden, and each component of the whole can become a small-scale rain garden in itself.

35.1 Restoring the water cycle and mitigating urbanization

In developed areas, natural depressions where storm water would pool, are filled in. The surface of the ground is often leveled or paved. Storm water is directed into storm drains which often may cause overflows of combined sewer systems or poisoning, erosion or flooding of waterways receiving the storm water runoff.[6][7][8] Redirected storm water is often warmer than the groundwater normally feeding a stream, and has been linked to upset in some aquatic ecosystems primarily through the reduction of dissolved oxygen (DO). Storm water runoff is also a source of a wide variety of pollutants washed off hard or compacted surfaces during rain events. These pollutants include volatile organic compounds, pesticides, herbicides, hydrocarbons and trace metals[9] Rain gardens are designed to capture the initial flow of storm water and reduce the accumulation of toxins flowing directly into natural waterways through ground filtration. They also reduce energy consumption. For example, "the cumulative storage capacity of these rain gardens exceeds a conventional stormwater's system's by 10 times."[10] The National Science Foundation, the United States Environmental Protection Agency, and a number of research institutions are presently studying the impact of augmenting rain gardens with materials capable of capture or chemical reduction of the pollutants to benign compounds.

Rain garden, SUNY College of Environmental Science and Forestry, Syracuse, New York

Rain gardens are often located near a building's roof drainpipe (with or without rainwater tanks). Most rain gardens are designed to be an endpoint of drainage with a capacity to percolate all incoming water through a series of soil or gravel layers beneath the surface plantings. A French drain may be used to direct a portion of the rainwater to an overflow location for heavier rain events. By reducing peak stormwater discharge, rain gardens extend hydraulic lag time and somewhat mimic the natural water cycle displaced by urban development and allow for groundwater recharge. While rain gardens always allow for restored groundwater recharge, and reduced stormwater volumes, they may also increase pollution unless remediation materials are included in the design of the filtration layers .[11]

The primary challenge of rain garden design centers on calculating the types of pollutants and the acceptable loads of pollutants the rain garden's filtration system can handle during storm-water events. This challenge is specifically acute when a rain event occurs after a longer dry period. The initial storm water is often highly contaminated with the accumulated pollutants from dry periods. Rain garden designers have previously focused on finding robust native plants and encouraging adequate biofiltration, but recently have begun augmenting filtration layers with media specifically suited to chemically reduce redox of incoming pollutant streams.

Rain gardens are beneficial for many reasons: improve water quality by filtering runoff, provide localized flood control, are aesthetically pleasing, and provide interesting planting opportunities. They also encourage wildlife and biodiversity, tie together buildings and their surrounding environments in attractive and environmentally advantageous ways, and provide significant partial solutions to important environmental problems that affect us all.

A rain garden provides a way to use and optimize any rain that falls, reducing or avoiding the need for irrigation. They allow a household or building to deal with excessive rainwater runoff without burdening the public storm water systems. Rain gardens differ from retention basins, in that the water will infiltrate the ground within a day or two. This creates the advantage that the rain garden does not allow mosquitoes to breed.

35.2 History

The first rain gardens were created to mimic the natural water retention areas that occurred naturally before development of an area. The rain gardens for residential use were developed in 1990 in Prince George's County, Maryland, when Dick Brinker, a developer building a new housing subdivision had the idea to replace the traditional best management practices (BMP) pond with a bioretention area. He approached Larry Coffman, the county's Associate Director for Programs and Planning in the Department of Environmental Resources, with the idea.[12] The result was the extensive use of rain gardens in Somerset, a residential subdivision which has a 300–400 sq ft (28–37 m^2) rain garden on each house's property.[13] This system proved to be highly cost-effective. Instead of a system of curbs, sidewalks, and gutters, which would have cost nearly $400,000, the planted drainage swales cost $100,000 to install.[12] This was also much more cost effective than building BMP ponds that could handle 2-, 10-, and 100-year storm events.[12] Flow monitoring done in later years showed that the rain gardens have resulted in a 75–80% reduction in stormwater runoff during a regular rainfall event.[13]

This is also referred to as Water Sensitive Urban Design (WSUD) in Australia, Sustainable Urban Drainage Systems or SUDS in the United Kingdom, and low impact development (LID) in the United States, and is cited by the EPA on their website as a success on the Stormwater Case Studies section of their website.[14] This webpage has many links to information on Prince George's County's literature on implementing LID in a community.

Some *de facto* rain gardens predate their recognition by professionals as a significant LID tool. Any shallow garden depression implemented to capture and retain rain water within the garden so as to drain adjacent land without running off a property is at conception a rain garden — particularly if vegetation is maintained with recognition of its role in this function. Vegetated roadside swales, now promoted as "bioswales", remain the conventional drainage system in many parts of the world from long before extensive networks of concrete sewers became the conventional engineering practice in the industrialized world.

What is new about such technology is the emerging rigor of increasingly quantitative understanding of how such tools may make sustainable development possible. This is as true for wealthy developed communities retrofitting bioretention into built stormwater management systems, as for developing communities seeking a faster and more sustainable development path.

35.3   Characteristics

A home rain garden recently planted

A rain garden requires an area where water can collect and infiltrate, and plants to maintain infiltration rates, diverse microbe communities, and water holding capacity. Transpiration by growing plants accelerates soil drying between storms. This includes any plant extending roots to the garden area.

Simply adjusting the landscape so that downspouts and paved surfaces drain into existing gardens may be all that is needed because the soil has been well loosened and plants are well established. However, many plants do not tolerate saturated roots for long and often more water runs off one's roof than people realize. Often the required location and storage capacity of the garden area must be determined first. Rain garden plants are then selected to match the situation, not the other way around.

35.3.1   Soil and drainage

When an area's soils are not permeable enough to allow water to drain and filter properly, the soil should be re-

placed and an underdrain installed. This bioretention mixture should typically contain 60% sand, 20% compost, and 20% topsoil, and there is a current trend to replace compost with biochar. Existing soil must be removed and replaced. Do not combine the sandy soil (bioretention) mixture with a surrounding soil that does not have high sand content. Otherwise, the clay particles will settle in between the sand particles and form a concrete-like substance, as demonstrated in a 1983 study.[15] Deep plant roots also create additional channels for storm water to filter into the ground. Microbial populations feed off plant root secretions and break down carbon (such as in mulch or desiccated plant roots) to aggregate soil particles which increases infiltration rates. A five-year USGS study[16] indicates that rain gardens in urban clay soils can be effective without the use of underdrains or replacement of native soils with the bioretention mix. Pre-installation infiltration rates should be at least .25 in/hour, however. Type D soils will require an underdrain paired with the sandy soil mix in order to drain properly.

Sometimes a drywell with a series of gravel layers near the lowest spot in the rain garden will help facilitate percolation. However, a drywell placed at the lowest spot can become clogged with silt prematurely turning the garden into an infiltration basin defeating its purpose. Depression-focused recharge of polluted water into wells poses a serious threat and should be avoided. Similarly plans to install a rain garden near a septic system should be reviewed by a qualified engineer. The more polluted the water, the longer it must be retained in the soil for purification. This is often achieved by installing several smaller rain garden basins with soil deeper than the seasonal high water table. In some cases lined bioretention cells with subsurface drainage are used to retain smaller amounts of water and filter larger amounts without letting water percolate as quickly.

Rain gardens are at times confused with bioswales. Swales slope to a destination, while rain gardens do not; however, a bioswale may end with a rain garden. Drainage ditches may be handled like bioswales and even include rain gardens in series, saving time and money on maintenance. Part of a garden that nearly always has standing water is a water garden, wetland, or pond, and not a rain garden. Using the proper terminology ensures that the proper methods are used to achieve the desired results.

35.3.2   Plant selection

Plants selected for use in a rain garden should tolerate both saturated and dry soil. Using native plants is generally encouraged. This way the rain garden may contribute to urban habitats for native butterflies, birds, and beneficial insects.

Well planned plantings require minimal maintenance to survive, and are compatible with adjacent land use. Trees under power lines, or that up-heave sidewalks when soils become moist, or whose roots seek out and clog drainage tiles can cause expensive damage.

Trees generally contribute most when located close enough to tap moisture in the rain garden depression, yet do not excessively shade the garden. That said, shading open surface waters can reduce excessive heating of habitat. Plants tolerate inundation by warm water for less time because heat drives out dissolved oxygen, thus a plant tolerant of early spring flooding may not survive summer inundation.

35.4 Rain garden projects

35.4.1 Australia

- Healthy Waterways Raingardens Program promotes a simple and effective form of stormwater treatment, and aims to raise peoples' awareness about how good stormwater management contributes to healthy waterways. The program encourages people to build rain gardens at home, and has achieved its target is to see 10,000 rain gardens built across Melbourne by 2013.[17]

- Melbourne Water's database of Water Sensitive Urban Design projects, including 57 case studies relating to rain gardens/bioretention systems. Melbourne Water is the Victorian State Government agency responsible for managing Melbourne's water supply catchments.[18]

- Water By Design is a capacity building program that supports the uptake of Water Sensitive Urban Design, including rain gardens, in South East Queensland. It was established by the South East Queensland Healthy Waterways Partnership in 2005, as an integral component of the SEQ Healthy Waterways Strategy.[19]

35.4.2 United Kingdom

- The Wildfowl and Wetlands Trust's London Wetland Centre includes a rain garden designed by Nigel Dunnett.[20]

- Islington London Borough Council commissioned sustainable drainage consultants Robert Bray Associates to design a pilot rain garden in the Ashby Grove development which was completed in 2011. This raingarden is fed from a typical modest domestic roof catchment area of 30m² and is designed to demonstrate how simple and cost effective domestic rain gardens are to install. Monitoring apparatus was built into

the design to allow Middlesex University to monitor water volumes, water quality and soil moisture content. The rain garden basin is 300mm deep and has a storage capacity of 2.17m³ which is just over the volume required to store runoff from the roof catchment in a 1 in 100 storm plus 30% allowance for climate change.[21][22]

- The Day Brook Rain Garden Project has introduced a number of rain gardens into an existing residential street in Sherwood, Nottingham[23]

35.4.3 United States of America

- The 12,000 rain garden campaign for Puget Sound is coordinating efforts to build 12,000 rain gardens in the Puget Sound Basin of Western Washington by 2016. The 12,000 rain gardens website provides information and resources for the general public, landscape professionals, municipal staff, and decision makers. By providing access to the best current guidance, easy-to-use materials, and a network of trained "Rain Garden Mentor" Master Gardeners, this campaign seeks to capture and cleanse over 200 Million gallons of polluted runoff each year, and thereby significantly improve Puget Sound's water quality.[24]

- Maplewood, Minnesota has implemented a policy of encouraging residents to install rain gardens. Many neighborhoods had swales added to each property, but installation of a garden at the swale was voluntary. The project was a partnership between the City of Maplewood, University of Minnesota Department of Landscape Architecture, and the Ramsey Washington Metro Watershed District. A focus group was held with residents and published so that other communities could use it as a resource when planning their own rain garden projects.

- In Seattle, a prototype project, used to develop a plan for the entire city, was constructed in 2003. Called *SEA Street,* for Street Edge Alternatives, it was a drastic facelift of a residential street. The street was changed from a typical linear path to a gentle curve, narrowed, with large rain gardens placed along most of the length of the street. The street has 11% less impervious surface than a regular street. There are 100 evergreen trees and 1100 shrubs along this 3-block stretch of road, and a 2-year study found that the amount of stormwater which leaves the street has been reduced by 99%.[25]

- 10,000 Rain Gardens is a public initiative in the Kansas City, Missouri metro area. Property owners are encouraged to create rain gardens, with an eventual goal of 10,000 individual gardens.

- The West Michigan Environmental Action Council has established Rain Gardens of West Michigan as an outreach water quality program.[26] Also in Michigan, the Southeastern Oakland County Water Authority has published a pamphlet to encourage residents to add a rain garden to their landscapes in order to improve the water quality in the Rouge River watershed.[27] In Washtenaw County, homeowners can volunteer for the Water Resources Commissioner's Rain Garden program, in which volunteers are annually selected for free professional landscape design. The homeowners build the gardens themselves as well as pay for landscaping material. Photos of the gardens as well as design documents and drainage calculations are available online.[28]

- The city of Portland, Oregon, has established a Clean River Rewards program, to encourage residents to disconnect downspouts from the city's combined sewer system and create rain gardens. Workshops, discounts on storm water bills, and web resources are offered.[29]

- In Delaware, several rain gardens have been created through the work of the University of Delaware Water Resources Agency, and environmental organizations, such as the Appoquinimink River Association.[30]

- In New Jersey, the Rutgers Cooperative Extension Water Resources Program has already installed over 125 demonstration rain gardens in suburban and urban areas. The Water Resources Program has begun to focus on using rain gardens as green infrastructure in urban areas, such as Camden and Newark to help prevent localized flooding, combined sewer overflows, and to improve water quality. The Water Resources Program has also revised and produced a rain garden manual in collaboration with The Native Plant Society of New Jersey.[31]

35.5   See also

- Bioretention

- Climate-friendly gardening

- Constructed wetland

- Ecohydrology

- Green infrastructure

- Green roof

- Low impact development

- Microclimate

- Runoff Footprint

- Stormwater

- Surface runoff

- Sustainable urban drainage systems

- Urban runoff

- Water-energy nexus

- Water garden

35.6   References

[1] University of Rhode Island. Healthy Landscapes Program. "Rain Gardens: Enhancing your home landscape and protecting water quality."

[2] Dussaillant et al. Journal of Hydrologic Engineering

[3] Sandy Coyman; Keota Silaphone. "Rain Gardens in Maryland's Coastal Plain" (PDF). p. 2. Retrieved 11 October 2011.

[4] B.C. Wolverton, Ph.D., R.C. McDonald-McCaleb (1986). "Biotransformation of Priority Pollutants Using Biofilms and Vascular Plants." *Journal Of The Mississippi Academy Of Sciences.* Vol. XXXI, pp. 79-89.

[5] A. Dussaillant, Ph.D., et al. (2005). *Water Science & Technology: Water Supply journal.* Vol. 5, pp. 173-179.

[6] Kuichling, E. 1889. "The relation between the rainfall and the discharge of sewers in populous districts." *Trans. Am. Soc. Civ. Eng.* 20, 1–60.

[7] Leopold, L. B. 1968. "Hydrology for urban land planning: A guidebook on the hydrologic effects of urban land use." *Geological Survey Circular* 554. United States Geological Survey.

[8] Waananen, A. O. 1969. "Urban effects on water yield" in W. L. Moore and C. W. Morgan (eds), *Effects of Watershed Changes on Streamflow.* University of Texas Press, Austin and London.

[9] Novotny, V. and Olem, H. 1994. "Water Quality: Prevention, Identification, and Management of Diffuse Pollution." Van Nostrand Reinhold, New York.

[10] STRASSBERG, VALERIE; BRAD LANCASTER (June 2011). "Fighting water with water: Behavioral change versus climate change" (PDF). *American Water Works Association* **103** (6): 59. Retrieved 29 March 2012.

[11] Dietz, Michael E.; Clausen, John C. (2005). "A Field Evaluation of Raingarden Flow and Pollutant Treatment". *Water, Air, and Soil Pollution* **167** (1–4): 123–138. doi:10.1007/s11270-005-8266-8.

[12] U.S. Environmental Protection Agency (EPA), Washington, D.C. *Nonpoint Source News-Notes*. August/September 1995. Issue #42. "Urban Runoff"

[13] *Wisconsin Natural Resources* (magazine). "Rain Gardens Made One Maryland Community Famous." February 2003.

[14] EPA. "Stormwater Case Studies."

[15] http://www.hriresearch.org/Docs/Publications/JEH/JEH_1983/JEH_1983_1_3/JEH%201-3-77-80.pdf

[16] http://www.sustainablecitynetwork.com/topic_channels/water/article_b65fe518-3de0-11e0-82aa-00127992bc8b.html?utm_source=SCN+InBox+e-Newsletter&utm_campaign=2c869b70df-Newsletter_2-23-11_Tech&utm_medium=email

[17] http://melbournewater.com.au/raingardens

[18] http://wsud.melbournewater.com.au/content/case_studies/case_studies.asp

[19] http://www.waterbydesign.com.au

[20] http://www.wwt.org.uk/visit/london/

[21] "Robert Bray Associates Design Statement - Islington Council Public Records" (PDF). Islington Council.

[22] "Ashby Grove residential retrofit rain garden, London". Susdrain. Retrieved 2013-12-02.

[23] "Nottingham Green Streets – Retrofit Rain Garden Project". Susdrain. Retrieved 2013-08-04.

[24] http://www.12000raingardens.org

[25] City of Seattle, Washington. Seattle Public Utilities. "Street Edge Alternatives (SEA Streets) Project."

[26] Rain Gardens of West Michigan, Grand Rapids, MI. "Rain Gardens of West Michigan"

[27] Southeastern Oakland County Water Authority, Royal Oak, MI.

  • "Rain Gardens for the Rouge River: A Citizen's Guide to Planning, Design, & Maintenance for Small Site Rain Gardens"

[28] Washtenaw County, Michigan. "Rain Garden Virtual Tour"

[29] Clean River Rewards, Portland, Oregon. "Clean River Rewards."

[30] University of Delaware Cooperative Extension. "Rain Gardens in Delaware."

[31] http://water.rutgers.edu/Rain_Gardens/RGWebsite/demoraingardens.html

35.7 Further reading

• Dunnett, Nigel and Andy Clayden. *Rain Gardens: Sustainable Rainwater Management for the Garden and Designed Landscape.* Timber Press: Portland, 2007. ISBN 978-0-88192-826-6

• Liu, Jia, David J. Sample, Cameron Bell and Yuntao Guan. 2014. "Review and Research Needs of Bioretention Used for the Treatment of Urban Stormwater". Water, 6 (4): 1069–1099. "doi:10.3390/w6041069"

• Prince George's County. 1993. *Design Manual for Use of Bioretention in Stormwater Management.* Prince George's County, MD Department of Environmental Protection. Watershed Protection Branch, Landover, MD.

• Prince George's County, 2002. "Bioretention Manual". Department of Environmental Resources, Landover, MD.

• Michael L. Clar, Billy J. Barfield, and Thomas P. O'Connor. 2004. "Stormwater Best Management Practice Design Guide, Volume 2: Vegetative Biofilters." US EPA, National Risk Management Research Laboratory.

• Kraus, Helen, and Anne Spafford. *Rain Gardening in the South: Ecologically Designed Gardens for Drought, Deluge & Everything in Between.* Eno Publishers: Hillsborough, NC, 2009. ISBN 978-0-9820771-0-8

• Bray, B., Gedge, D., Grant, G., Leuthvilay, L. *UK Rain Garden Guide.* Published by RESET Development, London, 2012

35.8 External links

• Rain garden case study, Burnsville, MN (USA). 2004. *Land & Water:* 48(5).

• Water at the Grass Roots A brief introduction to Low Impact Development and rain gardens

• Details for construction of rain garden with a long plant list from Brooklyn Botanical Garden

• Rain Garden Network - Local Solutions to Local Stormwater Issues — Chicago, Illinois, United States

• Stormwater Tender project — Little Stringybark Creek, Victoria, Australia

• Rain Garden Design Templates for the Chesapeake Bay Watershed

- Wisconsin Department of Natural Resources — Rain Gardens

- Healthy Waterways Raingardens Program — Melbourne, Victoria, Australia

- UK Rain Garden Guide

Chapter 36

Reconciliation ecology

A simple form of reconciliation ecology: the construction of nest boxes increases densities of bluebirds in areas where natural tree cavities are scarce due to short-rotation forestry.[1]

Reconciliation ecology is the branch of ecology which studies ways to encourage biodiversity in human-dominated ecosystems. Michael Rosenzweig first articulated the concept in his book *Win-Win Ecology*,[2] based on the theory that there is not enough area for all of earth's biodiversity to be saved within designated nature preserves. Therefore, humans should increase biodiversity in human-dominated landscapes. By managing for biodiversity in ways that do not decrease human utility of the system, it is a "win-win" situation for both human use and native biodiversity. The science is based in the ecological foundation of human land-use trends and species-area relationships. It has many benefits beyond protection of biodiversity, and there are numerous examples of it around the globe. Aspects of reconciliation ecology can already be found in management legislation, but there are challenges in both public acceptance and ecological success of reconciliation attempts.

36.1   Theoretical basis

36.1.1   Human land use trends

Main article: List of environmental issues

Traditional conservation is based on "reservation and restoration"; reservation meaning setting pristine lands aside for the sole purpose of maintaining biodiversity, and restoration meaning returning human impacted ecosystems to their natural state. However, reconciliation ecologists argue that there is too great a proportion of land already impacted by humans for these techniques to succeed.

While it is difficult to measure exactly how much land has been transformed by human use, estimates range from 39 to 50%. This includes agricultural land, pastureland, urban areas, and heavily harvested forest systems.[3] An estimated 50% of arable land is already under cultivation.[4] Land transformation has increased rapidly over the last fifty years, and is likely to continue to increase.[5] Beyond direct transformation of land area, humans have impacted the global biogeochemical cycles, leading to human caused change in even the most remote areas.[6] These include addition of nutrients such nitrogen and phosphorus, acid rain, ocean acidification, redistribution of water resources, and increased carbon dioxide in the atmosphere. Humans have also changed species compositions of many landscapes that they do not dominate directly by introducing new species or harvesting native species. This new assemblage of species has been compared to previous mass extinctions and speciation events caused by formation of land bridges and colliding of continents.[7]

36.1.2   Species-area relationships

The need for reconciliation ecology was derived from patterns of species distribution and diversity. The most relevant of these patterns is the species-area curve which states that a larger geographic area will contain higher species diversity. This relationship has been supported by so large a body of research that some scholars consider it to be an

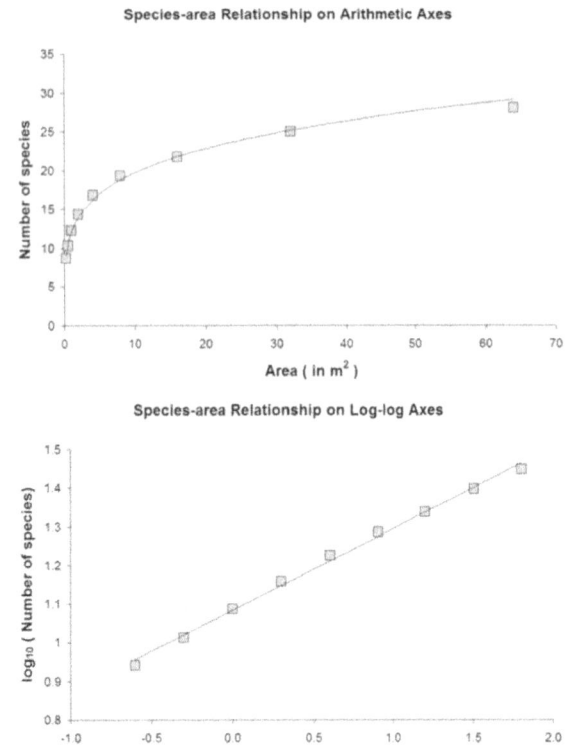

The species-area relationship for a contiguous habitat

ecological law.[8]

There are two main reasons for the relationship between number of species and area, both of which can be used as an argument for conservation of larger areas. The habitat heterogeneity hypothesis claims that a larger geographic area will have a greater variety of habitat types, and therefore more species adapted to each unique habitat type. Setting aside a small area will not encompass enough habitat variety to contain a large variety of species.[9] The equilibrium hypothesis draws from the theory of island biogeography as described by MacArthur and Wilson.[10] Large areas have large populations, which are less likely to go extinct through stochastic processes. The theory assumes that speciation rates are constant with area, and a lower extinction rate coupled with higher speciation leads to more species.

The species-area relationship has often been applied to conservation, often quantitatively. The simplest and most commonly used formula was first published by Frank W. Preston.[11] The number of species present in a given area increases in relationship to that area with the relationship $S = cA^z$ where S is the number of species, A is the area, and c and z are constants which vary with the system under study. This equation has frequently been used for designing reserve size and placement (see SLOSS debate).[12] The most common version of the equation used in reserve de-

sign is the formula for inter-island diversity, which has a z-value between 0.25-0.55,[13] meaning protecting 5% of the available habitat will preserve 40% of the species present. However, inter-provincial species area relationships have z-values closer to 1, meaning protecting 5% of habitat will only protect 5% of species diversity.[2]

Taken together, proponents of reconciliation ecology see the species-area relationship and human domination of a large percentage of the earth's area as a sign that we will not be able to set aside enough land to protect all of life's biodiversity. There can be negative effects of setting land aside because it means the remaining land is used more intensely.[4] For example, less land is required for crop production when high levels of inorganic fertilizer is applied, but these chemicals will affect nearby land set aside for natural ecosystems. The direct benefits of land transformation for the growing world population often make it ethically difficult to justify the tradeoff between biodiversity and human use.[14] Reconciled ecosystems are ones in which humans dominate, but natural biodiversity is encouraged to persist within the human landscape. Ideally, this creates a more sustainable socio-ecological system and does not necessitate a trade off between biodiversity and human use.

36.1.3 Beyond natural history

The life history of the great grey shrike is better understood as a result of focused natural history and reconciliation ecology.

How can understanding of species' natural history aid their effective conservation in human-dominated ecosystems? Humans often conduct activities that allow for the incorporation of other species, whether as a by-product or as a result of a focus on nature.[15] Traditional natural history can only inform how best to do this to a certain degree, because landscapes have been changed so dramatically. However, there is much more to learn through direct study of species' ecology in human-dominated ecosystems, through what is known as focused natural history. Rosenzweig [15] cites four examples: shrikes (Laniidae) thrived in altered landscapes when wooden fence post perches allowed them easy access to pouncing on prey, but inhospitable steel fence posts contributed to their decline. Replacing steel fence posts with wood fence posts reverses the shrikes' decline and allows humans to determine the reasons for the distribution and abundance of shrikes. Additionally, the cirl bunting (*Emberiza cirlus*) thrived on farms when fields alternated between harvests and hay, but declined where farmers began to plant winter grain crops, natterjack toads (*Bufo calamatus*) declined when reductions in sheep grazing ceased to alter ponds to their preferred shape and depth, and longleaf pine (*Pinus palustris*) declined in the Southeastern United States when lack of wildfires prevented its return after timbering.[15][16] Thus, applying focused natural history in human-dominated landscapes can contribute to conservation efforts.

36.2 Benefits

Reconciliation ecologists believe increasing biodiversity within human dominated landscapes will help to save global biodiversity. This is sometimes preferable to traditional conservation because it does not impair human use of the landscape and therefore may be more acceptable to stakeholders.[2] However, not only will it encourage biodiversity in the areas where it takes place, but many scholars cite other benefits of including biodiversity in human landscapes on both global conservation activities and human well-being.

36.2.1 Habitat connectivity benefits

Increasing wildlife habitat in human-dominated systems not only increases *in situ* biodiversity, it also aids in conservation of surrounding protected areas by increasing connectivity between habitat patches.[17][18] This may be especially important in agricultural systems where buffers, live fences, and other small habitat areas can serve as stops between major preserves.[19] This concept forms the basis of the subdiscipline countryside biogeography [14] which studies the potential of the matrix between preserves to provide habitat for species moving from preserve to preserve.

36.2.2 Educational benefits

Placing importance on native ecosystems and biodiversity within human landscapes increases human exposure to natural areas,[20] which has been shown to increase appreciation of nature. Studies have shown that students who participate in outdoor education programs show a greater un-

derstanding of their environment, greater willingness to act in order to save the environment, and even a greater enthusiasm for school and learning.[21][22] Green spaces have also been shown connect urban dwellers of all ages with nature, even when dominated by invasive species.[23] Reconnecting people with nature is especially important for conservation because there is a tendency for people to use the biodiversity present in the landscape they grew up in as a point of comparison for future trends (see Shifting baseline).[24]

36.2.3 Psychological benefits

The results of reconciliation ecology can also improve human well-being. E. O. Wilson has hypothesized that humans have an innate desire to be close to nature (see Biophilia),[25] and numerous studies have linked natural settings to decreased stress and faster recovery during hospital stays.[26]

36.3 Examples

Many examples of native plants and animals taking advantage of human dominated landscapes have been unintentional, but may be enhanced as part of reconciliation ecology. Others are intentional redesigns of human landscapes to better accommodate native biodiversity. These have been going on for many hundreds of years including examples within agricultural systems, urban and suburban systems, marine systems, and even industrial areas.

36.3.1 Historical examples

While Rosenzweig formalized the concept, humans have been encouraging biodiversity within human landscapes for millennia. In the Trebon Biosphere Reserve of the Czech Republic, a system of human-engineered aquaculture ponds built in the 1500s not only provides a profitable harvest of fish, but also provides habitat for a hugely diverse wetland ecosystem. Many cities in Europe take pride in their local population of storks, which nest on roofs or in church towers that replace the trees they would naturally nest in.[2] There are records of humans maintaining plants in pleasure gardens as early as ancient Mesopotamia, with an especially strong tradition of incorporating gardens into the architecture of human landscapes in China.[27]

36.3.2 Agricultural systems

Agroforestry provides many examples of reconciliation ecology at work. In tropical agroforestry systems, crops

Agroforestry in Burkina Faso allows sorghum crop to be grown under native tree species, preserving biodiversity.

such as coffee or fruit trees are cultivated under a canopy of shade trees, providing habitat for tropical forest species outside of protected areas.[28] For example, shade-grown coffee plantations typically have lower tree diversity than unmanaged forests, however they have much higher tree species diversity and richness than other agricultural methods.[29] Agriculture that mimics nature, encourages natural forest species along with the crops, and also takes pressure off nearby uncultivated forest areas where people are allowed to collect forest products.[28] The understory can also be managed with reconciliation ecology: allowing weeds to grow among crops (minimizing labor and preventing the invasion of noxious weed species) and leaving fallowlands alongside farmed areas can enhance understory plant richness with associated benefits for native insects and birds compared to other agricultural practices.[30]

The oil palm (*Elaeis guineensis*) provides another example of the potential of reconciliation ecology. It is one of the most important and rapidly expanding tropical crops,[31] so lucrative because it is used in many products throughout the world. Unfortunately, oil-palm agriculture is one of the main drivers of forest conversion in Southeast Asia and is devastating for native biodiversity, perhaps even more so than logging.[32] However, attempts are being made to foster the sustainability of this industry. As a monoculture, oil palm is subject to potentially devastating attacks from insect pests.[31][33] Many companies are attempting an integrated pest management approach which encourages the planting of species that support predators and parasitoids of these insect pests, as well as an active native bird community.[33] Experiments have shown that a functioning bird community, especially at higher densities, can serve to reduce insect herbivory on oil palms, promoting increased crop yields and profits.[33] Thus, oil palm plantation managers can participate in reconciliation ecology by promoting lo-

cal vegetation that is beneficial to insectivorous birds, including maintaining ground plants that serve as nesting sites, thereby protecting natural communities. Additionally, steps such as maintaining riparian buffer zones or natural forest patches can help to slow the loss of biodiversity within oil palm plantation landscapes.[32] By engaging in these environmentally friendly practices, fewer chemicals and less effort are required to maintain both plantation productivity and ecosystem services.[31][33]

There are many grazing practices that also encourage native biodiversity. In Rosenzweig's book he uses the example of a rancher in Arizona who intentionally deepened his cattle ponds in order to save a population of threatened leopard frogs (*Rana chiricahuensis*), with no detriment to the use of those tanks for cattle,[2] and a similar situation has occurred with the vulnerable California tiger salamander (*Ambystoma californiense*) in the Central Valley of California. Research has shown that without cattle grazing, many of the remaining vernal pools would dry too early for the salamanders to complete their life cycle under global climate change predictions.[34] In Central America, a large percentage of pastureland is fenced using live trees which are not only low maintenance for the farmer, but also provide habitat for birds, bats, and invertebrates which cannot persist in open pastureland.[35] Another example from Rosenzweig involves encouraging loggerhead shrikes (*Lanius ludovicianus*) to populate pastureland by placing perches around the pasture.[2] These are all simple, low-cost ways to encourage biodiversity without negatively impacting the human uses of the landscape.

36.3.3 Urban systems

Green roofs can help maintain species diversity in urban landscapes.

Urban ecology can be included under the umbrella of reconciliation ecology and it tackles biodiversity in cities, the most extreme of human-dominated landscapes. Cities occupy less than 3% of global surface area, but are responsible for a majority of carbon emissions, residential water use, and wood use.[36] Cities also have unique climatic conditions such as the urban heat island effect, which can greatly affect biodiversity.[37] There is a growing trend among city managers to take biodiversity into account when planning city development, especially in rapidly growing cities. Cities often have surprisingly high plant biodiversity due to their normally high degree of habitat heterogeneity and high numbers of gardens and green spaces cultivated to include a large variety of species.[37] However, these species are often not native, and a large part of the total urban biodiversity is usually made up of exotic species.[38]

Because cities are so highly impacted by human activities, restoration to the pristine state is not possible, however there are modifications that can be made to increase habitat without negatively impacting human needs. In urban rivers, addition of large woods and floating islands to provide habitat, modifications to walls and other structures to mimic natural banks, and buffer areas to reduce pollutants can all increase biodiversity without reducing the flood control and water supply services.[39] Urban green spaces can be re-designed to encourage natural ecosystems rather than manicured lawns, as is seen in the National Wildlife Federation's Backyard Wildlife Habitat program.[40] Peregrine falcons (*Falco peregrinus*), which were once endangered by pesticide use, are frequently seen nesting in tall urban buildings throughout North America, feeding chiefly on the introduced rock dove.[41] The steep walls of buildings mimic the cliffs peregrines naturally nest in and the rock doves replace the native prey species that were driven out of urban areas.

36.3.4 Industrial systems

In Florida, the Florida manatee (*Trichechus manatus latirostris*) uses warm water discharged from power plants as a refuge when the temperature of the Gulf of Mexico drops.[42] These warm areas replace the warm springs that manatees once naturally used in the winter. These springs have been drained or cut off from open water by human uses. American crocodiles (*Crocodylus acutus*) have a similar habitat in the cooling canals of the Turkey Point power plant, where an estimated 10% of the total North American population of the species lives.[2]

Wastewater treatment systems have shown potential for reconciliation ecology on numerous occasions. Man-made wetlands designed to remove nitrogen before runoff from agriculture enters the Everglades in Florida are used as breeding sites for a number of birds, including the endangered wood stork (*Mycteria americana*).[43] Stormwater

treatment ponds can provide important breeding habitat for amphibians, especially where natural wetlands have been drained by human development.[44]

36.3.5 Ocean systems

Coral reefs have been intensively impacted by human use, including overfishing and mining of the reef itself. One reconciliation approach to this problem is building artificial reefs that not only provide valuable habitat for aquatic species, but also protect nearby islands from storms when the natural structure has been mined away.[45] Even structures as simple as scrap metal and automobiles can be used as habitat, providing added benefits of freeing space in landfills.[46]

36.4 Legislation

Governmental intervention can aid in encouraging private landowners to create habitat or otherwise increase biodiversity on their land. The United States' Endangered Species Act requires landowners to halt any activities negatively affecting endangered species on their land, which is a disincentive for them to encourage endangered species to settle on their land in the first place.[2] To help mediate this problem, the US Fish and Wildlife Service has instituted safe harbor agreements whereby the landowner engages in restoration on their land to encourage endangered species, and the government agrees not to place further regulation on their activities should they want to reverse the restoration at a later date.[47] This practice has already led to an increase in aplomado falcons (*Falco femoralis*) in Texas and red-cockaded woodpeckers (*Picoides borealis*) in the Southeastern US.

Another example is the US Department of Agriculture's Conservation Reserve Program (CRP). The CRP was originally put in place to protect soil from erosion, but also has major implications for conservation of biodiversity. In the program, landowners take their land out of agricultural production and plant trees, shrubs, and other permanent, erosion controlling vegetation. Unintended, but ecologically significant consequences of this were the reduction of runoff, improved water quality, creation of wildlife habitat, and possible carbon sequestration.[48]

36.5 Challenges

While reconciliation ecology attempts to modify the human world to encourage biodiversity without negatively impacting human use, there are still difficulties in getting broad acceptance of the idea. For example, addition of large woods to urban river systems, which provides critical habitat structure for native fish and invertebrates may be seen as "untidy" and a sign of poor management by residents.[39] Similarly, many suburban areas do not allow long, unkempt lawns that provide useful wildlife habitat because of perceived damage to property values.[49] Many humans have negative feelings toward certain species, especially predators such as wolves, which are often based more on perceived risk than actual risk of loss or injury resulting from the animal.[50] Even with cooperation of the human element of the equation, reconciliation ecology can not help every species. Some animals, such as several species of waterfowl, show strong avoidance behaviors toward humans and any form of human disturbance.[51] No matter how nice an urban park is built, the proximity of humans will scare away some birds. Other species must maintain large territories, and barriers that abound in human habitats, such as roads, will stop them from coexisting with humans.[52] These animals will require undisturbed land set aside for them.

36.6 See also

- Sustainable agriculture
- Sustainability
- Sustainable development
- Community (ecology)
- Restoration ecology
- Human impacts on the nitrogen cycle
- Human impact of climate change
- Ecological footprint
- Species-area curve
- Conservation biology

36.7 References

[1] Twedt, D.J.; Henne-Kerr, J.L. (2001). "Artificial cavities enhance breeding bird densities in managed cottonwood forests". *Wildlife Society Bulletin* 29: 680–687.

[2] Rosenzweig, Michael (2003). *Win-win Ecology, How the Earth's species can survive in the midst of human enterprise*. Oxford, UK: Oxford University Press.

[3] Vitousek, P. M.; H. A. Mooney; J. Lubchenco; J. M. Melillo (1997). "Human Domination of Earth's Ecosystems". *Science* 277 (5325): 494–499. doi:10.1126/science.277.5325.494.

[4] Green, R. E.; S. J. Cornell; J. P. W. Scharlemann; A. Balmford (2005). "Farming and the fate of wild nature". *Science* 307 (5709): 550–557. doi:10.1126/science.1106049.

[5] Millennium Ecosystem Assessment (2005). *Ecosystems and Human Well-Being*. Washington, DC, USA: Island Press.

[6] Vitousek, P. M.; J. D. Aber; R. W. Howarth; G. E. Likens; P. A. Matson; D. W. Schindler; W. H. Schlesinger; D. G. Tilman (1997). "Human alterations of the global nitrogen cycle: sources and consequences". *Ecological Applications* 7: 737–750. doi:10.1890/1051-0761(1997)007[0737:haotgn]2.0.co;2.

[7] Mooney, H. A; E. E. Cleland (2001). "The evolutionary impact of invasive species". *Proceedings of the National Academy of Sciences of the United States of America* 98 (10): 5446–5451. doi:10.1073/pnas.091093398. PMC 33232. PMID 11344292.

[8] Lomolino, M. V. (2000). "Ecology's most general, yet protean pattern: the species-area relationship". *Journal of Biogeography* 27: 17–26. doi:10.1046/j.1365-2699.2000.00377.x.

[9] Rosenzweig, Michael (2003). "Reconciliation ecology and the future of species diversity". *Oryx* 37: 194–206. doi:10.1017/s0030605303000371.

[10] MacArthur, R. H. and E. O. Wilson (1967). *The Theory of Island Biogeography*. Princeton, USA: Princeton University Press.

[11] Preston, F. W. (1960). "Time and Space and the Variation of Species". *Ecology* 41: 612–627. doi:10.2307/1931793.

[12] Desmet, P; R. Cowling (2004). "Using the species-area relationship to set baseline targets for conservation". *Ecology and Society* 9: 11.

[13] Rosenzweig, Michael (1995). *Species diversity in space and time*. Cambridge, USA: Cambridge University Press.

[14] Daily, Gretchen (1997). *Countryside biogeography and the provision of ecosystem services.in Forum on Biodivesity*. National Research Council: National Academy Press.

[15] Rosenzweig, M.L. (2005). "Avoiding mass extinction: basic and applied challenges". *The American Midland Naturalist* 153: 195–208. doi:10.1674/0003-0031(2005)153[0195:amebaa]2.0.co;2.

[16] Noss, R.F.; Beier, P.; Covington, W.W.; Grumbine, R.E.; Lindenmayer, D.B.; Prather, J.W.; Schmiegelow, F.; Sisk, T.D.; Vosick, D.J. (2006). "Recommendations for integrating restoration ecology and conservation biology in ponderosa pine forests of the southwestern united states". *Restoration Ecology* 14: 4–10. doi:10.1111/j.1526-100x.2006.00099.x.

[17] Anand, M. O; J. Krishnaswamy; A. Kumar; A. Bali (2010). "Sustaining biodiversity conservation in human-modified landscapes in Western Ghats: Remnant forests

matter". *Biological Conservation* **143** (10): 2363–2374. doi:10.1016/j.biocon.2010.01.013.

[18] Lombard, A. T.; R. M. Cowling; J. H. J. Vlok; C. Fabricius (2010). "Designing conservation corridors in production landscapes: assessment methods, implementation issues, and lessons learned". *Ecology and Society* **15**: 7.

[19] Ulrich, R. S; R. F. Simons; B. D. Losito; E. Fiorito; M. A. Miles; M. Zelson (1991). "Stress recovery during exposures to natural and urban environments". *Journal of Environmental Psychology* **11** (3): 201–230. doi:10.1016/S0272-4944(05)80184-7.

[20] Miller, J. R (2005). "Biodiversity conservation and the extinction of experience". *Trends in Ecology & Evolution* **20** (8): 430–434. doi:10.1016/j.tree.2005.05.013.

[21] Bogner, F. X (1998). "The influence of short-term outdoor ecology education on long-term variables of environmental perspective". *The Journal of Environmental Education* **29** (4): 17. doi:10.1080/00958969809599124.

[22] Dillon, J. M; Rickinson, K. Teamey, M. Morris, M. Y. Choi, D. Sanders, and P. Benefield (2006). "The value of outdoor learning: evidence from research in the UK and elsewhere". *School Science Review* **87**: 107–111.

[23] Teillac-Deschamps, P; R. Lorrilliere, V. Servais, V. Delmas, A. Cadi, and A.-C. Prevot-Julliard (2009). "Management strategies in urban green spaces: Models based on an introduced exotic pet turtle". *Biological Conservation* **142** (10): 2258–2269. doi:10.1016/j.biocon.2009.05.004.

[24] Pauly, D (1995). "Anecdotes and the shifting baseline syndrome of fisheries". *Trends in Ecology & Evolution* **10** (10): 430. doi:10.1016/S0169-5347(00)89171-5.

[25] Wilson, E O (1984). *Biophilia*. Cambridge, USA: Harvard University Press.

[26] Ulrich, R. S.; R. F. Simons; B. D. Losito; E. Fiorito; M. A. Miles; M. Zelson (1991). "Stress recovery during exposures to natural and urban environments". *Journal of Environmental Psychology* **11** (3): 201–230. doi:10.1016/S0272-4944(05)80184-7.

[27] Chen, X.; J. Wu (2009). "Sustainable landscape architecture: implications of the Chinese philosophy of "unity of man with nature" and beyond". *Landscape Ecology* **24** (8): 1015–1026. doi:10.1007/s10980-009-9350-z.

[28] Bhagwat, S. A.; K. J. Willis; H. J. B. Birks; R. J. Whittaker (2008). "Agroforestry: a refuge for tropical biodiversity?". *Trends in Ecology and Evolution* **23** (5): 261–268. doi:10.1016/j.tree.2008.01.005.

[29] Correia, M; M. Diabate; P. Beavogui; K. Guilavogui; N. Lamanda; H. d. Foresta (2010). "Conserving forest tree diversity in Guinee Forestiere (Guinea, West Africa): the role of coffee-based agroforests". *Biodiversity Conservation* **19** (6): 1725–1747. doi:10.1007/s10531-010-9800-6.

[30] Bobo, K.S.; Waltert, M.; Sainge, N.M.; Njokagbor, J.; Fermon, H.; Muhlenberg, M. (2006). "From forest to farmland: species richness patterns of trees and understorey plants along a gradient of forest conversion in Southwestern Cameroon". *Biodiversity and Conservation* **15**: 4097–4117. doi:10.1007/s10531-005-3368-6.

[31] Koh, L.P. (2008). "Can oil palm plantations be made more hospitable for forest butterflies and birds?". *Journal of Applied Ecology* **45**: 1002–1009. doi:10.1111/j.1365-2664.2008.01491.x.

[32] Wilcove, D.S.; Koh, L.P. (2010). "Addressing the threats to biodiversity from oil-palm agriculture". *Conservation of Biodiversity* **19**: 999–1007. doi:10.1007/s10531-009-9760-x.

[33] Koh, L.P. (2008). "Birds defend oil palms from herbivorous insects". *Ecological Applications* **18**: 821–825. doi:10.1890/07-1650.1.

[34] Pyke, C. R.; J. Marty (2005). "Cattle Grazing Mediates Climate Change Impacts on Ephemeral Wetlands". *Conservation Bioloty* **19** (5): 1619–1625. doi:10.1111/j.1523-1739.2005.00233.x.

[35] Harvey, C. A.; C. Villanueva, J. Villacís, M. Chacón, D. Muñoz, M. López, M. Ibrahim, R. Gómez, R. Taylor, J. Martinez, A. Navas, Saenz, D. Sánchez, A. Medina, S. Vilchez, B. Hernández, A. Perez, F. Ruiz, F. López, I. Lang, and F. L. Sinclair (2005). "Contribution of live fences to the ecological integrity of agricultural landscapes". *Agriculture, Ecosystems & Environment* **111**: 200–230. doi:10.1016/j.agee.2005.06.011.

[36] Brown, L. R. (2001). *Eco-Economy: building an economy for the Earth*. New York, USA: Norton.

[37] Grimm, N. B.; S. H. Faeth; N. E. Golubiewski; C. L. Redman; J. Wu; X. Bai; J. M. Briggs (2008). "Global Change and the Ecology of Cities". *Science* **319** (5864): 756–760. doi:10.1126/science.1150195. PMID 18258902.

[38] Wang, G; G. Jiang; Y. Zhou; Q. Liu; Y. Ji; S. Wang; S. Chen; H. Liu (2007). "Biodiversity conservation in a fast-growing metropolitan area in China: a case study of plant diversity in Beijing". *Biodiversity Conservation* **16** (14): 4025–4038. doi:10.1007/s10531-007-9205-3.

[39] Francis, R. A. (2009). "Perspectives on the potential for reconciliation ecology in urban riverscapes". *CAB Reviews: Perspectives in Agriculture, Veterinary Science, Nutrition and Natural Resources* **4**: 1–20. doi:10.1079/pavsnnr20094073.

[40] Tufts, C. and P. Loewer (1995). *Gardening for wildlife*. Emmaus, USA: Rodale Press.

[41] Cade, T; D. Bird (1990). "Peregrine falcons, Falco peregrinus, nesting in an urban environment: a review". *Canadian field-naturalist* **104**: 209–218.

[42] Laist, D. W.; J. E. Reynolds (2005). "Florida manatees, warm-water refuges, and an uncertain future". *Coastal Management* **33** (3): 279–295. doi:10.1080/08920750590952018.

[43] Frederick, P. C; S. M. McGhee. (1994). "Wading bird use of wastewater treament wetlands in central Florida, USA". *Colonial Waterbirds* (17): 50–59.

[44] Brand, A. B.; J. W. Snodgrass (2009). "Value of artificial habitats for amphibian reproduction in altered landscapes". *Conservation Biology* **24** (1): 295–301. doi:10.1111/j.1523-1739.2009.01301.x. PMID 19681986.

[45] Clark, S; A. J. Edwards (1998). "An evaluation of artificial reef structures as tools for marine habitat rehabilitation in the Maldives". *Aquatic Conservation: Marine and Freshwater Ecosystems* **9**: 5–21. doi:10.1002/(sici)1099-0755(199901/02)9:1<5::aid-aqc330>3.0.co;2-u.

[46] Brock, R. E; J. E. Norris (1989). "An analysis of four artificial reef designs in tropical waters". *Bulletin of Marine Science* **44**: 934–941.

[47] Wilcove, D. S; J. Lee (2004). "Using economic and regulatory incentives to restore endangered species: lessons learned from three new programs". *Conservation Biology* (18): 639–645. doi:10.1111/j.1523-1739.2004.00250.x.

[48] Dunn, C. P; F. Stearns; G. R. Guntenspergen; D. M. Sharpe (1993). "Ecological benefits of the Conservation Reserve Program". *Conservation Biology* **7**: 132–139. doi:10.1046/j.1523-1739.1993.07010132.x.

[49] Robbins, P; A. Polderman; T. Birkenholtz (2001). "Lawns and Toxins: An Ecology of the City". *Cities* **18** (6): 369–380. doi:10.1016/S0264-2751(01)00029-4.

[50] Berger, K. M (2006). "Carnivore-Livestock Conflicts: Effects of Subsidized Predator Control and Economic Correlates on the Sheep Industry". *Conservation Biology* **20** (3): 751–761. doi:10.1111/j.1523-1739.2006.00336.x. PMID 16909568.

[51] Gill, J. A; K. Norris; W. J. Sutherland (2001). "Why behavioral responses may not reflect the population consequences of human disturbance". *Biological Conservation* **97** (2): 265–268. doi:10.1016/S0006-3207(00)00002-1.

[52] Riley, S. P. D; J. P. Pollinger; R. M. Savajot; E. C. York; C. Bromley; T. Fuller; R. K. Wayne (2006). "A southern California freeway is a physical and social barrier to gene flow in carnivores". *Molecular Ecology* **15** (7): 1733–1741. doi:10.1111/j.1365-294X.2006.02907.x. PMID 16689893.

36.8   External links

- The Careful Foot, a site on reconciliation ecology

- Michael Rosenzweig

Chapter 37

Riparian buffer

37.1 Benefits

Riparian buffers act to intercept sediment, nutrients, pesticides, and other materials in surface runoff and reduce nutrients and other pollutants in shallow subsurface water flow.[1] They also serve to provide habitat and wildlife corridors in primarily agricultural areas. They can also be key in reducing erosion by providing stream bank stabilization.

37.1.1 Water quality benefits

- Intercepting sediments/nutrients - Key to counteract eutrophication in downstream lakes and ponds which can be detrimental to aquatic habitats because of large fish kills that occur upon large-scale eutrophication.

- Intercepting pesticides - Riparian buffers keep chemicals that can be harmful to aquatic life out of the water. Some pesticides can be especially harmful if they bioaccumulate in the organism, with the chemicals reaching harmful levels once they are ready for human consumption.

- Bank stabilization - This is important because erosion can be a major problem in agricultural regions when cut (eroded) banks can take land out of production. Erosion can also lead to sedimentation and siltation of downstream lakes, ponds, and reservoirs. Siltation can greatly reduce the life span of reservoirs and the dams that create the reservoirs.

A riparian buffer of vegetation lining a farm creek in Story County, Iowa.

37.1.2 Habitat benefits

- Provide habitat - Riparian buffers can act as crucial habitat for a large number of species, especially those who have lost habitat due to agricultural land being put into production.

- Increase biodiversity - By adding this vegetated area of land near a water source it becomes a prime location for species that may have left the area due

A **riparian buffer** is a vegetated area (a "buffer strip") near a stream, usually forested, which helps shade and partially protect a stream from the impact of adjacent land uses. It plays a key role in increasing water quality in associated streams, rivers, and lakes, thus providing environmental benefits. With the decline of many aquatic ecosystems due to agricultural production, riparian buffers have become a very common conservation practice aimed at increasing water quality and reducing pollution.

to non-conservation land use to re-establish. With this re-establishment the number of native species and biodiversity in general can be increased.

- Buffers acting as corridors - Buffers also serve a major role in wildlife habitat. The habitat provided by the buffers also double as corridors for species that have had their habitat fragmented by various land uses.

- Shading water - The large trees in the first zone of the riparian buffer provide shade and therefore cooling for the water, increasing productivity and increasing habitat quality for aquatic species.

37.1.3 Economic benefits

- Increase land value - Often people who purchase land for recreational use are willing to pay more if there is more wooded area located on the land.

- Produce profitable alternative crops - Vegetation such as Black Walnut and Hazelnut, which can be profitably harvested, can be incorporated into the riparian buffer.

- Increase lease fees for hunting - The increased habitat means that the land will be more sought-after for hunting purposes.

37.2 Buffer design

Ground level view of riparian buffer between Munson Pond (off camera left) and an agricultural operation (off camera right)

A riparian buffer is usually split into three different zones, each having its own specific purpose for filtering runoff and interacting with the adjacent aquatic system. Buffer design is a key element in the effectiveness of the buffer. It is generally recommended that native species be chosen to plant in these three zones, with the general width of the buffer being 50 feet (15 m) on each side of the stream.[2]

Zone 1. This zone should function mainly to shade the water source and act as a bank stabilizer. The zone should include large native tree species that grow fast and can quickly act to perform these tasks. Although this is usually the smallest of the three zones and absorbs the fewest contaminants, most of the contaminants have been eliminated by Zone 2 and Zone 3.[3]

Zone 2. Usually made up of native shrubs, this zone provides a habitat for wildlife, including nesting areas for bird species. This zone also acts to slow and absorb contaminants that Zone 3 has missed. The zone is an important transition between grassland and forest.[3]

Zone 3. This zone is important as the first line of defense against contaminants. It consists mostly of native grasses and serves primarily to slow water runoff and begin to absorb contaminants before they reach the other zones. Although these grass strips should be one of the widest zones, they are also the easiest to install.[3]

Streambed Zone. The streambed zone of the riparian area is linked closely to Zone 1. Zone 1 provides fallen limbs, trees, and tree roots that in turn slow water flow, reducing erosional processes associated with increased water flow and flooding. This woody debris also increases habitat and cover for various aquatic species

The National Agroforestry Center has developed a Filter Strip Design tool (AgBufferBuilder), which is a GIS-based computer program for designing vegetative filter strips around agricultural fields that utilizes terrain analysis to account for spatially non-uniform runoff.

37.3 Species selection (example using both native Nebraska and introduced species)

In Zone 1: Cottonwood, Bur Oak, Hackberry, Swamp White Oak, Siberian Elm, Honeylocust, Silver Maple, Black Walnut, and Northern Red Oak.[4]

In Zone 2: manchurian apricot, Silver Buffaloberry, Caragana, Black Cherry, Chokecherry, Sandcherry, Peking Cotoneaster, Midwest Crabapple, Golden Currant, Elderberry, Washington Hawthorn, American Hazel, Amur Honeysuckle, Common Lilac, Amur Maple, American Plum, and Skunkbush Sumac.[4]

In Zone 3: Western Wheatgrass, Big Bluestem, Sand Bluestem, Sideoats Grama, Blue Grama, Hairy Grama, Buffalo Grass, Sand Lovegrass, Switchgrass, Little Bluestem, Indiangrass, Prairie Cordgrass, Prairie Dropseed, Tall Dropseed, Needleandthread, Green Needlegrass.

37.4 Managing forests in riparian area

Logging is sometimes recommended as a management practice in riparian buffers, usually to provide economic incentive. However, some studies have shown that logging can harm wildlife populations, especially birds. A study by the University of Minnesota found that there was a correlation between the harvesting of timber in riparian buffers and a decline in bird populations.[5] Therefore, logging is generally discouraged as an environmental practice, and left to be done in designated logging areas.

37.5 Conservation incentives

The Conservation Reserve Program (CRP), a farming assistance program in the United States, provides many incentives to landowners to encourage them to install riparian buffers around water systems that have a high chance of non-point water pollution and are highly erodible. For example, the Nebraska system of Riparian Buffer Payments offers the following:

- Annual rental payments = $62 to $116 per acre/year (up to $96 per acre per year for certain marginal pasture land)

- Sign up bonus: up-front payment of $10 per acre/year for the life of the contract (Example: $150/ac for a fifteen-year contract)

- Payments for 90% of costs for trees, shrubs, grass seed, site preparation and planting

- 10 to 15 year contracts

- Eligible areas:

  - Up to 180 feet (55 m) average width along each side of a continuous or intermittent stream

  - Strips of grass, shrubs, and/or trees can be used

  - Planting done by the local Natural Resource Districts (NRDs)

These incentives are offered to agriculturists to compensate them for their economic loss of taking this land out of production. If the land is highly erodible and produces little economic gain, it can sometimes be more economic to take advantage of these CRP programs.[6]

37.6 Effectiveness

Riparian buffers have undergone much scrutiny about their effectiveness, resulting in thorough testing and monitoring. A study done by the University of Georgia, conducted over a nine-year period, monitored the amounts of fertilizers that reached the watershed from the source of the application. It found that these buffers removed at least 60% of the nitrogen in the runoff, and at least 65% of the phosphorus from the fertilizer application. The same study showed that the effectiveness of the Zone 3 was much greater than that of both Zone 1 and 2 at removing contaminants.[7]

37.7 Long-term sustainability

After the initial installation of the riparian buffer, relatively little maintenance needs to be performed to keep the buffer in good condition. Once the trees and grasses reach maturity, they regenerate naturally and make a more effective buffer. The sustainability of the riparian buffer makes it extremely attractive to landowners, since they do relatively little work and still receive payments. Riparian buffers have the potential to be the most effective ways to protect aquatic biodiversity, water quality and manage water resources in developing countries that lack the funds to install water treatment and supply systems in midsize and small towns.

37.8 See also

- Agricultural wastewater treatment

- Agroforestry

- Ecoscaping

- Erosion control

- Nonpoint source pollution

37.9 References

[1] U.S. Natural Resources Conservation Service (NRCS). (2006). "National Conservation Practice Standard: Riparian Forest Buffer." Code 391. January 2006.

[2] Dosskey, M., Schultz, D., & Isenhart, T. (1997). "Riparian Buffers for Agricultural Land." *Agroforestry Notes,* No. 3, January 1997. National Agroforestry Center, USDA Forest Service, Lincoln, NE.

[3] Maryland Cooperative Extension. "Riparian Forest Buffer Design, Establishment, and Maintenance." University of Maryland, 1998.

[4] Nebraska Association of Resources Districts (2003). "Conservation Trees for Nebraska."

[5] Journal of Wildlife Management; Apr 2005, Vol. 69 Issue 2, p689-698, 10p

[6] University of Nebraska Cooperative Extension. "Benefits of Riparian Forest Buffers (Streamside Plantings of Trees, Shrubs and Grasses)." University Press, Lincoln, NE.

[7] Durham, Sharon. "Riparian Buffers Effective." Southeast Farm Press. 4 Feb 2004. p26

37.10   External links

- National Agroforestry Center (USDA)

- Filter Strip Design Tool (USDA)

- Extensive Riparian Buffer bibliography

Chapter 38

Rural–urban fringe

The rural-urban fringe of Bacchus Marsh, Victoria, Australia

The **rural–urban fringe**, also known as the **outskirts** or the **urban hinterland**, can be described as the "**landscape interface** between town and country",[1] or also as the transition zone where urban and rural uses mix and often clash.[2] Alternatively, it can be viewed as a landscape type in its own right, one forged from an interaction of urban and rural land uses.

38.1   Definition

Its definition shifts depending on the global location, but typically in Europe, where urban areas are intensively managed to prevent urban sprawl and protect agricultural land the urban fringe will be characterised by certain land uses which have either purposely moved away from the urban area, or require much larger tracts of land. As examples:

- Roads, especially motorways and bypasses
- Waste transfer stations, recycling facilities and landfill sites
- Park and ride sites,
- Airports,
- Large hospitals,
- Power, water and sewerage facilities.
- Factories
- Large out-of-town shopping facilities e.g. large supermarkets

Despite these 'urban' uses, the fringe remains largely open with the majority of the land agricultural, woodland or other rural use. However the quality of the countryside around urban areas tends to be low with severance between area of open land and bad maintained woodlands and hedgerows.

In recent years there has been a growing interest in how the full environmental and social potential of the urban fringe can be unlocked. In England in 2005, the Countryside Agency (now part of Natural England) together with Groundwork, a community and environmental regeneration body, produced a vision for the 'countryside in and around towns' that sets out ten 'functions' for a multi-functional urban fringe.[3] The realisation of this vision would provide a high quality environment right on the urban doorstep and provide the adjacent town or city with a host of 'ecosystem services'. It is estimated that within England the urban fringe covers as much as 20% of the land area. Such an extensive resource must be managed and used more intelligently and sustainably if the country as a whole is to develop and function sustainably.

In the United States the urban fringe generally consists of contiguous territory having a density of at least 1,000 persons per square mile. The urban fringe also includes outlying territory of such density if it was connected to the core of the contiguous area by road and is within 1 1/2 road miles of that core, or within 5 road miles of the core but separated by water or other undevelopable territory. Other territory with a population density of fewer than 1,000 people per square mile is included in the urban fringe if it eliminates an enclave or closes an indentation in the boundary of the urbanized area.[4]

38.2   See also

- Boomburbs

- Commuter town

- Edge city

- Edge effect

- Exurb

- Habitat

- Habitat destruction

- Microdistrict

- Natural landscape

- Penurbia

- Prime farmland

- Restoration ecology

- Suburb

- wildland–urban interface

38.3   References

[1]

[2] Griffiths, Michael. B. (2010) 'Lamb Buddha's Migrant Workers: Self-assertion on China's Urban Fringe'. Journal of Current Chinese Affairs (China Aktuell), 39, 2, 3–37

[3] Countryside In and Around Towns (Countryside Agency and Groundwork UK, 2005)

[4] URBAN AND RURAL DEFINITIONS/ US Census Bureau. 1995

38.4   External links

- Countryside Agency of England's online research library of urban rural fringe

- 'Case Studies' of the Urban Rural fringe for students

- Kay's Geography: Kingston Park - retail change at the edge

Chapter 39

Solvatten

SOLVATTEN is a combined portable water treatment and solar water heater system that has been designed for use at the household level in the developing world. The device uses natural UV radiation to treat water, and units are capable of rendering highly contaminated water drinkable (as defined by the WHO safe drinking water standards.[1]) in a few hours, provided there is sufficient sunlight.

SOLVATTEN incorporates three water treatment processes: filtration, pasteurisation and UV sterilisation.[2] Under optimal conditions, the device can eliminate all pathogenic material in 10-litres of water within 2 hours,[3] allowing for multiple batches of water to be treated in a given day.

The device is typically used in situations where water resources are scarce and prone to contamination, but it has also been applied in disaster relief scenarios.[4] The device and its inventor, Petra Wadström, have both achieved recognition through a series of national and international awards.

39.1 Design and Usage

Each SOLVATTEN unit consists of two five-litre containers that have transparent surfaces and can be filled with water. The transparent surfaces of the containers are made from a plastic that allows the penetration of a large portion of UV light, specifically UV-B which is highly effective at destroying microorganisms. Water is poured in through an opening that houses a filter of 35 microns to remove larger particles[5]

Once filled, the unit is placed in direct sunlight, which simultaneously heats the water and exposes it to ultraviolet radiation. The combination of these is an effective means of purifying water and, depending on conditions, the water will be free of pathogenic material in 2–6 hours. The water is heated to between 55 and 75 °C,[2] making it suitable for a number of other household and hygiene purposes, such as hand washing, bathing and domestic cleaning.

A study has shown that the use of SOLVATTEN is associated with health improvements in communities that are dependent upon contaminated water resources. It showed that it eliminated 100% of pathogenic material from water sourced from contaminated wells, whilst its use was also associated with as 60% reduction in diarrhoeal infections in children under five years old.[6]

The study, conducted in Mali, showed that households using SOLVATTEN reduced their combustion of firewood and charcoal by an average of 0.5 and 0.9 Kg/day respectively.[6] This had additional economic benefits for the user, with average savings of 15.8 USD/month.[6] A separate research exercise calculated the social return on investment (SROI) of SOLVATTEN, which accounts for extra-financial value (e.g. environmental and social value not reflected in conventional financial accounts) relative to resources invested. In that study, the SROI was calculated as being 1:26, meaning that for every unit of currency invested in SOLVATTEN, 26 were created[3]

39.2 Awards and recognition

SOLVATTEN was designed by the Swedish inventor and environmentalist Petra Wadström. Wadström has been recognised as one of Sweden's most prestigious modern entrepreneurs and environmentalists, whilst the SOLVATTEN device itself has been acknowledged as a significant innovation. Below is a list of some of the awards received by both Wadström and her invention.

39.3 References

[1] Climate Solver: Solar safe water system

[2] Lundström, H. & Hagström, E. A field study in Kenya of insolation parameters to make water drinkable in the household water treatment unit SOLVATTEN. Uppsala University, Sweden

[3] Jönsson, J., Wikman, A. & Wätthammar, T. (2011) Social Return on Investment, SROI, the value added for families before and after using Solvatten in the Bungoma district in Western Kenya. Swedish University of Agricultural Sciences (SLU), Uppsala

[4] Solvatten use following Typhoon Yolanda (in Swedish)

[5] Solvatten product brochure

[6] Diop, B. S. (2012) Sanitary, economic and environmental impacts of Solvatten in the West African peri-urban context: case of Dialakorodji, Bamako (Mali). University of Dakar, Senegal.

[7] Sweden's top 100 environmentalists 2014

[8] Sweden's top 100 women 2014

[9] Polhempriset 2013

[10] The Energy Globe Awards 2013

[11] Soroptomists Sweden

[12] SACC NY Deloitte Green Award 2011

[13] Änglamarks Prize 2011

[14] International Green Awards 2011

[15] WWF Climate Solver 2009

[16] Skapa Priset 2008

Chapter 40

Strategic environmental assessment

Strategic environmental assessment (SEA) is a systematic decision support process, aiming to ensure that environmental and possibly other sustainability aspects are considered effectively in policy, plan and programme making. In this context, following Fischer (2007)[1] SEA may be seen as:

- a structured, rigorous, participative, open and transparent environmental impact assessment (EIA) based process, applied particularly to plans and programmes, prepared by public planning authorities and at times private bodies,

- a participative, open and transparent, possibly non-EIA-based process, applied in a more flexible manner to policies, prepared by public planning authorities and at times private bodies, or

- a flexible non-EIA based process, applied to legislative proposals and other policies, plans and programmes in political/cabinet decision-making.

Effective SEA works within a structured and tiered decision framework, aiming to support more effective and efficient decision-making for sustainable development and improved governance by providing for a substantive focus regarding questions, issues and alternatives to be considered in policy, plan and programme (PPP) making.

SEA is an evidence-based instrument, aiming to add scientific rigour to PPP making, by using suitable assessment methods and techniques. Ahmed and Sanchez Triana (2008) developed an approach to the design and implementation of public policies that follows a continuous process rather than as a discrete intervention.[2]

40.1 History

The European Union Directive on Environmental Impact Assessments (85/337/EEC, known as the *EIA Directive*) only applied to certain projects.[3] This was seen as deficient as it only dealt with specific effects at the local level whereas many environmentally damaging decisions had already been made at a more strategic level (for example the fact that new infrastructure may generate an increased demand for travel).

The concept of strategic assessments originated from regional development / land use planning in the developed world. In 1981 the *U.S. Housing and Urban Development Department* published the *Area-wide Impact Assessment Guidebook*. In Europe the *Convention on Environmental Impact Assessment in a Transboundary Context* the so-called *Espoo Convention* laid the foundations for the introduction of SEA in 1991. In 2003, the Espoo Convention was supplemented by a Protocol on Strategic Environmental Assessment.

The European SEA Directive 2001/42/EC required that all member states of the European Union should have ratified the Directive into their own country's law by 21 July 2004.[4]

Countries of the EU started implementing the land use aspects of SEA first, some took longer to adopt the directive than others, but the implementation of the directive can now be seen as completed. Many EU nations have a longer history of strong Environmental Appraisal including Denmark, the Netherlands, Finland and Sweden. The newer member states to the EU have hurried in implementing the directive.

40.2 Relationship with environmental impact assessment

For the most part, an SEA is conducted before a corresponding EIA is undertaken. This means that information on the environmental impact of a plan can cascade down through the tiers of decision making and can be used in an EIA at a later stage. This should reduce the amount of work that needs to be undertaken. A handover procedure is fore-seen.

40.3   Aims and structure of SEA

The SEA Directive only applies to plans and programmes, not policies, although policies within plans are likely to be assessed and SEA can be applied to policies if needed and in the UK certainly, very often is.

The structure of SEA (under the Directive) is based on the following phases:

- "Screening", investigation of whether the plan or programme falls under the SEA legislation,

- "Scoping", defining the boundaries of investigation, assessment and assumptions required,

- "Documentation of the state of the environment", effectively a *baseline* on which to base judgments,

- "Determination of the likely (non-marginal) environmental impacts", usually in terms of Direction of Change rather than firm figures,

- Informing and consulting the public,

- Influencing "Decision taking" based on the assessment and,

- Monitoring of the effects of plans and programmes after their implementation.

The EU directive also includes other impacts besides the environmental, such as material assets and archaeological sites. In most western European states this has been broadened further to include economic and social aspects of sustainability.

SEA should ensure that plans and programmes take into consideration the environmental effects they cause. If those environmental effects are part of the overall decision taking it is called *Strategic Impact Assessment*.

40.4   SEA in the European Union

SEA is a legally enforced assessment procedure required by Directive 2001/42/EC (known as the SEA Directive).[4] The SEA Directive aims at introducing systematic assessment of the environmental effects of strategic land use related plans and programs. It typically applies to regional and local, development, waste and transport plans, within the European Union. Some plans, such as finance and budget plans or civil defence plans are exempt from the SEA Directive, it also only applies to plans that are required by law, which interestingly excludes national government's plans and programs, as their plans are 'voluntary', whereas local and regional governments are usually required to prepare theirs.

40.4.1   United Kingdom

SEA within the UK is complicated by different Regulations, guidance and practice between England, Scotland, Wales and Northern Ireland. In particular the SEA Legislation in Scotland (and in Northern Ireland, which specifically refers to the Regional Development Strategy) contains an expectation that SEA will apply to strategies as well as plans and programmes. In the UK, SEA is inseparable from the term 'sustainability', and an SEA is expected to be carried out as part of a wider Sustainability Appraisal (SA), which was already a requirement for many types of plan before the SEA directive and includes social, and economic factors in addition to environmental. Essentially an SA is intended to better inform decision makers on the sustainability aspects of the plan and ensure the full impact of the plan on sustainability is understood.

The United Kingdom in its strategy for sustainable development, *A Better Quality of Life* (May 1999), explained sustainable development in terms of four objectives. These are:

- social progress which recognises the needs of everyone

- effective protection of the environment

- prudent use of natural resources

- maintenance of high and stable levels of economic growth and employment.

These headline objectives are usually used and applied to local situations in order to asses the impact of the plan or program.

40.5   SEA Internationally

40.5.1   The pan-European region

The Protocol on Strategic Environmental Assessment was negotiated by the member States of the UNECE (in this instance Europe, Caucasus and Central Asia). It required ratification by 16 States to come into force, which it did in July 2010. It is now open to all UN Member States. Besides its potentially broader geographical application (global), the Protocol differs from the corresponding European Union Directive in its non-mandatory application to policies and legislation - not just plans and programmes. The Protocol also places a strong emphasis on the consideration of health, and there are other more subtle differences between the two instruments.

40.5.2 New Zealand

SEA in New Zealand is part of an integrated planning and assessment process and unlike the US is not used in the manner of Environmental impact assessment. The Resource Management Act has, as a principal objective, the aim of sustainable management. SEA is increasingly being considered for transportation projects.

40.5.3 The OECD DAC - SEA in development co-operation

Development assistance is increasingly being provided through strategic-level interventions, aimed to make aid more effective. SEA meets the need to ensure environmental considerations are taken into account in this new aid context. Applying SEA to development co-operation provides the environmental evidence to support more informed decision making, and to identify new opportunities by encouraging a systematic and thorough examination of development options.

The OECD Development Assistance Committee (DAC) Task Team on SEA has developed guidance on how to apply SEA to development co-operation. The document explains the benefits of using SEA in development co-operation and sets out key steps for its application, based on recent experiences.

40.6 See also

- Environmental impact assessment

- Hydropower Sustainability Assessment Protocol

- Millennium Ecosystem Assessment (MEA)

- Strategic Environmental Assessment (Denmark) (SEA)

- UN-Habitat Guidelines for Strategic Environmental Assessment

- Strategic Environmental Assessment Good Practices Guide - Portugal

40.7 References

[1] Fischer, T. B. (2007). Theory and Practice of Strategic Environmental Assessment, Earthscan, London.

[2] Ahmed, Kulsum; Sánchez-Triana, Ernesto. 2008. Strategic Environmental Assessment for Policies : An Instrument for Good Governance. © Washington, DC : World Bank. https://openknowledge.worldbank.org/handle/10986/6461

[3] Council Directive 85/337/EEC on the Assessment of the Effects of Certain Public and Private Projects on the Environment (1985-06-27) from Eur-Lex

[4] Directive 2001/42/EC of the European Parliament and of the Council (2001-06-27) from Eur-Lex

- nssd.net - *Strategic Environmental Assessment: A rapidly evolving approach.*

40.8 External links

40.8.1 Organisations

- International Association for Impact Assessment

- World Bank

- European Union

- United Nations Economic Commission for Europe

- United Nations University's Open Educational Resource on SEA: Contains a Course Module, Wiki and Instructional Guide

- UK government guidance

- The Strategic Environmental Assessment Task Team Network

- The OECD DAC Network on Environment and Development Co-operation

- Strategic Environmental Assessment Information Service

- Environmental Protection Agency Ireland SEA Workshop Series

- Netherlands Commission for Environmental Assessment

40.8.2 EC projects

- BEACON - Strategic Environmental Assessment of transport plans and programmes

Chapter 41

Strategic Environmental Assessment (Denmark)

The **Strategic Environmental Assessment** (SEA), is a process in Denmark for assessing the environmental effects of proposed government projects and programmes. Established in 1993 by an administrative order of Denmark's Prime Minister's Office,[1] this requirement was not initially enshrined in law, but was supported by a government circular which required an SEA to be carried out on "government proposals with major environmental effects".[2] The SEA process was limited only to government proposals and did not extend to plans and programmes. SEAs were required to focus on the impacts proposals would have on physical, ecological, cultural, health and risk factors.[3] In 1995, the SEA requirement was extended to new parliamentary acts in addition to government proposals at the national level.

41.1 Initial administrative order

The administrative order broadly outlined four steps in the SEA process.[4]

1. Screening – using the checklist contained in the guidance (water, air, climate, surface of the earth, soil, flora and fauna, landscape, resources, waste, historical buildings, population health, safety and transport of harmful substances) proposals that were likely to have a significant environmental impact had to be identified.

2. Scoping – the cumulative effects of a bill or policy had to be identified.

3. Assessment – analysis of the effects that had been identified as significant in the previous stages. Crucially the guidance stated that it was not possible to give an overall description, so a list of factors to be referenced were included in the guidance.

4. Report – a separate report of the environmental effects had to be included, attached to the bill that was to be put before parliament. The report had to be non-technical and

easy to understand – it also had to be made available to the public.

41.1.1 Legal Framework

The SEA Directive (2001/42/EC) was integrated into the Danish planning system in 2004. The SEA Directive now ensures that SEA has been extended to plans and programmes, rather than just government proposals. The Act on the Environmental Assessment of Plans and Programmes (Danish SEA Act) was passed into law in May 2004 and "is strictly in accordance with the requirements of the [2001/42/EC] directive".[5]

The SEA Act allows the Minister for the Environment a certain amount of discretion (as the administrative order of 1993 did), with the Minister able to determine whether or not a private or publicly owned company is required to carry out a SEA on their plan or programme, when they are acting in the capacity of a public authority (for example, the provision of utilities).[6]

It is also important to note that the SEA Act can be over-ruled, as long as a plan or programme complies with the procedural and substantive requirements of the SEA Directive. This helps to avoid duplication with "land use plans that have been prepared in accordance with the [Danish] Planning Act".[7]

41.1.2 Guidance Documents

Despite the integration of the SEA Directive, SEA guidance was restricted to a publication called Strategic Environmental Assessment of Bills and Other Proposals: Examples and Experience issued in 1995 by the Ministry for the Environment. New guidance was prepared in 2002, but never published due to a change in government.[8] The 1995 guidance issues a six step approach with regards to the type of information that should be included within an SEA[9][10]

- Formulating the problem and describing the process of the proposal and the alternatives that have been considered

- Identifying the relevant environmental effects

- Describing the extent of the likely environmental effects

- Identifying mitigation measures and monitoring follow up programmes

- assessing and weighing the environmental effects in relation to policy objectives

- A statement with the summary of the main findings

In 2006, The Environment Ministry released up to date guidance that related to the SEA requirements for plans and programmes dating from 2004. Instruction number 9664 entitled 'Vejledning om miljøvurdering af planer og programmer' replaced the earlier guidance dating from 1995. The new guide was later supplemented, in 2007, by examples of SEAs that had been conducted.[11][12]

41.1.3 SEA steps required

SEA steps that are required are defined by the SEA Act and supplemented by the guidance.

The table sets out the specific SEA process as defined by 2006 and 2007 guidance from the Danish Ministry of the Environment. The process is broadly split into nine main stages:

41.2 Examples of SEAs produced so far

Examples of Danish SEAs carried out following the introduction of the EU SEA Directive are "so far few".[13] Below are some examples of SEA that have been produced prior to the introduction of the EU SEA Directive.

41.2.1 North Jutland Regional Development Plan (1995-1997)

This was the very first SEA to be carried out in Denmark and was carried out before the introduction of the EU SEA Directive. As a result of this pilot SEA, the major concern amongst stakeholders was reconciling negative environmental impacts that resulted from planning activities.

Overall, while stakeholders felt that no new knowledge was gained from carrying out an SEA, the planning process became more transparent. Politicians felt that there was more information on which they could make an informed decision about the choices presented to them. It was also decided that consultation should be carried out during the scoping process, rather than at later stages of the SEA process.[14]

41.2.2 Regional groundwater plan for Vejle County, Eastern Jutland (2002)

This SEA was carried out as a consequence of the regional planning process (which included the requirement for SEAs) and is notable for the alternatives suggested. Groundwater extraction takes places beneath polluted urban areas and two strategies were devised – 'cleaning up' or extracting water from rural areas. The resulting analysis found there was no clear best solution and therefore other factors, such as economic considerations, should be taken into account to aid the decision making process.[15]

41.2.3 Regional mineral extraction plan for Vejle County (2003)

Extraction of minerals controlled under the plan were assessed under six criteria – transportation, economy, changes in landscape, landscape dynamics, pollution suffered by local residents and stress factors for local residents. This SEA clearly influenced the decision making process as the decision on the best alternative was deferred until suitable mitigation measures could be found.[16]

41.3 Strengths and Weaknesses of the SEA system

41.3.1 Strengths

Before the implementation of the SEA Directive into the Danish planning system, there had been a long history of consideration of the environmental effects that a planning process may have. Therefore, the transposition of the SEA Directive into Danish legislation conforms fully, and the supporting guidance (albeit late) ensures that SEAs conducted should be directive compliant. This historical approach has ensured that the SEA process has been institutionalised into the planning process. This ensures that environmental considerations are the subject of political debates and that parliamentary bills are prepared adequately with the SEA Directive and Danish SEA Act in mind.

The SEA process in Denmark is transparent and SEAs have a discernible influence upon the decision making process (Jones et al., 2005). There is a long tradition of public participation in the Danish planning system and this philosophy is no exception with regards to the SEA system. This is coupled with a feeling among planners and stakeholders, that despite the costs of SEA, overall it is a worthwhile and productive process.

41.3.2 Weaknesses

The major weakness in the Danish SEA system largely relates to government ministries – this is ironic given that they are responsible for the integration of the SEA Directive into Danish legislation.

- There is a lack of satisfactory scoping procedure when ministries prepare bills and conduct environmental assessments.

- There is an administrative failure of ministries excluding negative effects and including only positive effects.

- Political adaptation of environmental assessments often takes place within ministries, sometimes with the result that information on the environmental impact of bills is not submitted to Parliament.

- Further consideration should be given to how the public could play a more attentive role in the SEA process to help overcome issues of stakeholder dominance and political short-sightedness.

- The development of SEA at the plan and programme level should be monitored to evaluate its contribution on the policy level and to identify how the practice on both levels could complement and stimulate each other.[17]

Monitoring systems currently in place need to be strengthened. Currently monitoring of environmental impacts is done only through informal communication instead of the Minister for the Environment issuing specific rules.[18]

41.4   See also

- Environmental impact assessment

- Strategic Environmental Assessment

- Social Impact Assessment

- Healthy development measurement tool

41.5   References

[1] Dalal-Clayton and Sadler, Strategic Environmental Assessment A Sourcebook and Reference Guide to International Experience, Earthscan, London, 2005.

[2] Therival and Partidario (1996), The Practice of Strategic Environmental Assessment, Earthscan, London, p25.

[3] Therival and Partidario (1996), The Practice of Strategic Environmental Assessment, Earthscan, London.

[4] Dalal-Clayton and Sadler, Strategic Environmental Assessment A Sourcebook and Reference Guide to International Experience, Earthscan, London, 2005, pg 67-68.

[5] Elling cited in: Jones et al., Strategic Environmental Assessment and Land Use Planning: An International Evaluation, Earthscan, London, 2005, pg66.

[6] Elling cited in: Jones et al., Strategic Environmental Assessment and Land Use Planning: An International Evaluation, Earthscan, London, 2005.

[7] Elling cited in: Jones et al., Strategic Environmental Assessment and Land Use Planning: An International Evaluation, Earthscan, London, 2005, pg66.

[8] Dalal-Clayton and Sadler, Strategic Environmental Assessment A Sourcebook and Reference Guide to International Experience, Earthscan, London, 2005.

[9] Dalal-Clayton and Sadler, Strategic Environmental Assessment A Sourcebook and Reference Guide to International Experience, Earthscan, London, 2005.

[10] http://www.blst.dk/Planlaegning/Miljoekonsekvenser/ MiljoevurderingPlaner/, Danish Ministry for the Environment (2008), Miljøvurdering af planer og programmer (SEA Legislation & Guidance), Retrieved 18/03/08.

[11] https://www.retsinformation.dk/Forms/R0710.aspx?id= 12953, Danish Ministry for the Environment (2008 (2)), Vejledning om miljøvurdering af planer og programmer, Retrieved 19/03/08.

[12] http://www.blst.dk/NR/rdonlyres/ 46372868-F4B7-4900-A743-063A553CB6E5/49265/ Eksempelsamlingendeligudgave.pdf, Danish Ministry for the Environment (2008 (3)), Eksempelsamling for miljøvurdering af planer og programmer September 2007, Retrieved 19/03/08.

[13] http://www.danidadevforum.um.dk/en/menu/Topics/ EnvironmentalSustainability/ToolsAndReferences/ Guidelines/StrategicEnvironmentalAssessment/, Danish Ministry of Foreign Affairs (2007), Strategic Environmental Assessment, Retrieved 25/03/08.

[14] Jones et al. (2005), Strategic Environmental Assessment and Land Use Planning: An International Evaluation, Earthscan, London

[15] Jones et al. (2005), Strategic Environmental Assessment and Land Use Planning: An International Evaluation, Earthscan, London

[16] Jones et al. (2005), Strategic Environmental Assessment and Land Use Planning: An International Evaluation, Earthscan, London

[17] http://www.unece.org/env/eia/documents/PolicySEA/ SEA%20of%20Policies%20volume.pdf, RECCEE (2005), Strategic Environmental Assessment at the Policy Level: Recent Progress, Current Status and Future Progress, Retrieved 25/03/08

[18] Jones et al. (2005), Strategic Environmental Assessment and Land Use Planning: An International Evaluation, Earthscan, London

http://sea.linddal.net/docs/seanote.pdf, DIDC (2007), An introduction to Strategic Environmental Assessment (SEA) for Danish International Development Cooperation, Retrieved 25/03/08.

Chapter 42

Sustainability appraisal

In United Kingdom planning law, a **sustainability appraisal** is an appraisal of the economic, environmental, and social effects of a plan from the outset of the preparation process to allow decisions to be made that accord with sustainable development.

Since 2001, sustainability appraisals have had to be in conformity with the EU directive on strategic environmental assessment.

The Hydropower Sustainability Assessment Protocol is a sector specific sustainability appraisal method.

42.1   See also

- Equality impact assessment

Chapter 43

Sustainable architecture

Hanging gardens of One Central Park, Sydney

Energy-plus-houses at Freiburg-Vauban in Germany

Sustainable architecture is architecture that seeks to minimize the negative environmental impact of buildings by efficiency and moderation in the use of materials, energy, and development space. Sustainable architecture uses a conscious approach to energy and ecological conservation in the design of the built environment.[1]

The idea of sustainability, or ecological design, is to ensure that our actions and decisions today do not inhibit the opportunities of future generations.[2]

43.1 Sustainable energy use

Main articles: Low-energy house and Zero-energy building
Energy efficiency over the entire life cycle of a build-

K2 sustainable apartments in Windsor, Victoria, Australia by DesignInc (2006) features passive solar design, recycled and sustainable materials, photovoltaic cells, wastewater treatment, rainwater collection and solar hot water.

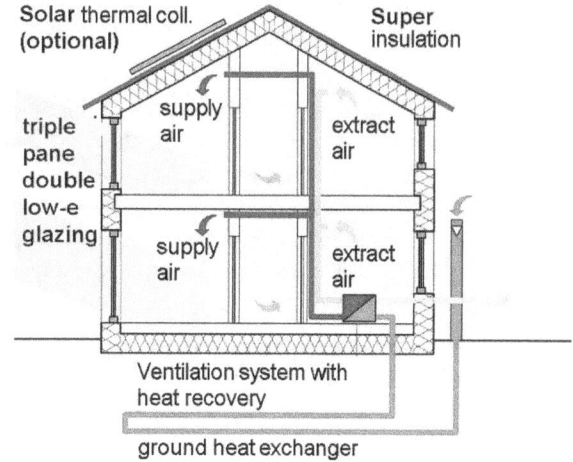

The passivhaus standard combines a variety of techniques and technologies to achieve ultra-low energy use.

ing is the most important goal of sustainable architecture. Architects use many different passive and active techniques

Following its destruction by a tornado in 2007, the town of Greensburg, Kansas (USA) elected to rebuild to highly stringent LEED Platinum environmental standards. Shown is the town's new art center, which integrates its own solar panels and wind generators for energy self-sufficiency.

to reduce the energy needs of buildings and increase their ability to capture or generate their own energy.[3] One of the keys to exploit local environmental resources and influence energy-related factors such as daylight, solar heat gains and ventilation is the use of site analysis.

43.1.1 Heating, ventilation and cooling system efficiency

Numerous passive architectural strategies have been developed over time. Examples of such strategies include the arrangement of rooms or the sizing and orientation of windows in a building,[3] and the orientation of facades and streets or the ratio between building heights and street widths for urban planning.[4]

An important and cost-effective element of an efficient heating, ventilating, and air conditioning (HVAC) system is a well-insulated building. A more efficient building requires less heat generating or dissipating power, but may require more ventilation capacity to expel polluted indoor air.

Significant amounts of energy are flushed out of buildings in the water, air and compost streams. Off the shelf, on-site energy recycling technologies can effectively recapture energy from waste hot water and stale air and transfer that energy into incoming fresh cold water or fresh air. Recapture of energy for uses other than gardening from compost leaving buildings requires centralized anaerobic digesters.

HVAC systems are powered by motors. Copper, versus other metal conductors, helps to improve the electrical energy efficiencies of motors, thereby enhancing the sustainability of electrical building components. *(For main article, see: Copper in energy-efficient motors).*

Site and building orientation have some major effects on a building's HVAC efficiency.

Passive solar building design allows buildings to harness the energy of the sun efficiently without the use of any active so-

lar mechanisms such as photovoltaic cells or solar hot water panels. Typically passive solar building designs incorporate materials with high thermal mass that retain heat effectively and strong insulation that works to prevent heat escape. Low energy designs also requires the use of solar shading, by means of awnings, blinds or shutters, to relieve the solar heat gain in summer and to reduce the need for artificial cooling. In addition, low energy buildings typically have a very low surface area to volume ratio to minimize heat loss. This means that sprawling multi-winged building designs (often thought to look more "organic") are often avoided in favor of more centralized structures. Traditional cold climate buildings such as American colonial saltbox designs provide a good historical model for centralized heat efficiency in a small-scale building.

Windows are placed to maximize the input of heat-creating light while minimizing the loss of heat through glass, a poor insulator. In the northern hemisphere this usually involves installing a large number of south-facing windows to collect direct sun and severely restricting the number of north-facing windows. Certain window types, such as double or triple glazed insulated windows with gas filled spaces and low emissivity (low-E) coatings, provide much better insulation than single-pane glass windows. Preventing excess solar gain by means of solar shading devices in the summer months is important to reduce cooling needs. Deciduous trees are often planted in front of windows to block excessive sun in summer with their leaves but allow light through in winter when their leaves fall off. Louvers or light shelves are installed to allow the sunlight in during the winter (when the sun is lower in the sky) and keep it out in the summer (when the sun is high in the sky). Coniferous or evergreen plants are often planted to the north of buildings to shield against cold north winds.

In colder climates, heating systems are a primary focus for sustainable architecture because they are typically one of the largest single energy drains in buildings.

In warmer climates where cooling is a primary concern, passive solar designs can also be very effective. Masonry building materials with high thermal mass are very valuable for retaining the cool temperatures of night throughout the day. In addition builders often opt for sprawling single story structures in order to maximize surface area and heat loss. Buildings are often designed to capture and channel existing winds, particularly the especially cool winds coming from nearby bodies of water. Many of these valuable strategies are employed in some way by the traditional architecture of warm regions, such as south-western mission buildings.

In climates with four seasons, an integrated energy system will increase in efficiency: when the building is well insulated, when it is sited to work with the forces of nature, when heat is recaptured (to be used immediately or stored), when the heat plant relying on fossil fuels or electricity is greater than 100% efficient, and when renewable energy is used.

43.1.2 Renewable energy generation

BedZED (Beddington Zero Energy Development), the UK's largest and first carbon-neutral eco-community: the distinctive roofscape with solar panels and passive ventilation chimneys

Solar panels

Main article: Solar PV

Active solar devices such as photovoltaic solar panels help to provide sustainable electricity for any use. Electrical output of a solar panel is dependent on orientation, efficiency, latitude, and climate—solar gain varies even at the same latitude. Typical efficiencies for commercially available PV panels range from 4% to 28%. The low efficiency of certain photovoltaic panels can significantly affect the payback period of their installation.[5] This low efficiency does not mean that solar panels are not a viable energy alternative. In Germany for example, Solar Panels are commonly installed in residential home construction.

Roofs are often angled toward the sun to allow photovoltaic panels to collect at maximum efficiency. In the northern hemisphere, a true-south facing orientation maximizes yield for solar panels. If true-south is not possible, solar panels can produce adequate energy if aligned within 30° of south. However, at higher latitudes, winter energy yield will be significantly reduced for non-south orientation.

To maximize efficiency in winter, the collector can be angled above horizontal Latitude +15°. To maximize efficiency in summer, the angle should be Latitude −15°. However, for an annual maximum production, the angle of the panel above horizontal should be equal to its latitude.[6]

Wind turbines

Main article: Wind power

The use of undersized wind turbines in energy production in sustainable structures requires the consideration of many factors. In considering costs, small wind systems are generally more expensive than larger wind turbines relative to the amount of energy they produce. For small wind turbines, maintenance costs can be a deciding factor at sites with marginal wind-harnessing capabilities. At low-wind sites, maintenance can consume much of a small wind turbine's revenue.[7] Wind turbines begin operating when winds reach 8 mph, achieve energy production capacity at speeds of 32-37 mph, and shut off to avoid damage at speeds exceeding 55 mph.[7] The energy potential of a wind turbine is proportional to the square of the length of its blades and to the cube of the speed at which its blades spin. Though wind turbines are available that can supplement power for a single building, because of these factors, the efficiency of the wind turbine depends much upon the wind conditions at the building site. For these reasons, for wind turbines to be at all efficient, they must be installed at locations that are known to receive a constant amount of wind (with average wind speeds of more than 15 mph), rather than locations that receive wind sporadically.[8] A small wind turbine can be installed on a roof. Installation issues then include the strength of the roof, vibration, and the turbulence caused by the roof ledge. Small-scale rooftop wind turbines have been known to be able to generate power from 10% to up to 25% of the electricity required of a regular domestic household dwelling.[9] Turbines for residential scale use are usually between 7 feet (2 m) to 25 feet (8 m) in diameter and produce electricity at a rate of 900 watts to 10,000 watts at their tested wind speed.[10] Building integrated wind turbine performance can be enhanced with the addition of an aerofoil wing on top of a roof mounted turbine.[11]

See also: Design feasibilty of Wind turbine systems

Solar water heating

Main article: Solar thermal power

Solar water heaters, also called solar domestic hot water systems, can be a cost-effective way to generate hot water for a home. They can be used in any climate, and the fuel they use—sunshine—is free.[12]

There are two types of solar water systems- active and passive. An active solar collector system can produce about 80 to 100 gallons of hot water per day. A passive system will have a lower capacity.[13]

There are also two types of circulation, direct circulation systems and indirect circulation systems. Direct circulation systems loop the domestic water through the panels. They should not be used in climates with temperatures below freezing. Indirect circulation loops glycol or some other fluid through the solar panels and uses a heat exchanger to heat up the domestic water.

The two most common types of collector panels are Flat-Plate and Evacuated-tube. The two work similarly except that evacuated tubes do not convectively lose heat, which greatly improves their efficiency (5%−25% more efficient). With these higher efficiencies, Evacuated-tube solar collectors can also produce higher-temperature space heating, and even higher temperatures for absorption cooling systems.[14]

Electric-resistance water heaters that are common in homes today have an electrical demand around 4500 kW·h/year. With the use of solar collectors, the energy use is cut in half. The up-front cost of installing solar collectors is high, but with the annual energy savings, payback periods are relatively short.[14]

Heat pumps

Air-source heat pumps (ASHP) can be thought of as reversible air conditioners. Like an air conditioner, an ASHP can take heat from a relatively cool space (e.g. a house at 70 °F) and dump it into a hot place (e.g. outside at 85 °F). However, unlike an air conditioner, the condenser and evaporator of an ASHP can switch roles and absorb heat from the cool outside air and dump it into a warm house.

Air-source heat pumps are inexpensive relative to other heat pump systems. However, the efficiency of air-source heat pumps decline when the outdoor temperature is very cold or very hot; therefore, they are only really applicable in temperate climates.[14]

For areas not located in temperate climates, ground-source (or geothermal) heat pumps provide an efficient alternative. The difference between the two heat pumps is that the ground-source has one of its heat exchangers placed underground—usually in a horizontal or vertical arrangement. Ground-source takes advantage of the relatively constant, mild temperatures underground, which means their efficiencies can be much greater than that of an air-source heat pump. The in-ground heat exchanger generally needs a considerable amount of area. Designers have placed them in an open area next to the building or underneath a parking lot.

Energy Star ground-source heat pumps can be 40% to 60% more efficient than their air-source counterparts. They are also quieter and can also be applied to other functions like domestic hot water heating.[14]

In terms of initial cost, the ground-source heat pump system costs about twice as much as a standard air-source heat pump to be installed. However, the up-front costs can be more than offset by the decrease in energy costs. The reduction in energy costs is especially apparent in areas with typically hot summers and cold winters.[14]

Other types of heat pumps are water-source and air-earth. If the building is located near a body of water, the pond or lake could be used as a heat source or sink. Air-earth heat pumps circulate the building's air through underground ducts. With higher fan power requirements and inefficient heat transfer, Air-earth heat pumps are generally not practical for major construction.

43.2   Sustainable building materials

See also: Green building

Some examples of sustainable building materials include recycled denim or blown-in fiber glass insulation, sustainably harvested wood, Trass, Linoleum,[15] sheep wool, concrete (high and ultra high performance[16] roman self-healing concrete[17]), panels made from paper flakes, baked earth, rammed earth, clay, vermiculite, flax linnen, sisal, seegrass, expanded clay grains, coconut, wood fibre plates, calcium sand stone, locally obtained stone and rock, and bamboo, which is one of the strongest and fastest growing woody plants, and non-toxic low-VOC glues and paints.

43.2.1   Recycled materials

Sustainable architecture often incorporates the use of recycled or second hand materials, such as reclaimed lumber and recycled copper. The reduction in use of new materials creates a corresponding reduction in embodied energy (energy used in the production of materials). Often sustainable architects attempt to retrofit old structures to serve new needs in order to avoid unnecessary development. Architectural salvage and reclaimed materials are used when appropriate. When older buildings are demolished, frequently any good wood is reclaimed, renewed, and sold as flooring. Any good dimension stone is similarly reclaimed. Many other parts are reused as well, such as doors, windows, mantels, and hardware, thus reducing the consumption of new goods. When new materials are employed, green designers look for materials that are rapidly replenished, such as bamboo, which can be harvested for commercial use after only 6 years of growth, sorghum or wheat straw, both of

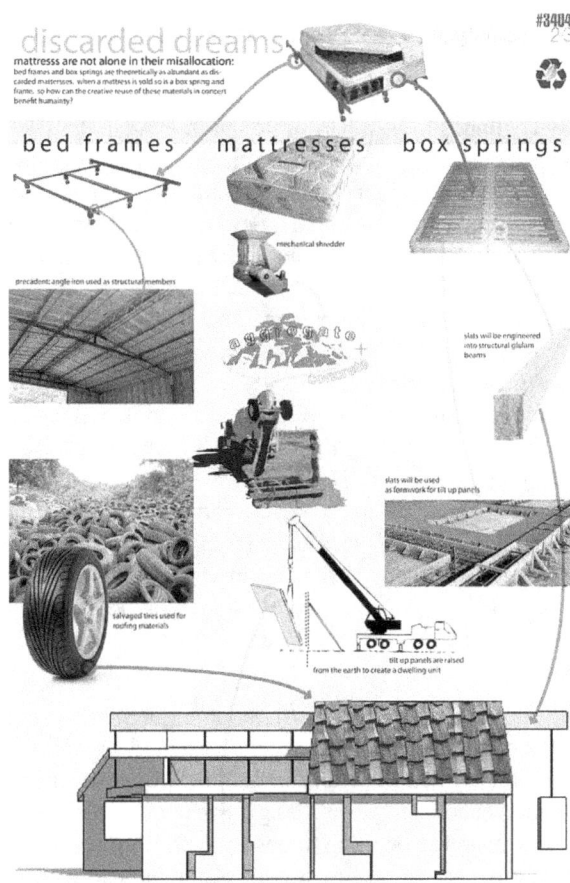

Recycling items for building

which are waste material that can be pressed into panels, or cork oak, in which only the outer bark is removed for use, thus preserving the tree. When possible, building materials may be gleaned from the site itself; for example, if a new structure is being constructed in a wooded area, wood from the trees which were cut to make room for the building would be re-used as part of the building itself.

43.2.2 Lower volatile organic compounds

Low-impact building materials are used wherever feasible: for example, insulation may be made from low VOC (volatile organic compound)-emitting materials such as recycled denim or cellulose insulation, rather than the building insulation materials that may contain carcinogenic or toxic materials such as formaldehyde. To discourage insect damage, these alternate insulation materials may be treated with boric acid. Organic or milk-based paints may be used.[18] However, a common fallacy is that "green" materials are always better for the health of occupants or the environment. Many harmful substances (including formaldehyde, arsenic, and asbestos) are naturally occur-

ring and are not without their histories of use with the best of intentions. A study of emissions from materials by the State of California has shown that there are some green materials that have substantial emissions whereas some more "traditional" materials actually were lower emitters. Thus, the subject of emissions must be carefully investigated before concluding that natural materials are always the healthiest alternatives for occupants and for the Earth.[19]

Volatile organic compounds (VOC) can be found in any indoor environment coming from a variety of different sources. VOCs have a high vapor pressure and low water solubility, and are suspected of causing sick building syndrome type symptoms. This is because many VOCs have been known to cause sensory irritation and central nervous system symptoms characteristic to sick building syndrome, indoor concentrations of VOCs are higher than in the outdoor atmosphere, and when there are many VOCs present, they can cause additive and multiplicative effects.

Green products are usually considered to contain fewer VOCs and be better for human and environmental health. A case study conducted by the Department of Civil, Architectural, and Environmental Engineering at the University of Miami that compared three green products and their non-green counterparts found that even though both the green products and the non-green counterparts both emitted levels of VOCs, the amount and intensity of the VOCs emitted from the green products were much safer and comfortable for human exposure.[20]

43.2.3 Materials sustainability standards

Despite the importance of materials to overall building sustainability, quantifying and evaluating the sustainability of building materials has proven difficult. There is little coherence in the measurement and assessment of materials sustainability attributes, resulting in a landscape today that is littered with hundreds of competing, inconsistent and often imprecise eco-labels, standards and certifications. This discord has led both to confusion among consumers and commercial purchasers and to the incorporation of inconsistent sustainability criteria in larger building certification programs such as LEED. Various proposals have been made regarding rationalization of the standardization landscape for sustainable building materials.[21]

43.3 Waste management

Waste takes the form of spent or useless materials generated from households and businesses, construction and demolition processes, and manufacturing and agricultural industries. These materials are loosely categorized as municipal solid waste, construction and demolition (C&D) debris, and industrial or agricultural by-products.[22] Sustainable architecture focuses on the on-site use of waste management, incorporating things such as grey water systems for use on garden beds, and composting toilets to reduce sewage. These methods, when combined with on-site food waste composting and off-site recycling, can reduce a house's waste to a small amount of packaging waste.This is the new techniques of sustainable architecture .

43.4 Building placement

One central and often ignored aspect of sustainable architecture is building placement. Although the ideal environmental home or office structure is often envisioned as an isolated place, this kind of placement is usually detrimental to the environment. First, such structures often serve as the unknowing frontlines of suburban sprawl. Second, they usually increase the energy consumption required for transportation and lead to unnecessary auto emissions. Ideally, most building should avoid suburban sprawl in favor of the kind of light urban development articulated by the New Urbanist movement. Careful mixed use zoning can make commercial, residential, and light industrial areas more accessible for those traveling by foot, bicycle, or public transit, as proposed in the Principles of Intelligent Urbanism. The study of Permaculture, in its holistic application, can also greatly help in proper building placement that minimizes energy consumption and works with the surroundings rather than against them, especially in rural and forested zones.

43.5 Sustainable building consulting

A sustainable building consultant may be engaged early in the design process, to forecast the sustainability implications of building materials, orientation, glazing and other physical factors, so as to identify a sustainable approach that meets the specific requirements of a project.

Norms and standards have been formalized by performance-based rating systems e.g. LEED[23] and Energy Star for homes.[24] They define benchmarks to be met and provide metrics and testing to meet those benchmarks. It is up to the parties involved in the project to determine the best approach to meet those standards.

43.6 Changing pedagogies

Critics of the reductionism of modernism often noted the abandonment of the teaching of architectural history as a

causal factor. The fact that a number of the major players in the shift away from modernism were trained at Princeton University's School of Architecture, where recourse to history continued to be a part of design training in the 1940s and 1950s, was significant. The increasing rise of interest in history had a profound impact on architectural education. History courses became more typical and regularized. With the demand for professors knowledgeable in the history of architecture, several PhD programs in schools of architecture arose in order to differentiate themselves from art history PhD programs, where architectural historians had previously trained. In the US, MIT and Cornell were the first, created in the mid-1970s, followed by Columbia, Berkeley, and Princeton. Among the founders of new architectural history programs were Bruno Zevi at the Institute for the History of Architecture in Venice, Stanford Anderson and Henry Millon at MIT, Alexander Tzonis at the Architectural Association, Anthony Vidler at Princeton, Manfredo Tafuri at the University of Venice, Kenneth Frampton at Columbia University, and Werner Oechslin and Kurt Forster at ETH Zürich.[25]

The term "sustainability" in relation to architecture has so far been mostly considered through the lens of building technology and its transformations. Going beyond the technical sphere of "green" design, invention and expertise, some scholars are starting to position architecture within a much broader cultural framework of the human interrelationship with nature. Adopting this framework allows tracing a rich history of cultural debates about our relationship to nature and the environment, from the point of view of different historical and geographical contexts.[26]

43.7 Sustainable urbanism and architecture

Concurrently, the recent movements of New Urbanism and New Classical Architecture promote a sustainable approach towards construction, that appreciates and develops smart growth, architectural tradition and classical design.[27][28] This in contrast to modernist and globally uniform architecture, as well as leaning against solitary housing estates and suburban sprawl.[29] Both trends started in the 1980s. The Driehaus Architecture Prize is an award that recognizes efforts in New Urbanism and New Classical Architecture, and is endowed with a prize money twice as high as that of the modernist Pritzker Prize.[30]

43.8 Criticism

There are conflicting ethical, engineering, and political orientations depending on the viewpoints.[31]

There is no doubt Green Technology has made its headway into the architectural community, the implementation of given technologies have changed the ways we see and perceive modern day architecture. While green architecture has been proven to show great improvements of ways of living both environmentally and technologically the question remains, is all this sustainable? Many building codes have been demeaned to international standards. "LEED" (Leadership in Energy & Environmental Design) has been criticized for exercising flexible codes for building to follow. Contractors do this to save as much money as they possibly can. For example, a building may have solar paneling but if the infrastructure of the building's core doesn't support that over a long period of time improvements would have to be made on a constant basis and the building itself would be vulnerable to disasters or enhancements. With companies cutting paths to make shortcuts with sustainable architecture when building their structures it fuels to the irony that the "sustainable" architecture isn't sustainable at all. Sustainability comes in reference to longevity and effectiveness.

Ethics and Politics also play into sustainable architecture and its ability to grow in urban environment. Conflicting viewpoints between engineering techniques and environmental impacts still are popular issues that resonate in the architectural community. With every revolutionary technology or innovation there comes criticisms of legitimacy and effectiveness when and how it is being utilized. Many of the criticisms of sustainable architecture do not reflect every aspect of it but rather a broader spectrum across the international community.

43.9 See also

- Alternative natural materials
- BREEAM
- Deconstruction (building)
- Earthship
- Ecological design
- Ecological footprint
- Energy-plus-house
- Fab Tree Hab: 100% Ecological Home
- Haute qualité environnementale French standard for green building - HQE
- Low-energy house
- New Urbanism
- Organic architecture
- Passive house
- Principles of Intelligent Urbanism
- Renewable heat
- Solar chimney
- Straw-bale construction
- Superinsulation
- Sustainable design
- Sustainable development
- Sustainable flooring
- Sustainable landscape architecture
- Sustainable preservation
- Sustainable refurbishment
- Windcatcher
- Zero-energy building

43.10 References

[1] "Sustainable Architecture and Simulation Modelling", Dublin Institute of Technology,

[2] Doerr Architecture, Definition of Sustainability and the Impacts of Buildings

[3] M. DeKay & G.Z. Brown, Sun Wind & Light, architectural design strategies, 3rd ed. *Wiley*, 2014

[4] M. Montavon, Optimization of Urban Form by the Evaluation of the Solar Potential, *EPFL*, 2010

[5] shamilton. "Module Pricing". Solarbuzz. Retrieved 2012-11-07.

[6] G.Z. Brown, Mark DeKay. Sun, Wind & Light. 2001

[7] Brower, Michael; *Cool Energy, The Renewable Solution to Global Warming*; Union of Concerned Scientists, 1990

[8] Gipe, Paul; *Wind Power: Renewable Energy for Farm and Business*; Chelsea Green Publishing, 2004

[9] The Sunday Times, "Home wind turbines dealt killer blow", April 16, 2006

[10] "Wind turbine, a powerful investment", Rapid City Journal, February 20, 2008

[11] Factors enhancing aerofoil wings for wind energy harnessing in buildings,7 November 2013 http://bse.sagepub.com/content/early/2013/11/07/0143624413509097.abstract?papetoc

[12] U.S. Department of Energy, Energy Efficiency and Renewable Energy, Solar Water Heaters, March 24, 2009

[13] "Solar Water Heaters". Toolbase.org. Retrieved 2012-11-07.

[14] John Randolph and Gilbert M. Masters, 2008. "Energy for Sustainability: Technology, Planning, Policy," Island Press, Washington, DC.

[15] Duurzaam en Gezond Bouwen en Wonen by Hugo Vanderstadt,

[16] Time:Cementing the future

[17] Roman concrete self-healing

[18] Information on low-emitting materials may be found at www.buildingecology.com/iaq_links.php IAQ links

[19] Building Emissions Study accessed at California Integrated Waste Management web site

[20] James, J.P., Yang, X. Indoor and Built Environment, Emissions of Volatile Organic Compounds from Several Green and Non-Green Building Materials: A Comparison, January 2004. Retrieved: 2008-04-30.

[21] J. Contreras, M. Lewis, H. Roth, "Toward a Rational Framework for Sustainable Building Materials Standards", Standards Engineering, Vol. 63, No. 5, p. 1, September/October 2011

[22] John Ringel., University of Michigan, Sustainable Architecture, Waste Prevention

[23] United States Green Building Council

[24] Energystar.gov

[25] Mark Jarzombek, "The Disciplinary Dislocations of Architectural History," *Journal of the Society of Architectural Historians* 58/3 (September 1999), p. 489. See also other articles in that issue by Eve Blau, Stanford Anderson, Alina Payne, Daniel Bluestone, Jeon-Louis Cohen and others.

[26] McGrath, Brian (2013). *Urban Design Ecologies: AD Reader*. John Wiley & Sons, Inc. pp. 220–237. ISBN 978-0-470-97405-6.

[27] Charter of the New Urbanism

[28] "Beauty, Humanism, Continuity between Past and Future". Traditional Architecture Group. Retrieved 23 March 2014.

[29] Issue Brief: Smart-Growth: Building Livable Communities. American Institute of Architects. Retrieved on 2014-03-23.

[30] "Driehaus Prize". Together, the $200,000 Driehaus Prize and the $50,000 Reed Award represent the most significant recognition for classicism in the contemporary built environment.. Notre Dame School of Architecture. Retrieved 23 March 2014.

[31] Mark Jarzombek, "Sustainability - Architecture: between Fuzzy Systems and Wicked Problems" (PDF), *Blueprints* **21** (1): 6–9

43.11   External links

- World Green Building Council

- El Paso Solar Energy Association Information page about passive solar water heating

- Energy Recovery Council

- Passivhaus Institut German institute for passive buildings

- U.S. EPA - Landfill Research Bioreactor landfill research supports sustainable waste management initiatives

Chapter 44

Sustainable design standards

Design standards, **reference standards** and **performance standards** are familiar throughout business and industry, virtually for anything that is definable. Sustainable design, taken as reducing our impact on the earth and making things better at the same time, is in the process of becoming defined. There are also lots of well organized specific methodologies that are used by different communities of people for different purposes.

44.1 Design standards

One of the better known is the Leadership in Energy and Environmental Design (LEED) green building rating system, which uses a diverse group of hard measures of environmental quality and impacts to define a holistic approach to sustainable building and assign ratings to individual projects.

Sustainable design is really just a more determined effort to consider the whole range of impacts on our environment in making any decision. A more complete design guide, guided more by whole project impact measures, is the model offered by the U.S. cooperating agencies in the "Whole Building Design Guide".

Green construction codes and standards are beginning to emerge on the national code stage. The standards go beyond energy standards such as ASHRAE 90.1 and the International Energy Conservation Code (IECC) to cover additional areas such as site sustainability, water efficiency, indoor environmental quality and materials and resources. The first is ASHRAE 189.1, Standard for the Design of High-Performance, Green Buildings Except Low-Rise Residential Buildings, published by ASHRAE in January 2010 in conjunction with the U.S. Green Building Council and the Illuminating Engineering Society. Standard 189.1 provides criteria by which a building can be judged as "green," written in model code language that jurisdictions can use to develop a green building construction code.

Several organizations have developed their own ways of setting goals for energy reductions, such as Architecture 2030 and for qualifying performance toward them such as Cradle to Cradle.

44.2 Design methods

Developing real methods for how to discover the design opportunities that would allow you to meet or exceed the standards was one of the objectives of the environmental design movement in architectural schools in the 1960s and '70s, but though some of the issues introduced then are still an important part of the process, not much actually changed about the methods of design. Now with the combination of many more interactive tools and much higher stakes in the outcome, and long gestating rethinking about natural systems in general, a dramatic new revolution in methodology seems inevitable.

BIM (building information modeling) allows designers to work with many remote consultants on the same data file that represents all the decisions being made by the team. That same file is available to the climate and energy and environmental impact analysis and cost analysis tools and consultants, ... and of course to the prospective contractors and the regulators. Along with this new integrated access to the model there in needed a new way to integrate the conversation of so many people, each with some interest in reviewing each other's comments on the progress with the central design model. That is likely to involve development of wiki tools for the process. One such very early impliemention of a Wiki SD tool called "4Dsustainability" organizes the project design evolution around the general learning process of **how you define the problem** by exploring its environment, and following that through the project.

The main difference between sustainable design methods and conventional design is incorporating the entire environment of the project's stakeholders on the design team, essentially, and that calls for new ways to explore connections and for more people and perspectives to be taken into account. Other methods that recognize this requirement are the "AIA SDAT" (sustainable design assessment team) program and the "Scenarios for sustainability" process design tools.

44.3   References

Chapter 45

Sustainable development

Wind powers 5 MW wind turbines on a wind farm 28 km off the coast of Belgium.

Sustainable development (SD) is a process for meeting human development goals while maintaining the ability of natural systems to continue to provide the natural resources and ecosystem services upon which the economy and society depend. While the modern concept of sustainable development is derived most strongly from the 1987 Brundtland Report, it is rooted in earlier ideas about sustainable forest management and twentieth century environmental concerns .

Sustainable development is the organizing principle for sustaining finite resources necessary to provide for the needs of future generations of life on the planet. It is a process that envisions a desirable future state for human societies in which living conditions and resource-use continue to meet human needs without undermining the "integrity, stability and beauty" of natural biotic systems.[1]

45.1   Definition

Sustainability can be defined as the practice of reserving resources for future generation without any harm to the nature and other components of it .[2][3] Sustainable development ties together concern for the carrying capacity of natural systems with the social, political, and economic challenges faced by humanity. Sustainability science is the study of the concepts of sustainable development and environmental science. There is an additional focus on the present generations' responsibility to regenerate, maintain and improve planetary resources for use by future generations.[4]

45.2   Dimensions

See also: Planetary boundaries and Triple bottom line

Sustainable development has been described in terms of three dimensions, domains or pillars. In the three-dimension model, these are seen as "economic, environmental and social" or "ecology, economy and equity";[5] this has been expanded by some authors to include a fourth pillar of culture,[6][7] institutions or governance.[5]

45.2.1   Ecology

See also: Ecological engineering

The ecological sustainability of human settlements is part

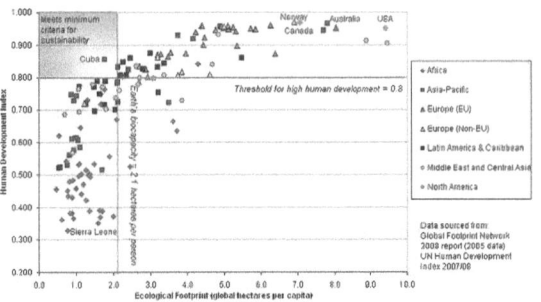

Relationship between ecological footprint and Human Development Index (HDI)

The Blue Marble, photographed from Apollo 17 in 1972, quickly became an icon of environmental conservation.[11]

of the relationship between humans and their natural, social and built environments.[8] Also termed human ecology, this broadens the focus of sustainable development to include the domain of human health. Fundamental human needs such as the availability and quality of air, water, food and shelter are also the ecological foundations for sustainable development;[9] addressing public health risk through investments in ecosystem services can be a powerful and transformative force for sustainable development which, in this sense, extends to all species.[10]

Environment

See also: Environmental engineering and Environmental technology

Environmental sustainability concerns the natural environment and how it endures and remains diverse and productive. Since natural resources are derived from the environment, the state of air, water, and the climate are of particular concern. The IPCC Fifth Assessment Report outlines current knowledge about scientific, technical and socioeconomic information concerning climate change, and lists options for adaptation and mitigation.[12] Environmental sustainability requires society to design activities to meet human needs while preserving the life support systems of the planet. This, for example, entails using water sustainably, utilizing renewable energy, and sustainable material supplies (e.g. harvesting wood from forests at a rate that maintains the biomass and biodiversity).

An unsustainable situation occurs when natural capital (the sum total of nature's resources) is used up faster than it can be replenished. Sustainability requires that human activity only uses nature's resources at a rate at which they can be

replenished naturally. Inherently the concept of sustainable development is intertwined with the concept of carrying capacity. Theoretically, the long-term result of environmental degradation is the inability to sustain human life. Such degradation on a global scale should imply an increase in human death rate until population falls to what the degraded environment can support. If the degradation continues beyond a certain tipping point or critical threshold it would lead to eventual extinction for humanity.

Integral elements for a sustainable development are research and innovation activities. A telling example is the European environmental research and innovation policy, which aims at defining and implementing a transformative agenda to greening the economy and the society as a whole so to achieve a truly sustainable development. Research and innovation in Europe is financially supported by the programme Horizon 2020, which is also open to participation worldwide.[13] An promising direction towards sustainable development is to design systems that are flexible and reversible [14][15]

Agriculture

See also: Sustainable agriculture

Sustainable agriculture consists of environmentally-friendly methods of farming that allow the production of crops or livestock without damage to human or natural systems. It involves preventing adverse effects to soil, water, biodiversity, surrounding or downstream resources—as well as to those working or living on the farm or in neighboring areas. The concept of sustainable agriculture extends intergenerationally, passing on a conserved or improved natural resource, biotic, and economic base rather than one which has been depleted or polluted.[16] Elements of sustainable agriculture include permaculture, agroforestry, mixed farming, multiple cropping, and crop rotation.[17]

Numerous sustainability standards and certification systems have been established in recent years, offering consumer choices for sustainable agriculture practices. These include Organic certification, Rainforest Alliance, Fair Trade, UTZ Certified, Bird Friendly, and the Common Code for the Coffee Community (4C).[18][19]

Energy

Main articles: Smart grid and Sustainable energy

Sustainable energy is the sustainable provision of energy that is clean and lasts for a long period of time. Unlike the fossil fuel that most of the countries are using, renewable energy only produces little or even no pollution.[20] The most common types of renewable energy in US are solar and wind energy, solar energy are commonly used on public parking meter, street lights and the roof of buildings.[21] Wind energy has expanded quickly, generating 12,000 MW in 2013. The largest wind power station is in Texas and California.[22][23] Household energy consumption can also be improved in a sustainable way, like using electronics with Energy Star logos which conserve water and energy. Most of California's fossil fuel infrastructures are sited in or near low-income communities, and have traditionally suffered the most from California's fossil fuel energy system. These communities are historically left out during the decision-making process, and often end up with dirty power plants and other dirty energy projects that poison the air and harm the area. These toxicants are major contributors to health problems in the communities. As renewable energy becomes more common, fossil fuel infrastructures is replaced by renewables, providing better social equity to these community.[24]

Transportation

See also: Sustainable transport

Transportation is a large contributor to greenhouse gas emissions. It is said that one-third of all gasses produced are due to transportation.[25] Some western countries are making transportation more sustainable in both long-term and short-term implementations.[26] An example is the modifications in available transportation in Freiburg, Germany.

The city has implemented extensive methods of public transportation, cycling, and walking, along with large areas where cars are not allowed.[25]

Since many western countries are highly automobile-orientated areas, the main transit that people use is personal vehicles. About 80% of their travel involves cars.[25] Therefore, California, deep in the automobile-oriented west, is one of the highest greenhouse gases emitters in the country. The federal government has to come up with some plans to reduce the total number of vehicle trips in order to lower greenhouse gases emission. Such as:

- Improve public transit through the provision of larger coverage area in order to provide more mobility and accessibility, new technology to provide a more reliable and responsive public transportation network.[27]

- Encourage walking and biking through the provision of wider pedestrian pathway, bike share station in commercial downtown, locate parking lot far from the shopping center, limit on street parking, slower traffic lane in downtown area.

- Increase the cost of car ownership and gas taxes through increased parking fees and tolls, encouraging people to drive more fuel efficient vehicles. They can produce social equity problem, since lower people usually drive older vehicles with lower fuel efficiency. Government can use the extra revenue collected from taxes and tolls to improve the public transportation and benefit the poor community.[28]

Other states and nations have built efforts to translate knowledge in behavioral economics into evidence-based sustainable transportation policies.

45.2.2 Economics

See also: Ecological economics

It has been suggested that because of rural poverty and overexploitation, environmental resources should be treated as important economic assets, called natural capital.[29] Economic development has traditionally required a growth in the gross domestic product. This model of unlimited personal and GDP growth may be over.[30] Sustainable development may involve improvements in the quality of life for many but may necessitate a decrease in resource consumption.[31] According to ecological economist Malte Faber, ecological economics is defined by its focus on nature, justice, and time. Issues of intergenerational equity, irreversibility of environmental change, uncertainty

A sewage treatment plant that uses solar energy, located at Santuari de Lluc monastery, Majorca.

of long-term outcomes, and sustainable development guide ecological economic analysis and valuation.[32]

As early as the 1970s, the concept of sustainability was used to describe an economy "in equilibrium with basic ecological support systems."[33] Scientists in many fields have highlighted *The Limits to Growth*,[34][35] and economists have presented alternatives, for example a 'steady state economy';[36] to address concerns over the impacts of expanding human development on the planet. In 1987 the economist Edward Barbier published the study *The Concept of Sustainable Economic Development*, where he recognized that goals of environmental conservation and economic development are not conflicting and can be reinforcing each other.[37]

A World Bank study from 1999 concluded that based on the theory of genuine savings, policymakers have many possible interventions to increase sustainability, in macroeconomics or purely environmental.[38] A study from 2001 noted that efficient policies for renewable energy and pollution are compatible with increasing human welfare, eventually reaching a golden-rule steady state.[39] The study, *Interpreting Sustainability in Economic Terms*, found three pillars of sustainable development, interlinkage, intergenerational equity, and dynamic efficiency.[40]

A meta review in 2002 looked at environmental and economic valuations and found a lack of "sustainability policies".[41] A study in 2004 asked if we consume too much.[42] A study concluded in 2007 that knowledge, manufactured and human capital(health and education) has not compensated for the degradation of natural capital in many parts of the world.[43] It has been suggested that intergenerational equity can be incorporated into a sustainable development and decision making, as has become common in economic valuations of climate economics.[44] A meta review in 2009 identified conditions for a strong case to act

on climate change, and called for more work to fully account of the relevant economics and how it affects human welfare.[45] According to John Baden[46] "the improvement of environment quality depends on the market economy and the existence of legitimate and protected property rights." They enable the effective practice of personal responsibility and the development of mechanisms to protect the environment. The State can in this context "create conditions which encourage the people to save the environment."[47]

Business

See also: Corporate sustainability

The most broadly accepted criterion for corporate sustainability constitutes a firm's efficient use of natural capital. This eco-efficiency is usually calculated as the economic value added by a firm in relation to its aggregated ecological impact.[48] This idea has been popularised by the World Business Council for Sustainable Development (WBCSD) under the following definition: "Eco-efficiency is achieved by the delivery of competitively priced goods and services that satisfy human needs and bring quality of life, while progressively reducing ecological impacts and resource intensity throughout the life-cycle to a level at least in line with the earth's carrying capacity." (DeSimone and Popoff, 1997: 47)[49]

Similar to the eco-efficiency concept but so far less explored is the second criterion for corporate sustainability. Socio-efficiency[50] describes the relation between a firm's value added and its social impact. Whereas, it can be assumed that most corporate impacts on the environment are negative (apart from rare exceptions such as the planting of trees) this is not true for social impacts. These can be either positive (e.g. corporate giving, creation of employment) or negative (e.g. work accidents, mobbing of employees, human rights abuses). Depending on the type of impact socio-efficiency thus either tries to minimize negative social impacts (i.e. accidents per value added) or maximise positive social impacts (i.e. donations per value added) in relation to the value added.

Both eco-efficiency and socio-efficiency are concerned primarily with increasing economic sustainability. In this process they instrumentalize both natural and social capital aiming to benefit from win-win situations. However, as Dyllick and Hockerts[50] point out the business case alone will not be sufficient to realise sustainable development. They point towards eco-effectiveness, socio-effectiveness, sufficiency, and eco-equity as four criteria that need to be met if sustainable development is to be reached.

Income

At the present time, sustainable development as well as solidarity or Catholic social teaching can impact reduce the poverty. Because over many thousands of years the 'stronger' (economically or physically) used to defeat/eliminate the weaker, nowadays, no matter what we call the reason for this decision – within Catholic social teaching, social solidarity, and sustainable development – the stronger helps the weaker. This aid may take the form of in-kind or material, refer to the present or the future. 'The Stronger', should offer real help and not, as demonstrated by the frequent experience – strive for the elimination or annihilation of another entity. Sustainable development reduce poverty through economic (among other things, a balanced budget), environmental (living conditions) and also social (including equality of income) dimensions.[51]

Architecture

See also: Sustainable architecture

In sustainable architecture the recent movements of New Urbanism and New Classical architecture promote a sustainable approach towards construction, that appreciates and develops smart growth, architectural tradition and classical design.[52][53] This in contrast to modernist and International Style architecture, as well as opposing to solitary housing estates and suburban sprawl, with long commuting distances and large ecological footprints.[54] Both trends started in the 1980s. (It should be noted that sustainable architecture is predominantly relevant to the economics domain while architectural landscaping pertains more to the ecological domain.)

45.2.3 Politics

See also: Environmental politics, Environmental governance and Sustainability metrics and indices

A study concluded that social indicators and, therefore, sustainable development indicators, are scientific constructs whose principal objective is to inform public policy-making.[55] The International Institute for Sustainable Development has similarly developed a political policy framework, linked to a sustainability index for establishing measurable entities and metrics. The framework consists of six core areas, international trade and investment, economic policy, climate change and energy, measurement and assessment, natural resource management, and the role of communication technologies in sustainable development.

The United Nations Global Compact Cities Programme has defined sustainable political development is a way that broadens the usual definition beyond states and governance. The political is defined as the domain of practices and meanings associated with basic issues of social power as they pertain to the organisation, authorisation, legitimation and regulation of a social life held in common. This definition is in accord with the view that political change is important for responding to economic, ecological and cultural challenges. It also means that the politics of economic change can be addressed. They have listed seven subdomains of the domain of politics:[56]

1. Organization and governance

2. Law and justice

3. Communication and critique

4. Representation and negotiation

5. Security and accord

6. Dialogue and reconciliation

7. Ethics and accountability

This accords with the Brundtland Commission emphasis on development that is guided by human rights principles (see above).

45.2.4 Culture

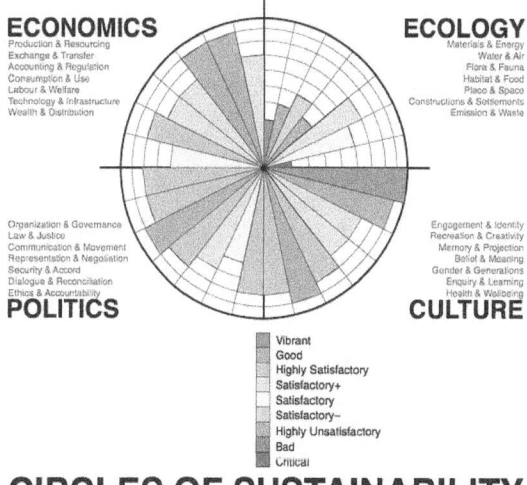

Framing of sustainable development progress according to the Circles of Sustainability, used by the United Nations.

Working with a different emphasis, some researchers and institutions have pointed out that a fourth dimension should be added to the dimensions of sustainable development, since the triple-bottom-line dimensions of economic, environmental and social do not seem to be enough to reflect the complexity of contemporary society. In this context, the Agenda 21 for culture and the United Cities and Local Governments (UCLG) Executive Bureau lead the preparation of the policy statement "Culture: Fourth Pillar of Sustainable Development", passed on 17 November 2010, in the framework of the World Summit of Local and Regional Leaders – 3rd World Congress of UCLG, held in Mexico City. although some which still argue that economics is primary, and culture and politics should be included in 'the social'. This document inaugurates a new perspective and points to the relation between culture and sustainable development through a dual approach: developing a solid cultural policy and advocating a cultural dimension in all public policies. The Circles of Sustainability approach distinguishes the four domains of economic, ecological, political and cultural sustainability.[57][58]

Other organizations have also supported the idea of a fourth domain of sustainable development. The Network of Excellence "Sustainable Development in a Diverse World",[59] sponsored by the European Union, integrates multidisciplinary capacities and interprets cultural diversity as a key element of a new strategy for sustainable development. The Fourth Pillar of Sustainable Development Theory has been referenced by executive director of IMI Institute at UNESCO Vito Di Bari[60] in his manifesto of art and architectural movement Neo-Futurism, whose name was inspired by the 1987 United Nations' report Our Common Future. The Circles of Sustainability approach used by Metropolis defines the (fourth) cultural domain as practices, discourses, and material expressions, which, over time, express continuities and discontinuities of social meaning.[56]

45.3 Themes

45.3.1 Progress

See also: Sustainable development goals

The United Nations Conference on Sustainable Development (UNCSD), also known as Rio 2012, Rio+20, or Earth Summit 2012, was the third international conference on sustainable development, which aimed at reconciling the economic and environmental goals of the global community. An outcome of this conference was the development of the Sustainable Development Goals that aim to promote sustainable progress and eliminate inequalities around the world. However, few nations met the World Wide Fund for Nature's definition of sustainable development criteria es-

tablished in 2006.[61] Although some nations are more developed than others, all nations are constantly developing because each nation struggles with perpetuating disparities, inequalities and unequal access to fundamental rights and freedoms. [62]

45.3.2 Measurement

Main articles: Ecological footprint and Sustainability measurement

In 2007 a report for the U.S. Environmental Protection

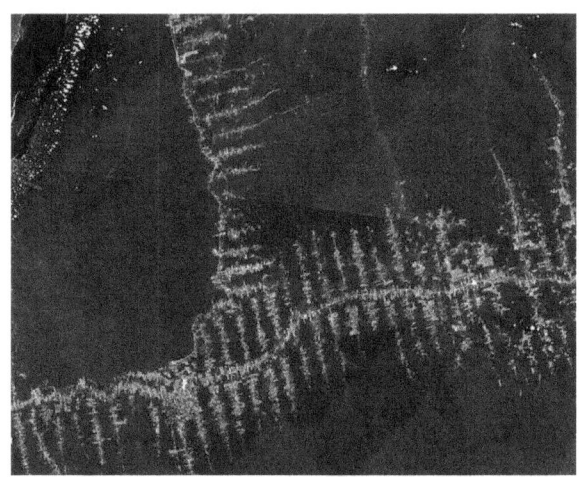

Deforestation and increased road-building in the Amazon Rainforest are a concern because of increased human encroachment upon wilderness areas, increased resource extraction and further threats to biodiversity.

Agency stated: "While much discussion and effort has gone into sustainability indicators, none of the resulting systems clearly tells us whether our society is sustainable. At best, they can tell us that we are heading in the wrong direction, or that our current activities are not sustainable. More often, they simply draw our attention to the existence of problems, doing little to tell us the origin of those problems and nothing to tell us how to solve them."[63] Nevertheless, a majority of authors assume that a set of well defined and harmonised indicators is the only way to make sustainability tangible. Those indicators are expected to be identified and adjusted through empirical observations (trial and error).[64]

The most common critiques are related to issues like data quality, comparability, objective function and the necessary resources.[65] However a more general criticism is coming from the project management community: How can a sustainable development be achieved at global level if we cannot monitor it in any single project?[66][67]

The Cuban-born researcher and entrepreneur Sonia Bueno suggests an alternative approach that is based upon the integral, long-term cost-benefit relationship as a measure and monitoring tool for the sustainability of every project, activity or enterprise.[68][69] Furthermore, this concept aims to be a practical guideline towards sustainable development following the principle of conservation and increment of value rather than restricting the consumption of resources.

Reasonable qualifications of sustainability are seen U.S. Green Building Council's (USGBC) Leadership in Energy and Environmental Design (LEED). This design incorporates some ecological, economic, and social elements. The goals presented by LEED design goals are sustainable sites, water efficiency, energy and atmospheric emission reduction, material and resources efficiency, and indoor environmental quality. Although amount of structures for sustainability development is many, these qualification has become a standard for sustainable building.

45.3.3 Natural capital

Deforestation of native rain forest in Rio de Janeiro City for extraction of clay for civil engineering (2009 picture).

The sustainable development debate is based on the assumption that societies need to manage three types of capital (economic, social, and natural), which may be non-substitutable and whose consumption might be irreversible.[70] Daly (1991),[36] for example, points to the fact that natural capital can not necessarily be substituted by economic capital. While it is possible that we can find ways to replace some natural resources, it is much more unlikely that they will ever be able to replace eco-system services, such as the protection provided by the ozone layer, or the climate stabilizing function of the Amazonian forest. In fact natural capital, social capital and economic capital are often complementarities. A further obstacle to substitutability lies also in the multi-functionality of many natural resources. Forests, for example, not only provide the raw material for paper (which can be substituted quite easily), but they also maintain biodiversity, regulate water flow, and absorb CO2.

Another problem of natural and social capital deterioration lies in their partial irreversibility. The loss in biodiversity, for example, is often definite. The same can be true for cultural diversity. For example, with globalisation advancing quickly the number of indigenous languages is dropping at alarming rates. Moreover, the depletion of natural and social capital may have non-linear consequences. Consumption of natural and social capital may have no observable impact until a certain threshold is reached. A lake can, for example, absorb nutrients for a long time while actually increasing its productivity. However, once a certain level of algae is reached lack of oxygen causes the lake's ecosystem to break down suddenly.

45.3.4 Business-as-usual

Before flue-gas desulfurization was installed, the air-polluting emissions from this power plant in New Mexico contained excessive amounts of sulfur dioxide.

If the degradation of natural and social capital has such important consequence the question arises why action is not taken more systematically to alleviate it. Cohen and Winn[71] point to four types of market failure as possible explanations: First, while the benefits of natural or social capital depletion can usually be privatized, the costs are often externalized (i.e. they are borne not by the party responsible but by society in general). Second, natural capital is often undervalued by society since we are not fully aware of the real cost of the depletion of natural capital. Information asymmetry is a third reason—often the link between cause and effect is obscured, making it difficult for actors to make informed choices. Cohen and Winn close with the realization that contrary to economic theory many firms are not perfect optimizers. They postulate that firms often do not optimize resource allocation because they are caught in a "business as usual" mentality.

45.4 Historical Development

Main articles: Sustainability and History of sustainability

Sustainable development has its roots in ideas about sustainable forest management which were developed in Europe during the seventeenth and eighteenth centuries.[72][73] In response to a growing awareness of the depletion of timber resources in England, John Evelyn argued that "sowing and planting of trees had to be regarded as a national duty of every landowner, in order to stop the destructive over-exploitation of natural resources" in his 1662 essay *Sylva*. In 1713 Hans Carl von Carlowitz, a senior mining administrator in the service of Elector Frederick Augustus I of Saxony published *Sylvicultura oeconomica*, a 400-page work on forestry. Building upon the ideas of Evelyn and French minister Jean-Baptiste Colbert, von Carlowitz developed the concept of managing forests for sustained yield.[72] His work influenced others, including Alexander von Humboldt and Georg Ludwig Hartig, leading in turn to the development of a science of forestry. This in term influenced people like Gifford Pinchot, first head of the US Forest Service, whose approach to forest management was driven by the idea of wise use of resources, and Aldo Leopold whose land ethic was influential in the development of the environmental movement in the 1960s.[72][73]

Following the publication of Rachel Carson's *Silent Spring* in 1962, the developing environmental movement drew attention to the relationship between economic growth and development and environmental degradation. Kenneth E. Boulding in his influential 1966 essay *The Economics of the Coming Spaceship Earth* identified the need for the economic system to fit itself to the ecological system with its limited pools of resources.[73] One of the first uses of the term sustainable in the contemporary sense was by the Club of Rome in 1972 in its classic report on the *Limits to Growth*, written by a group of scientists led by Dennis and Donella Meadows of the Massachusetts Institute of Technology. Describing the desirable "state of global equilibrium", the authors wrote: "We are searching for a model output that represents a world system that is sustainable without sudden and uncontrolled collapse and capable of satisfying the basic material requirements of all of its people."[4]

In 1980 the International Union for the Conservation of Nature published a world conservation strategy that included one of the first references to sustainable development as a global priority.[74] Two years later, the United Nations World Charter for Nature raised five principles of conservation by which human conduct affecting nature is to be guided and judged.[75] In 1987 the United Nations World Commission on Environment and Development re-

leased the report *Our Common Future*, commonly called the Brundtland Report. The report included what is now one of the most widely recognized definitions of sustainable development.[76][77]

Sustainable development is development that meets the needs of the present without compromising the ability of future generations to meet their own needs. It contains within it two key concepts:

- The concept of 'needs', in particular, the essential needs of the world's poor, to which overriding priority should be given; and

- The idea of limitations imposed by the state of technology and social organization on the environment's ability to meet present and future needs.

— World Commission on Environment and Development, *Our Common Future* (1987)

In 1992, the UN Conference on Environment and Development published in 1992 the Earth Charter, which outlines the building of a just, sustainable, and peaceful global society in the 21st century. The action plan Agenda 21 for sustainable development identified information, integration, and participation as key building blocks to help countries achieve development that recognizes these interdependent pillars. It emphasises that in sustainable development everyone is a user and provider of information. It stresses the need to change from old sector-centered ways of doing business to new approaches that involve cross-sectoral co-ordination and the integration of environmental and social concerns into all development processes. Furthermore, Agenda 21 emphasises that broad public participation in decision making is a fundamental prerequisite for achieving sustainable development.[78] Under the principles of the United Nations Charter the Millennium Declaration identified principles and treaties on sustainable development, including economic development, social development and environmental protection. Broadly defined, sustainable development is a systems approach to growth and development and to manage natural, produced, and social capital for the welfare of their own and future generations. The term sustainable development as used by the United Nations incorporates both issues associated with land development and broader issues of human development such as education, public health, and standard of living.

A 2013 study concluded that sustainability reporting should be reframed through the lens of four interconnected domains: ecology, economics, politics and culture.[79]

45.5 See also

- Applied sustainability
- Conservation biology
- Conservation development
- Ecological modernization
- Ecologically sustainable development
- Environmental issue
- Environmental justice
- Green development
- Micro-sustainability
- Outline of sustainability
- Social sustainability
- Sustainable coffee
- Sustainable fishery
- Sustainable forest management
- Sustainable land management
- Sustainable living
- Sustainable yield
- Sustainopreneurship
- Weak and strong sustainability
- Zero-carbon city

45.6 References

[1] Aldo Leopold, *A Sand County Almanac*, 1949.

[2] Melvin K. Hendrix, *Sustainable Backyard Polyculture: Designing for ecological resiliency* Smashwords ebook edition. 2014.

[3] Lynn R. Kahle, Eda Gurel-Atay, Eds (2014). *Communicating Sustainability for the Green Economy*. New York: M.E. Sharpe. ISBN 978-0-7656-3680-5.

[4] Finn (2009), pp. 3–8

[5] United Nations (2014). *Prototype Global Sustainable Development Report* (Online unedited ed.). New York: United Nations Department of Economic and Social Affairs, Division for Sustainable Development.

[6] James, Paul; with Magee, Liam; Scerri, Andy; Steger, Manfred B. (2015). *Urban Sustainability in Theory and Practice: Circles of Sustainability*. London: Routledge.

[7] Circles of Sustainability Urban Profile Process and Scerri, Andy; James, Paul (2010). "Accounting for sustainability: Combining qualitative and quantitative research in developing 'indicators' of sustainability". *International Journal of Social Research Methodology* **13** (1): 41–53. doi:10.1080/13645570902864145.

[8] http://citiesprogramme.com/aboutus/our-approach/circles-of-sustainability; Scerri, Andy; James, Paul (2010). "Accounting for sustainability: Combining qualitative and quantitative research in developing 'indicators' of sustainability". *International Journal of Social Research Methodology* **13** (1): 41–53. doi:10.1080/13645570902864145..

[9] White, F; Stallones, L; Last, JM. (2013). *Global Public Health: Ecological Foundations*. Oxford University Press. ISBN 978-0-19-975190-7.

[10] Bringing human health and wellbeing back into sustainable development. In: IISD Annual Report 2011-12. http://www.iisd.org/pdf/2012/annrep_2011_2012_en.pdf

[11] Blewitt (2015), p. 7

[12] IPCC Fifth Assessment Report (2014). "Climate Change 2014: Impacts, Adaptation and Vulnerability" (PDF). Geneva (Switzerland): IPCC.

[13] See Horizon 2020 – the EU's new research and innovation programme http://europa.eu/rapid/press-release_MEMO-13-1085_en.htm

[14] . 2012 http://www.tandfonline.com/doi/abs/10.1080/09613218.2012.702565. Missing or empty |title= (help)

[15] Zhang, S.X.; V. Babovic (2012). "A real options approach to the design and architecture of water supply systems using innovative water technologies under uncertainty" (PDF). *Journal of Hydroinformatics*.

[16] Networld-Project (1998-02-09). "Environmental Glossary". Green-networld.com. Retrieved 2011-09-28.

[17] Ben Falk, *The resilient farm and homestead: An innovative permaculture and whole systems design approach*. Chelsea Green, 2013. pp. 61-78.

[18] Manning, Stephen; Boons, Frank; Von Hagen, Oliver; Reinecke, Juliane (2012). "National Contexts Matter: The Co-Evolution of Sustainability Standards in Global Value Chains". *Ecological Economics* **83**: 197–209.

[19] Reinecke, Juliane; Manning, Stephen; Von Hagen, Oliver (2012). "The Emergence of a Standards Market: Multiplicity of Sustainability Standards in the Global Coffee Industry". *Organization Studies* **33** (5/6): 789–812.

[20] Fainstein, Susan S. 2000. "New Directions in Planning Theory," Urban Affairs Review 35:4 (March)

[21] Bedsworf, Louise W. and Ellen Hanak. 2010. "Adaptation to Climate Change, "Journal of the American Planning Association, 76:4.

[22] Barbour, Elissa and Elizabeth A. Deakin. 2012. "Smart Growth Planning for Climate Protection" 78:1

[23] America Wind Power Association. US Gov. Web. http://www.awea.org/

[24] Campbell, Scott. 1996. "Green Cities, Growing Cities, Just Cities?: Urban planning and the Contradictions of Sustainable Development," Journal of the American Planning Association

[25] Buehler, Ralph; Pucher, John (2011). "Sustainable Transport in Freiburg: Lessons from Germany's Environmental Capital". *International Journal of Sustainable Transportation*. doi:10.1080/15568311003650531.

[26] Barbour, Elissa and Elizabeth A. Deakin. 2012. "Smart Growth Planning for Climate Protection"

[27] [Murthy, A.S. Narasimha Mohle, Henry. Transportation Engineering Basics (2nd Edition). (American Society of Cilil Engineers 2001). At <http://site.ebrary.com/lib/alltitles/docDetail.action?docID=10447877&p00=transportation%20improvement>]

[28] Levine, Jonathan. 2013. "Urban Transportation and Social Equity: Transportation Planning Paradigms that Impede Policy Reform," in Naomi Carmon and Susan S. Fainstein, eds. Policy, Planning and people: promoting Justice in Urban Development (Penn)

[29] Barbier, Edward B. (2006). *Natural Resources and Economic Development*. //books.google.com/books?id=fYrEDA-VnyUC&pg=PA45: Cambridge University Press. pp. 44–45. ISBN 9780521706513. Retrieved April 8, 2014.

[30] Korowitz, David (2012), *Ignorance by Consensus*, Foundation for the Economics of Sustainability

[31] Brown, L. R. (2011). *World on the Edge*. Earth Policy Institute. Norton. ISBN 978-0-393-08029-2.

[32] Malte Faber. (2008). How to be an ecological economist. *Ecological Economics* **66**(1):1-7. Preprint.

[33] Stivers, R. 1976. The Sustainable Society: Ethics and Economic Growth. Philadelphia: Westminster Press.

[34] Meadows, D.H., D.L. Meadows, J. Randers, and W.W. Behrens III. 1972. The Limits to Growth. Universe Books, New York, NY. ISBN 0-87663-165-0

[35] Meadows, D.H.; Randers, Jørgen; Meadows, D.L. (2004). *Limits to Growth: The 30-Year Update*. Chelsea Green Publishing. ISBN 978-1-931498-58-6.

[36] Daly, H. E. 1973. Towards a Steady State Economy. San Francisco: Freeman. Daly, H. E. 1991. Steady-State Economics (2nd ed.). Washington, D.C.: Island Press.

[37] Barbier, E. (1987). "The Concept of Sustainable Economic Development". *Environmental Conservation* **14** (2): 101–110. doi:10.1017/S0376892900011449.

[38] Hamilton, K.; Clemens, M. (1999). "Genuine savings rates in developing countries". *World Bank Economy Review* **13** (2): 333–356. doi:10.1093/wber/13.2.333.

[39] Ayong Le Kama, A. D. (2001). "Sustainable growth renewable resources, and pollution". *Journal of Economic Dynamics and Control* **25** (12): 1911–1918. doi:10.1016/S0165-1889(00)00007-5.

[40] Stavins, R.; Wagner, A.; Wagner, G. "Interpreting Sustainability in Economic Terms: Dynamic Efficiency Plus Intergenerational Equity". *Economic Letters* **79** (3): 339–343. doi:10.1016/S0165-1765(03)00036-3.

[41] Pezzey, John C. V.; Michael A., Toman (2002). "The Economics of Sustainability: A Review of Journal Articles" (PDF). Resources for the future. Retrieved April 8, 2014.

[42] Arrow, K. J.; Dasgupta, P.; Goulder, L.; Daily, G.; Ehrlich, P. R.; Heal, G. M.; Levin, S.; Maler, K-G.; Schneider, S.; Starrett, D. A.; Walker, B. (2004). "Are we consuming too much?". *Journal of Economic Perspectives* **18** (3): 147–172. doi:10.1257/0895330042162377. JSTOR 3216811.

[43] Dasgupta, P. (2007). "The idea of sustainable development". *Sustainability Science* **2** (1): 5–11. doi:10.1007/s11625-007-0024-y.

[44] Heal, G. (2009). "Climate Economics: A Meta-Review and Some Suggestions for Future Research". *Review of Environmental Economics and Policy* **3** (1): 4–21. doi:10.1093/reep/ren014.

[45] Heal, Geoffrey (2009). "Climate Economics: A Meta-Review and Some Suggestions for Future Research". *Review of Environmental Economics and Policy* (Oxford Journals). doi:10.1093/reep/ren014. Retrieved April 8, 2104. Check date values in: |access-date= (help)

[46] chairman of the Foundation for Research on Economics and the Environment (FREE)

[47] « L'économie politique du développement durable », John Baden, document de l'ICREI

[48] Schaltegger, S. & Sturm, A. 1998. Eco-Efficiency by Eco-Controlling. Zürich: vdf.

[49] DeSimone, L. & Popoff, F. 1997. Eco-efficiency: The business link to sustainable development. Cambridge: MIT Press.

[50] Dyllick, T. & Hockerts, K. 2002. Beyond the business case for corporate sustainability. Business Strategy and the Environment, 11(2): 130–141.

[51] S. Adamiak, D. Walczak, Catholic social teaching, sustainable development and social solidarism in the context of social security, Copernican Journal of Finance & Accounting, Vol 3, No 1, p. 12,17.

[52] Charter of the New Urbanism

[53] "Beauty, Humanism, Continuity between Past and Future". Traditional Architecture Group. Retrieved 23 March 2014.

[54] Issue Brief: Smart-Growth: Building Livable Communities. American Institute of Architects. Retrieved on 2014-03-23.

[55] Paul-Marie Boulanger (2008). "Sustainable development indicators: a scientific challenge, a democratic issue. "S.A.P.I.EN.S." "'1'" (1)". Sapiens.revues.org. Retrieved 2011-09-28.

[56] http://citiesprogramme.com/archives/resource/ circles-of-sustainability-urban-profile-process Liam Magee, Andy Scerri, Paul James, James A. Thom, Lin Padgham, Sarah Hickmott, Hepu Deng, Felicity Cahill (2013). "Reframing social sustainability reporting: Towards an engaged approach". *Environment, Development and Sustainability* (Springer). doi:10.1007/s10668-012-9384-2.

[57] United Cites and Local Governments, "Culture: Fourth Pillar of Sustainable Development".

[58] http://www.uclg.org/en/node/21824

[59] "Sus.Div". Sus.Div. Retrieved 2011-09-28.

[60] "Agreement between UNESCO and the City of Milan concerning the International Multimedia Institute (IMI) - Appointment of Executive Director — UNESCO Archives ICA AtoM catalogue". Atom.archives.unesco.org. 1999-10-08. Retrieved 2014-01-17.

[61] "Living Planet Report 2006" (PDF). World Wide Fund for Nature, Zoological Society of London, Global Footprint Network. 24 October 2006. p. 19. Retrieved 18 August 2012.; World failing on sustainable development

[62] Nussbaum, Martha (2011). *Creating Capabilities: The Human Development Approach.* Cambridge, Massachusetts and London, England: The Belknap Press of Harvard University Press. p. 16. ISBN 978-0-674-05054-9.

[63] Joy E. Hecht, Can Indicators and Accounts Really Measure Sustainability? Considerations for the U.S. Environmental Protection

[64] KM.FAO.org "An adaptive learning process for developing and applying sustainability indicators with local communities". *Ecological economics* 59 (2006) 406-418

[65] "Annette Lang, Ist Nachhaltigkeit messbar?, Uni Hannover, 2003" (PDF). Retrieved 2011-09-28.

[66] "Project Management T-kit, Council of Europe and European Commission, Strasbourg, 2000" (PDF). Retrieved 2011-09-28.

[67] "Do global targets matter?, The Environment Times, Poverty Times #4, UNEP/GRID-Arendal, 2010". Grida.no. Retrieved 2011-09-28.

[68] "Sostenibilidad en la construcción. Calidad integral y rentabilidad en instalaciones hidro-sanitarias, Revista de Arquitectura e Ingeniería, Matanzas, 2009". Empai-matanzas.co.cu. 2009-01-17. Retrieved 2011-09-28.

[69] "Transforming the water and waste water infrastructure into an efficient, profitable and sustainable system, Revista de Arquitectura e Ingeniería, Matanzas, 2010" (PDF). Retrieved 2014-05-14.

[70] Dyllick, T. & Hockerts, K. 2002. Beyond the business case for corporate sustainability. Business Strategy and the Environment, 11(2): 130–141

[71] Cohen, B. & Winn, M. I. 2007. Market imperfections, opportunity and sustainable entrepreneurship. Journal of Business Venturing, 22(1): 29–49.

[72] Ulrich Grober: Deep roots — A conceptual history of "sustainable development" (Nachhaltigkeit), Wissenschaftszentrum Berlin für Sozialforschung, 2007

[73] Blewitt (2015), pp. 6–16

[74] *World Conservation Strategy: Living Resource Conservation for Sustainable Development* (PDF). International Union for Conservation of Nature and Natural Resources. 1980.

[75] *World Charter for Nature*, United Nations, General Assembly, 48th Plenary Meeting, October 28, 1982

[76] Brundtland Commission (1987). "Report of the World Commission on Environment and Development". United Nations.

[77] Smith, Charles; Rees, Gareth (1998). *Economic Development, 2nd edition.* Basingstoke: Macmillan. ISBN 0-333-72228-0.

[78] Will Allen. 2007."Learning for Sustainability: Sustainable Development."

[79] Liam Magee, Andy Scerri, Paul James, James A. Thom, Lin Padgham, Sarah Hickmott, Hepu Deng, Felicity Cahill (2013). "Reframing social sustainability reporting: Towards an engaged approach". *Environment, Development and Sustainability* (University of Melbourne). doi:10.1007/s10668-012-9384-2.

45.7 Literature cited

• Blewitt, John (2015). *Understanding Sustainable Development* (Second ed.). Routledge.

• Finn, Donovan (2009). *Our Uncertain Future: Can Good Planning Create Sustainable Communities?.* Ph.D. dissertation. University of Illinois at Urbana-Champaign.

45.8   Further reading

- Ahmed, Faiz (2008). *An Examination of the Development Path Taken by Small Island Developing States* (PDF). (pp. 17–26)

- Atkinson, G., S. Dietz, and E. Neumayer (2009). *Handbook of Sustainable Development*. Edward Elgar Publishing, ISBN 1848444729.

- Bakari, Mohamed El-Kamel. "Globalization and Sustainable Development: False Twins?." New Global Studies 7.3: 23-56. ISSN (Online) 1940-0004, ISSN (Print) 2194-6566, DOI: 10.1515/ngs-2013-021, November 2013.

- Bertelsmann Stiftung, ed. (2013). Winning Strategies for a Sustainable Future. Reinhard Mohn Prize 2013. Verlag Bertelsmann Stiftung, Gütersloh. ISBN 978-3-86793-491-6.

- Beyerlin, Ulrich. Sustainable Development, *Max Planck Encyclopedia of Public International Law*

- Cook, Sarah and Esuna Dugarova (2014). "Rethinking Social Development for a Post-2015 World". *Development* **57** (1): 30–35. doi:10.1057/dev.2014.25.

- Danilov-Danil'yan, Victor I., Losev, K.S., Reyf, Igor E. *Sustainable Development and the Limitation of Growth: Future Prospects for World Civilization*. Transl. Vladimir Tumanov. Ed. Donald Rapp. New York: Springer Praxis Books, 2009. at Google Books

- Edwards, A.R., and B. McKibben (2010). *Thriving Beyond Sustainability: Pathways to a Resilient Society*. New Society Publishers, ISBN 0865716412.

- Huesemann, M.H., and J.A. Huesemann (2011). *Technofix: Why Technology Won't Save Us or the Environment*, Chapter 6, "Sustainability or Collapse?", and Chapter 13, "The Design of Environmentally Sustainable and Socially Appropriate Technologies", New Society Publishers, ISBN 0865717044.

- James, Paul; Nadarajah, Yaso; Haive, Karen; Stead, Victoria (2012). *Sustainable Communities, Sustainable Development: Other Paths for Papua New Guinea*. Honolulu: University of Hawaii Press.

- James, Paul; with Magee, Liam; Scerri, Andy; Steger, Manfred B. (2015). *Urban Sustainability in Theory and Practice: Circles of Sustainability*. London: Routledge.

- Jarzombek, Mark, "Sustainability — Architecture: between Fuzzy Systems and Wicked Problems," Blueprints 21/1 (Winter 2003), pp. 6–9.

- Li, Rita Yi Man. , *Building Our Sustainable Cities"* Illinois, Published by Common Ground Publishing.

- Raudsepp-Hearne, C; Peterson, GD; Tengö, M; Bennett, EM; Holland, T; Benessaiah, K; MacDonald, GM; Pfeifer, L (2010). "Untangling the Environmentalist's Paradox: Why is Human Well-Being Increasing as Ecosystem Services Degrade?". *BioScience* **60** (8): 576–589. doi:10.1525/bio.2010.60.8.4.

- Rogers, P., K.F. Jalal, and J.A. Boyd (2007). *An Introduction to Sustainable Development*. Routledge, ISBN 1844075214.

- Sianipar, C. P. M., Dowaki, K., Yudoko, G., & Adhiutama, A. (2013). Seven Pillars of Survivability: Appropriate Technology with a Human Face. European Journal of Sustainable Development, 2(4), 1-18. ISSN 2239-5938.

- Van der Straaten, J., and J.C van den Bergh (1994). *Towards Sustainable Development: Concepts, Methods, and Policy*. Island Press, ISBN 1559633492.

- Wallace, Bill (2005). *Becoming part of the solution : the engineer's guide to sustainable development*. Washington, DC: American Council of Engineering Companies. ISBN 0-910090-37-8.

45.9   External links

- Global Sustainable Development, an undergraduate degree program offered by the University of Warwick.

- Circles of Sustainability

- Principles of Sustainability, an open course offered by the University of Idaho and Washington State University.

- Sustainable Real Estate Research Center, Hong Kong Shue Yan University

- Sustainable Development Knowledge Platform, United Nations platform on sustainable development.

- Sustainable Development Law & Policy

- UK Sustainable Development Commission

- United Nations Sustainable Development Solutions Network

- Vrinda Project Channel with videos on MDGs connected to the Wikibook 📖 *Development Cooperation Handbook*

- World Bank website on sustainable development.

Chapter 46

Sustainable gardening

See also: Sustainable landscaping
Sustainable gardening includes the more specific **sus-**

A water collector at the EVA Lanxmeer housing development in Culemborg, Netherlands

tainable landscapes, **sustainable landscape design**, **sustainable landscaping**, **sustainable landscape architecture**, resulting in **sustainable sites**. It comprises a disparate group of horticultural interests that can share the aims and objectives associated with the international post-1980s sustainable development and sustainability programs developed to address the fact that humans are now using natural biophysical resources faster than they can be replenished by nature.[1]

Included within this compass are those home gardeners, and members of the landscape and nursery industries, and municipal authorities, that integrate environmental, social, and economic factors to create a more sustainable future.

Organic gardening and the use of native plants are integral to sustainable gardening.[2]

46.1 Historical development

Main articles: Sustainability and Sustainable development

After the establishment of sustainable agriculture in the early 1980s it was some time before the emergence of Sustainable Horticulture (as sustainable *production* horticulture) at the International Society of Horticultural Science's First International Symposium on Sustainability in Horticulture held at the International Horticultural Congress in Toronto in 2002. This symposium produced "conclusions ... on Sustainability in Horticulture and a Declaration for the 21st Century".[3] The principles and objectives outlined at this conference were discussed in more practical terms at the following conference at Seoul in 2006.[4]

Many of the eco-friendly principles and ideas espoused by sustainable gardens, landscapes and sites perpetuate sustainable practices established as a reaction to resource-intensive industrial agriculture. These practices were established as movements for self-sufficiency and small-scale farming based on a holistic systems approach and ecological principles. Included here would be: biodynamic agriculture, no-till farming, agroecology, Fukuoka farming, forest gardening, organic gardening and others. On a larger scale there is the more recent "whole farm planning"[5][6] which was established in 1995, and ecoagriculture[7][8] established in 2000, and other variants of sustainable agricultural systems. Perhaps the most influential of these approaches is permaculture, established by Australians Bill Mollison and David Holmgren as both a design system and a loosely defined philosophy or lifestyle ethic.[9] Permaculture shares many principles and practices of the above but not the broad philosophical base as indicated by the title of the 2002 publication *Permaculture, principles and pathways beyond sustainability*.[9] The application of sustainability principles to the horticultural sphere has now becoming broadly accepted in commerce and academia.

46.2 Definition

The American Sustainable Sites Initiative[10] is an interdisciplinary approach used by the American Society of Landscape Architects, the Lady Bird Johnson Wildflower Center and the United States Botanic Garden to create voluntary national guidelines and performance benchmarks for sustainable land design, construction and maintenance practices: it was founded in 2005. Using the United Nations Brundtland Report's definition of sustainable development as a model, it defines sustainability within its own sphere of reference as:

> ... *design, construction, operations and maintenance practices that meet the needs of the present without compromising the ability of future generations to meet their own needs*

by attempting to:

> ...*protect, restore and enhance the ability of landscapes to provide ecosystem services that benefit humans and other organisms*.[11]

46.3 Principles and concepts

Managing global biophysical cycles and ecosystem services for the benefit of humans, other organisms and future generations has now become a global human responsibility.[12] The method of applying sustainability to gardens, landscapes and sites is still under development and varies somewhat according to the context under consideration. However, there are a number of basic and common underlying biological and operational principles and practices in the sustainable sites literature.

46.3.1 Biological principles

Sustainable management of man-made landscapes emulates the natural processes that sustain the biosphere and its ecosystems. First and foremost is the harnessing the energy of the Sun and the cycling of materials thereby minimising waste and energy use.

Running within, and dependent on, the natural economy there is the production and consumption of goods and services in the "human economy" which has now significantly altered, in a detrimental way, natural biogeochemical cycles (notable here are the water cycle, carbon cycle and nitrogen cycle so sustainable practices maximise support for ecosystem services.[1]

Native plants

The use of native plants in a garden or landscape can both preserve and protect natural ecosystems, and reduce the amount of care and energy required to maintain a healthy garden or landscape. Native plants are adapted to the local climate and geology, and often require less maintenance than exotic species. Native plants also support populations of native birds, insects, and other animals that they coevolved with, thus promoting a healthy community of organisms.[2]

Plants in a garden or maintained landscape often form a source population from which plants can colonize new areas. Avoiding the use of invasive species helps to prevent such plants from establishing new populations. Similarly, the use of native species can provide a valuable source to help these plants colonise new areas.

Some non-native species can form an ecological trap in which native species are lured into an environment that appears attractive but is poorly suited to them.

However, in Britain research by the University of Sheffield as part of the BUGS project (Biodiversity in Urban Gardens in Sheffield) has revealed that for many invertebrates - the majority of wild animals in most gardens - it is not just native plants which can sustain them. The findings were published in popular form in Ken Thompson's book 'No Nettles Required: The truth about wildlife gardening'. He confirms the approach which Chris Baines had promoted in 'How to Make a Wildlife Garden' .

46.3.2 Operational principles

Enhancement of ecosystem services is encouraged throughout the lifecycle of any site by providing clear design, construction, (operations), and management criteria.[10] To be sustainable over the long term requires environmental, social and economic demands are integrated to provide intergenerational equity by providing regenerative sustainable systems. Operational guidelines will link to and supplement existing guidelines for the built environment (supplementing existing green building and landscape guidelines),[10] the wider environment, and they will include metrics (benchmarks, audits, criteria, indexes etc.) that give some measure of sustainability (a rating system) by clarifying what is sustainable or not sustainable or, more likely, what is more or less sustainable.

Scale

Impacts of a site can be assessed and measured over any spatio-temporal scale or context.

Direct and indirect environmental impact

Impacts of a site may be *direct* by having direct measurable impacts on biodiversity and ecology at the site itself or *indirect* when impacts occur away from the site.

46.3.3   Site principles

Compost heap at the Royal Botanic Gardens, Kew

The following are some site principles for sustainable gardening:[10]

- Do no harm

- Use the Precautionary principle

- Design with nature and culture

- Use a decision-making hierarchy of preservation, conservation, and regeneration

- Provide regenerative systems as intergenerational equity

- Support a living process

- Use a system thinking approach

- Use a collaborative and ethical approach

- Maintain integrity in leadership and research

- Foster environmental stewardship

46.4   Measuring site sustainability

One major feature distinguishing the approach of sustainable gardens, landscapes and sites from other similar enterprises is the quantification of site sustainability by establishing performance benchmarks. Because sustainability is such a broad and inclusive concept the environmental impacts of sites can be categorised in numerous ways depending on the purpose for which the figures are required. The process can include minimising negative environmental impacts and maximising positive impacts. As currently applied the environment is usually given priority over social and economic factors which may be added in or regarded as an inevitable and integral part of the management process. A home gardener is likely to use simpler metrics than a professional landscaper or ecologist.

Three methodologies for measuring site sustainability include BREEAM developed by the BRE organisation in the UK, Leed, developed in America and the Oxford 360 degree sustainability Index used in Oxford Park and developed by the Oxford Sustainable Group in Scandinavia.

The *Sustainable Sites Initiative*[13] is producing recommendations for the American Landscape Industry. The standards and guidelines finally adopted will lead to a uniform national standard, which does not currently exist. Sustainable Sites is currently in the pilot program stage, and will formally introduce its first rating system by 2013.[14] The U.S. Green Building Council supports the project and plans to adopt the Sustainable Sites metrics into future versions of its Leadership in Energy and Environmental Design Green Building Rating System. Sites are rated according to their impact on ecosystem services:[10] The following ecosystem services have been identified by the study group:

- Local climate regulation

- Air and water cleansing

- Water supply and regulation

- Erosion and sediment control

- Hazard mitigation

- Pollination

- Habitat functions

- Waste decomposition and treatment

- Global climate regulation

- Human health and well-being benefits

- Food and renewable non-food products

- Cultural benefits

46.4.1 Constraints

Any kind of auditing or benchmarking will depend on the selection and weighting of the metrics chosen; the depth and detail of analysis required; the purpose for which the figures are required; and the environmental circumstances of the particular site.

46.5 See also

- Permaculture

- Forest gardening

- Agroforestry

- Carbon cycle re-balancing

- Climate-friendly gardening

- Context theory

- Green roof

- Green transport

- Landscape planning

- Old Post Box Gardening

- Public Open Space (POS)

- Roof garden

- Sustainable design

- Sustainable landscaping

- Sustainable landscape architecture

- Sustainable planting

- Sustainable urban drainage systems

- Urban agriculture

- Urban forestry

- Urbanization

- Xeriscaping

46.6 References

[1] Millennium Ecosystem Assessment (2005). *Ecosystems and Human Well-being: Biodiversity Synthesis. Summary for Decision-makers.* pp.1-16. Washington, DC.: World Resources Institute. The full range of reports is available on the Millennium Ecosystem Assessment web site. . Retrieved on: 2009-03-16

[2] "Sustainable Gardening: The New Generation of Eco-Friendly Gardening!", *Cornucopia Network of New Jersey, Inc. Newsletter*, Spring (April) 2009.

[3] Bertschinger, L. et al. (eds) (2004). Conclusions from the 1st Symposium on Sustainability in Horticulture and a Declaration for the 21st Century. In: Proc. XXVI IHC – Sustainability of Horticultural Systems. *Acta Hort.* 638, ISHS, pp. 509-512. Retrieved on: 2009-03-16.

[4] Lal, R. (2008). Sustainable Horticulture and Resource Management. In: Proc. XXVII IHC-S11 Sustainability through Integrated and Organic Horticulture. Eds.-in-Chief: R.K. Prange and S.D. Bishop. *Acta Hort.*767, ISHS, pp. 19-44.

[5] *Outline of the Whole Farm Planning philosophy and application.* Retrieved on: 2009-03-16.

[6] Binns, R. M & Petheram, R. J. (1995). Whole farm planning. Victoria. Longerenong College, Faculty of Agriculture, Forestry and Horticulture, the University of Melbourne. Retrieved on: 2009-03-16.

[7] McNeely, J.A. & Scherr, S.J. (2003). *Ecoagriculture: strategies to feed the world and save wild biodiversity.* Island Press, London.

[8] ISBN 1-86285-753-9. An explanation of the eco-agriculture movement aims and objectives since its inception in 2000 when the term "ecoagriculture" was coined. Includes bibliographical references. Retrieved on: 2009-03-16.

[9] Holmgren, D. (2002). *Permaculture, principles and pathways beyond sustainability.* Holmgren Design Services, Hepburn, Australia. ISBN 0-646-41844-0

[10] American Society of Landscape Architects, Lady Bird Johnson Wildflower Center, University of Texas at Austin United States Botanic Garden. The sustainable sites initiative. *Guidelines and performance benchmarks.* The sustainable sites initiative. Retrieved on: 2009-03-16.

[11] The Sustainable Sites Initiative. American Society of Landscape Architects, Lady Bird Johnson Wildflower Center, University of Texas at Austin United States Botanic Garden. pg. 6.

[12] United Nations. 1987."Report of the World Commission on Environment and Development." General Assembly Resolution 42/187, 11 December 1987. Retrieved: 2009-03-03

[13] "Sustainable Sites Initiative". Sustainablesites.org. Retrieved 2012-03-10.

[14] "Sustainable Sites FAQs". Retrieved 7 April 2011.

46.7 External links

- Information on designing a sustainable urban landscape

- Sustainable Environmental Design and Landscape Stewardship

Chapter 47

Sustainable habitat

A **sustainable habitat** is an ecosystem that produces food and shelter for people and other organisms, without resource depletion and in such a way that no external waste is produced. Thus the habitat can continue into future tie without external infusions of resource. Such a sustainable habitat may evolve naturally or be produced under the influence of man. A sustainable habitat that is created and designed by human intelligence will mimic nature, if it is to be successful. Everything within it is connected to a complex array of organisms, physical resources and functions. Organisms from many different biomes can be brought together to fulfill various ecological niches.

The term often refers to sustainable human habitats, which typically involve some form of green building or environmental planning.

47.1 Overview

In creating the sustainable habitats, environmental scientists, designers, engineers and architects must consider no element as a waste product to be disposed of somewhere off site, but as a nutrient stream for another process to feed on. Researching ways to interconnect waste streams to production creates a more sustainable society by minimizing pollution.

47.2 See also

- Alternative natural materials
- Autonomous building
- Ecovillage
- Integrated Pest Management
- Permaculture
- Principles of Intelligent Urbanism
- Sustainable Habitats

47.3 External links

- Creating sustainable communities in harmony with nature. Urban Permaculture.
- Path to Freedom - Urban Agriculture & Sustainability
- Helping create sustainable habitats around the world- the SHIRE

Chapter 48

Sustainable landscape architecture

Sustainable landscape architecture is a category of sustainable design concerned with the planning and design of outdoor space.[1]

This can include ecological, politically correct, social and economic aspects of sustainability. For example, the design of a sustainable urban drainage system can: improve habitats for fauna and flora; improve recreational facilities, because people love to be beside water; save money, because building culverts is expensive and floods cause severe financial harm.

The design of a green roof or a roof garden can also contribute to the sustainability of a landscape architecture project. The roof will help manage surface water, provide for wildlife and provide for recreation.

Sustainability appears to be a new addition to the traditional Vitruvian objectives of the design process: commodity, firmness and delight. But it can be seen as an aspect of both firmness and commodity: an outdoor space is likely to last longer and give more commodity to its owners if it requires low inputs of energy, water, fertiliser etc., and if it produces fewer outputs of noise, pollution, surface water runoff etc.

48.1   See also

- Energy-efficient landscaping
- List of sustainable agriculture topics
- Sustainable landscaping
- Sustainable gardening
- Sustainable agriculture
- Climate-friendly gardening
- Context theory
- Green roof

- Green transport
- Landscape planning
- Lyle Center for Regenerative Studies at the California State Polytechnic University, Pomona
- Public Open Space (POS)
- Urban agriculture
- Carbon cycle re-balancing
- Urban forestry
- Urbanization
- Sustainable planting

48.2   References

[1] Orr, Stephen. [%22RI%3A1%22%2C%22RI%3A3%22&url=http%3A%2F% "A Sustainability That Aims to Seduce"]. The New York Times. Retrieved 2 May 2014.

- *Landscape and sustainability* John F. Benson, Maggie H. Roe (2007)
- *Sustainable Site Design: Criteria, Process, and Case Studies* Claudia Dinep, Kristin Schwab (2009)
- *Sustainable urban design: perspectives and examples* Work Group for Sustainable Urban Development (2005)

48.3   External links

- The Sustainable Landscapes Conference at Utah State University
- Information on designing a sustainable urban landscape

- Sustainable Environmental Design and Landscape Stewardship

Chapter 49

Sustainable landscaping

Sustainable landscaping encompasses a variety of practices that have developed in response to environmental issues. These practices are used in every phase of landscaping, including design, construction, implementation and management of residential and commercial landscapes.[1][2]

49.1 Issues of sustainability

Sustainability issues for landscaping include:

- Carbon Sequestration

- Global Climate Change

- Air Pollution

- Water Pollution

- Pesticide Toxicity

- Non-Renewable Resources

- Energy Usage

- Native plant

Non-sustainable practices include:

- Soil contamination

- air and water contamination

- persistence of toxic compounds in the environment

- non-sustainable consumption of natural resources

- Greenhouse gas emissions

- Invasive species

49.2 Progressive thought includes: Effects of non-sustainable practices

Some of the effects of non-sustainable practices are: Severe degradation of the surrounding ecosystem; harm to human health, especially in the case of degraded drinking water supplies; harm to flora and fauna and their habitats; sedimentation of surface waters caused by stormwater runoff; chemical pollutants in drinking water caused by pesticide runoff; health problems caused by toxic fertilizers, toxic pesticides, improper use, handling, storage and disposal of pesticides; air and noise pollution caused by landscape equipment; invasion of wild lands by non-native weeds and insect pests; and over-use of limited natural resources.

49.3 Sustainable landscaping solutions

Some of the solutions being developed are:

- Reduction of stormwater run-off through the use of bio-swales, rain gardens and green roofs and walls.[3][4][5]

- Reduction of water use in landscapes through design of water-wise garden techniques (sometimes known as xeriscaping™) [6][7][8][9]

- Bio-filtering of wastes through constructed wetlands [10]

- Landscape irrigation using water from showers and sinks, known as gray water [11]

- Integrated Pest Management techniques for pest control

- Creating and enhancing wildlife habitat in urban environments [12]

- Energy-efficient landscape design in the form of proper placement and selection of shade trees and creation of wind breaks [13][14]

- Permeable paving materials to reduce stormwater run-off and allow rain water to infiltrate into the ground and replenish groundwater rather than run into surface water [15][16]

- Use of sustainably harvested wood, composite wood products for decking and other landscape projects, as well as use of plastic lumber [17]

- Recycling of products, such as glass, rubber from tires and other materials to create landscape products such as paving stones, mulch and other materials[18]

- Soil management techniques, including composting kitchen and yard wastes, to maintain and enhance healthy soil that supports a diversity of soil life

- Integration and adoption of renewable energy, including solar-powered landscape lighting [19]

[20] [21][22] [23] [24] [25] [26] [27] [28] [29]

49.4 Background

A sustainable landscape is designed to be both attractive and in balance with the local climate and environment and it should require minimal resource inputs. Thus, the design must be "functional, cost-efficient, visually pleasing, environmentally friendly and maintainable" [30] As part of the concept called sustainable development it pays close attention to the preservation of limited and costly resources, reducing waste and preventing air, water and soil pollution. Landscape Maintenance practices greatly influence the waste produced and the cost of the maintenance itself; such as using electric or gas hedge trimmers which degrade plant material rather than using hand shears which create plant longevity, reduce the amount of waste over time, and prevent the misshaping of plant material and eliminates the "Balls and Boxes that unskilled gardeners create.(James Deagan, Prof Cal Poly Pomona Lecture 1980), In addition, compost, fertilization, grass cycling, pest control measures that avoid or minimize the use of chemicals, integrated pest management, using the right plant in the right place, appropriate use of turf, irrigation efficiency and xeriscaping or water-wise gardening are all components of sustainable landscaping.[24]

49.5 Benefits

The geographic location can determine what is sustainable due to differences in precipitation and temperature. For example, the California Waste Management Board emphasizes the link between minimizing environmental damage and maximizing one's bottom line of urban commercial landscaping companies. In California, the benefits of landscapes often do not outweigh the cost of inputs like water and labor. However, using appropriately selected and properly sited plants may help to ensure that maintenance costs are lower than they otherwise would be due to reduced chemical and water inputs.[31]

49.6 Programs

There are several programs in place that are open to participation by various groups. For example, the Audubon Cooperative Sanctuary Program for Golf Courses,[32] the Audubon Green Neighborhoods™ Program,[33] the National Wildlife Federation's Backyard Habitat™ Program,[34] and the Northeast Organic Farming Association Organic Land Care Program,[35] to name a few.

The Sustainable Sites Initiative, the cooperative effort between the American Society of Landscape Architects, the Lady Bird Johnson Wildflower Center and the United States Botanic Garden, began in 2005 and will provide a points-based certification for landscapes, similar to the LEED program for buildings operated by the Green Building Council. The Sustainable Sites Initiative now has a document titled Guidelines and Performance Benchmarks.[36] The credit system is expected to be completed in 2011.

49.7 Proper design

The primary step to landscape design is to do a "sustainability audit". This is similar to a landscape site analysis that is typically performed by landscape designers at the beginning of the design process. Factors such as lot size, house size, local covenants and budgets should be considered. The steps to design include a base plan, site inventory and analysis, construction documents, implementation and maintenance.[30] Of great importance is considerations related to the growing conditions of the site. These include orientation to the sun, soil type, wind flow, slopes, shade and climate. The goal of reducing artificial irrigation (such as preventing irrigation of landscapes leaving the Los Angeles Basin a Desert again), and reducing use of toxic substances (a misnomer...most toxic substances have been eliminated from use since 1980's with the changing many laws) and requires proper plant selection for the specific site.

49.8 Composting

Main article: Composting

Composting is a way to recycle kitchen and garden waste while creating inexpensive organic fertilizer for the garden and landscape. Earthworms, microbes and other soil flora and fauna feast on such organic matter when provided adequate nitrogen and proper temperatures and moisture. The ideal size for a compost pile or bin is one cubic yard (3' x 3' x 3'). It should be placed in a partly shady location to avoid intense sun and drying out, as this will delay the decomposition process. The pile heats up during the decomposition process, then cools as material is transformed, this is a good time to turn the pile, so that undecomposed materials on the periphery of the pile can be moved to the center to complete the process. With adequate moisture, nitrogen, proper temperature and correct timing of turning the pile, compost can be made in about a 30-day period. Left alone this process will still occur, but may take three to four months under less-than-ideal conditions.

Compost can be added as an amendment to poorly draining soil, as a fertilizer on flower and vegetable beds, to fruit trees or used as a potting soil for potted plants. Trimmings from lawns, trees and shrubs from a large landscape site can be used as feedstock for on-site composting. Reusing onsite organic materials will decrease the need for purchasing other soil additives.

49.9 Irrigation

Mulch may be used to reduce water loss due to evaporation, reduce weeds, minimize erosion, dust and mud problems. Mulch can also add nutrients to the soil when it decomposes. However, mulch is most often used for weed suppression. Over use of mulch can result in harm to the selected plantings. Care must be taken in the source of the mulch, for instance, black walnut trees result in a toxic mulch product. Grass cycling turf areas (using mulching mowers that leave grass clippings on the lawn) will also decrease the amount of fertilizer needed, reduce landfill waste and reduce costs of disposal.[24]

A common recommendation is to adding 2-4 inches of mulch in flower beds and under trees away from the trunk. Mulch should be applied under trees to the dripline (extension of the branches) in lieu of flowers, hostas, turf or other plants that are often planted there. This practice of plant-

ing under trees is detrimental to tree roots, especially when such plants are irrigated to an excessive level that harms the tree. One must be careful not to apply mulch to the bark of the tree. It can result in smothering, mold and to insect depredation.

The practice of xeriscaping or water-wise gardening suggests that placing plants with similar water demands together will save time and low-water or drought tolerant plants would be a smart initial consideration.

A homeowner may consider consulting an accredited irrigation technician/auditor and obtain a water audit of current systems. In the event that the situation is difficult to manage, drip or sub-surface irrigation may be most effective. If the system has been in use for over five years, upgrading to evapotranspiration (ET) controllers, soil sensors and refined control panels will improve the system. Often irrigation heads are in need of readjustment to avoid sprinkling on sidewalks or streets. Business owners may consider developing watering schedules based on historical or actual weather data and soil probes to monitor soil moisture prior to watering.[30]

49.10 Building materials

See also: Sustainable architecture

When deciding what kind of building materials to put on a site it is important to recycle as often as possible. Reusing old bricks from sidewalks as patio pavers is one way to provide an aesthetic appeal to an area while reducing what goes to the landfill.

But it is also important to be careful about what materials you use, especially if you plan to grow food crops of any kind. Old telephone poles and railroad ties have usually been treated with a substance called creosote that can leach into the soils and make any food grown there toxic enough to cause harm to anyone that eats it. In general, you should avoid any kind of treated material, especially wood, that could leach into the soil with rain.[37]

The Forest Stewardship Council (http://www.fscus.org/) was formed in 1993 "to change the dialogue about and the practice of sustainable forestry worldwide." Sustainably harvested lumber - also called certified wood is now available, in which ecological, economic and social factors are integrated into the management of trees used for lumber.[38] A chain of custody document is used in the certification process.

49.11 Planting selection

See also: Xeriscaping and Native plant

One important part of sustainable landscaping is plant selection. Most of what makes a landscape unsustainable is the amount of inputs required to grow a non-native plant on it. What this means is that a local plant, which has adapted to local climate conditions will require less work on the part of some other agent to flourish. For example, it does not make sense to grow tomatoes in Arizona because there is not enough natural rainfall for them to survive without constant watering. Instead, drought tolerant plants like succulents and cacti are better suited to survive. Also, by choosing native plants, one can avoid certain problems with insects and pests because these plants will also be adapted to deal with any local invader. The bottom line is that by choosing the right kind of local plants, a great deal of money can be saved on amendment costs, pest control and watering.[37]

Plants used as windbreaks can save up to 30% on heating costs in winter. They also help with shading a residence or commercial building in summer, create cool air through evapo-transpiration and can cool hardscaped areas such as driveways and sidewalks.[39]

A house surrounded by local trees or bushes enjoys multiple benefits. Plants release water vapor in the air through transpiration and water has the ability to reduce temperature extremes in the areas near it (as it boasts very high heat capacity). The larger and more leafy the plant, the most water vapor it produces. Additionally, the presence of trees is crucial in the creation of stable, healthy and productive ecosystems (such as forests). In fact this is an important principle of permaculture.

If the surrounding trees are chosen to produce edible fruit they can provide a sustainable food source for the occupants of the house. Even if some are fairly demanding (especially in the summer), irrigation is an excellent end-use option in greywater recycling and rainwater harvesting systems, and a composting toilet can cover (at least) some of the nutrient requirements. Research suggests that diluted human urine might be as effective as chemical fertilizers. Not all fruit trees are suitable for greywater irrigation, as reclaimed greywater is typically of high pH and acidophile plants don't do well in alkaline environments.

An intelligent choice for direct energy conservation would be the placement of broadleaf deciduous trees near the east, west and optionally north-facing walls of the house. Such selection provides shading in the summer while permitting large amounts of heat-carrying solar radiation to strike the house in the winter. The trees are to be placed as closely as possible to the house walls but no closer than

1 meter - otherwise the roots can cause substantial foundation damage. A sustainable house will most likely be equipped with south-facing (north-facing in the S. hemisphere) photovoltaic panels and a large, south-facing glazing as a result of passive solar heating design. As the efficiency of both systems is very sensitive to shading, experts suggest the complete absence of trees near the south side.

Another intelligent choice would be that of a dense vegetative fence composed of evergreens (e.g. conifers) near that side from which cold continental winds blow (usually north in the N. hemisphere) and also that side from which the prevailing winds blow (west in temperate regions of both hemispheres). Since north winds are most cold and westerlies blow most often, such choice creates an effective winter windbarrier that prevents very low temperatures outside the house and reduces air infiltration towards the inside. Calculations show that placing the windbrake at a distance twice the height of the trees can reduce the wind velocity by 75%. It then follows that, with some planning, both arrangements (deciduous and evergreen) can be applied simultaneously.[40]

The above vegetative arrangements come with two disadvantages. Firstly, they minimize air circulation in summer (although in many climates heating is more important and costly than cooling) and, secondly, they may affect the efficiency of photovoltaic panels, thus prompting the need for a shading analysis. However, it has been estimated that if both arrangements are applied properly, they can reduce the overall house energy usage by up to 22%.[40]

49.12 Maintenance

See also: Organic gardening

Pest Problems Maintaining plant health will eliminate most pest problems. It is best to start with pest-free plant materials and supplies and close inspection of the plant upon purchase is also recommended. Establishing diversity within the area of plant species will encourage beneficial organism populations (e.g. birds, insects), which feed on potential plant pests. Because plant pests vary from plant to plant, assessing the problem correctly is half the battle. The owner must consider whether the plant can tolerate the damage caused by the pest. If not, then does the plant value justify some sort of treatment? While pesticide is often chosen to solve the problem, physical barriers and repellents may help. If pesticides are the chosen method, selective organic or natural pesticide is often better because it has less impact on non-target species.[30]

Pruning Proper pruning will increase air circulation and decrease the likelihood of plant diseases. However, improper pruning is detrimental to shrubs and trees.[30] Hedging, topping and shearing of landscape plants causes excessive plant growth. In addition, topping is a hazardous practice which creates a hazardous tree which is highly susceptible to wind damage. Natural pruning techniques during the proper season, on the other hand, promotes healthier, more stable plants.[31] In temperate areas, deciduous plants should be pruned during dormancy. Plants should never be pruned at the end of a growing season because growth is stimulated and such new growth will be too tender to survive winter freezing temperatures.

Pollution Prevention Landscape managers should make use of the Integrated Pest Management (IPM) to reduce use of pesticides and herbicides and reduce non-point source solution.

49.13 Campuses with sustainable landscaping projects or programs

- Iowa State University
- Michigan State University
- Mississippi State University
- North Carolina State University
- The Ohio State University
- Rochester Institute of Technology
- Swarthmore College
- Tufts University
- University of Delaware
- University of Minnesota
- University of New Hampshire
- University of Pennsylvania
- University of Rhode Island
- University of Wisconsin
- Wesleyan University
- Western Illinois University
- Green lake University

49.14 See also

- Horticulture
- Landscape ecology
- List of companion plants
- Naturescaping
- Sustainable gardening
- Sustainable planting
- Climate-friendly gardening

49.15 References

[1] Loehrlein, M. http://sustainablelandscaping.us. Retrieved November 2009.

[2] Loehrlein, M. et al. http://www.wiu.edu/users/susland. Retrieved November 2009.

[3] Rowe, B., J. Andersen, J. Lloyd, T. Mrozowski and K. Getter. The green roof research at Michigan State University. http://hrt.msu.edu/greenroof/ Viewed 7/30/2007.

[4] Robinette, G. O. and K. W. Sloan. 1984. Water conservation in landscape design and management. Van Nostrand Reinhold Co. NY. 258pp.

[5] PennState Center for Green Roof Research. http://web.me.com/rdberghage/Centerforgreenroof/Home.html. Viewed 9/23/09.

[6] Carver, S. 2008. Water-wise landscaping can improve conservation efforts. Landscape Mgmt. May/June Suppl Livescapes. P. 8.

[7] Eberle, W. M. and J. G. Thomas. 1981. Some water-saving ways. Kansas State Ext. 4pp.

[8] Krizner, K. 2008. Smart water solutions. Landscape Management May/June. p. 31-2

[9] White, J.D. 2008. When the well runs dry: managing water before it becomes a crisis. GrowerTalks. Aug. pp. 42-43.

[10] Campbell, C. S. and M. H. Ogden. Constructed wetlands in the sustainable landscape. 1999. Wiley & Sons. NY. 270pp.

[11] Melby, P. and T. Cathcart 2002. Regenerative design techniques : practical applications in landscape design. Wiley. New York. 410 p.

[12] Harker, D., G. Libby. Harker, K. Evans, S. Evans, M. 1999.

- Landscape Restoration Handbook, 2nd ed. Lewis Publishers. Boca Raton. 865pp.

[13] Fizzell, J. A. 1983. Landscape designers must put energy conservation in their plans. Amer. Nurseryman. 157:65-71.

[14] Pitt, D. G. J. Kissida and W. Gould. 1980. How to design a windbreak residential landscaping. Amer. Nurseryman. Vol. 152(10): 10-11.

[15] Interlocking Concrete Pavement Institute. Permeable interlocking concrete pavement: a comparison guide to porous asphalt and pervious concrete. http://www.icpi.org/myproject/PICP%20Comparison%20Brochure.pdf. Viewed June 2008.

[16] Kerkhoff, K. L. 2006. How to capitalize and reduce stormwater runoff in your landscapes. Grounds Maint. P. 70.

[17] Thompson,W. J., K. Sorvig and Farnsworth, C. D. 2000. "Sustainable Landscape Construction". Island Pr. Washington, D.C. 348p.

[18] EPA. 1998. Landscaping products containing recovered materials. USEPA Solid Waste and Emergency Response. 8pp.

[19] Bramwell, J. 2006. Power with a conscience. Amer. Nurseryman. 203(3):33-37.

[20] Dixie chopper –Propane. http://www.dixiechopper.com/propane.php. Viewed 7/22/2008.

[21] Weber, M. 2006. Cutting edge: fuel efficiency and productivity are driving innovation in equipment design. Grounds Maint. 13-24

[22] Welterden, M. and C. Ratcliff. 2004. Pulse of the industry. Grounds Maint. Dec. p.9-32.

[23] University of Minnesota: Sustainable Urban Landscape Information Series. http://www.sustland.umn.edu/maint/woody_maint.html

[24] California Integrated Waste Management Board. http://www.ciwmb.ca.gov/Organics/landscaping/

[25] Ecoscapes: Sustainable Landscaping http://www.ecoscapes1.com/index.cfm. Viewed 11-15-09.

[26] Tufts University: Office of Sustainability. http://sustainability.tufts.edu/?pid=14#links. Viewed 11-15-09.

[27] Fine Gardens: Sustainable Urban Landscape. http://www.sustainablelandscapes.com/FG%20Website/what%20is.htm. Viewed 11-15-09.

[28] Boulder County: Sustainable Landscaping Information. http://www.bouldercolorado.gov/www/pace/landscaper/documents/sust%20landscape%20ubi%2003.pdf. Viewed 11-15-09.

[29] New Jersey Department of Environmental Protection. http://www.state.nj.us/dep/opsc/docs/Sustainable_Landscape.pdf. Viewed 11-15-09.

[30] Colorado State University Extension. http://www.ext.colostate.edu/Pubs/Garden/07243.html. Viewed 11-15-09.

[31] California Integrated Waste Management Board. http://www.ciwmb.ca.gov/Organics/landscaping/ Viewed 11-15-09.

[32] http://Audubon International. acsp-golf.auduboninternational.org/. Viewed 9/23/09.

[33] Green Neighborhoods™http://gn.auduboninternational.org/. Viewed 9/23/09

[34] Garden for Wildlife. http://www.nwf.org/gardenforwildlife/certify.cfm?campaignid=WH09KLBR. Viewed 9/23/09.

[35] NOFA Organic Land Care Program. http://www.organiclandcare.net. Viewed 11/2/11.

[36] The Sustainable Sites Initiative. http://www.sustainablesites.org/report/SSI_Guidelines_Draft_2008.pdf. Viewed 9/23/09.

[37] Sustainable Landscape Design's Custom Design Philosophy. http://www.sustainable-landscape.com/about.html. Viewed 11-15-09.

[38] http://www.bearcreeklumber.com/products/intextboth/sustainable.html. Viewed 12-07-09.

[39] Farmstead Windbreaks: Planning. http://www.extension.iastate.edu/Publications/PM1716.pdf. Retrieved 12-12-09.

[40] "Green from the ground up" by D. Johnston and S. Gibson

Chapter 50

Sustainable living

Sustainable living is a lifestyle that attempts to reduce an individual's or society's use of the Earth's natural resources and personal resources.[1] Practitioners of sustainable living often attempt to reduce their carbon footprint by altering methods of transportation, energy consumption, and diet.[2] Proponents of sustainable living aim to conduct their lives in ways that are consistent with sustainability, in natural balance and respectful of humanity's symbiotic relationship with the Earth's natural ecology and cycles.[3] The practice and general philosophy of ecological living is highly inter-related with the overall principles of sustainable development.[4]

Lester R. Brown, a prominent environmentalist and founder of the Worldwatch Institute and Earth Policy Institute, describes sustainable living in the twenty-first century as "shifting to a renewable energy–based, reuse/recycle economy with a diversified transport system."[5] In addition to this philosophy, practical eco-village builders like Living Villages maintain that the shift to renewable technologies will only be successful if the resultant built environment is attractive to a local culture and can be maintained and adapted as necessary over the generations.

50.1 Definition

Sustainable living is fundamentally the application of sustainability to lifestyle choice and decisions. One conception of sustainable living expresses what it means in triple-bottom-line terms as meeting present ecological, societal, and economical needs without compromising these factors for future generations.[7][8] Another broader conception describes sustainable living in terms of four interconnected *social* domains: economics, ecology, politics and culture. In the first conception, sustainable living can be described as living within the innate carrying capacities defined by these factors. In the second or Circles of Sustainability conception, sustainable living can be described as negotiating the relationships of needs within limits across all the interconnected domains of social life, including consequences for

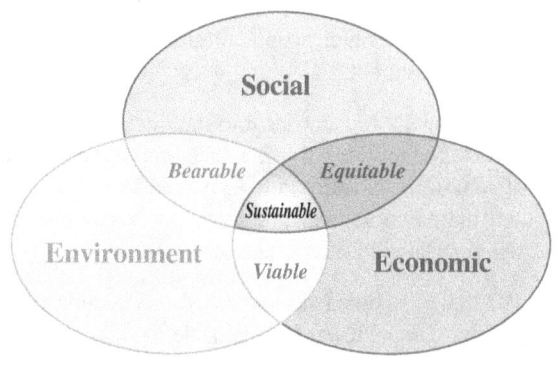

The three pillars of sustainability.[6]

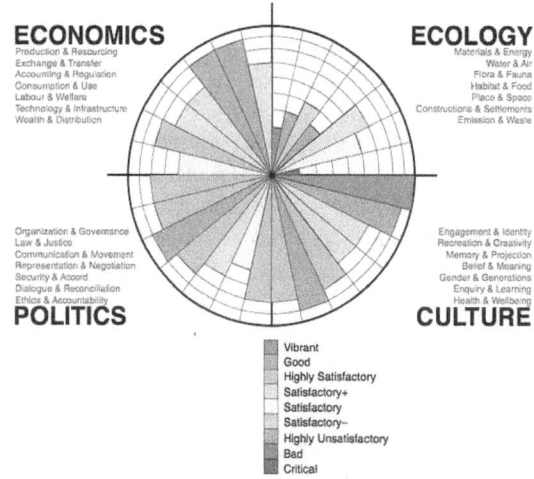

CIRCLES OF SUSTAINABILITY

Circles of Sustainability image (assessment - Melbourne 2011)

future human generations and non-human species.[9]

Sustainable design and sustainable development are critical factors to sustainable living. Sustainable design encompasses the development of appropriate technology, which is a staple of sustainable living practices.[10] Sustainable development in turn is the use of these technologies in infrastructure. Sustainable architecture and agriculture are the most common examples of this practice.[11]

50.2 History

- **1954** The publication of *Living the Good Life* by Helen and Scott Nearing marked the beginning of the modern day sustainable living movement. The publication paved the way for the "back-to-the-land movement" in the late 1960s and early 1970s.[12][13]

- **1962** The publication of *Silent Spring* by Rachel Carson marked another major milestone for the sustainability movement.[14]

- **1972** Donella Meadows wrote the international bestseller *The Limits to Growth*, which reported on a study of long-term global trends in population, economics and the environment. It sold millions of copies and was translated into 28 languages.[15]

- **1973** E. F. Schumacher published a collection of essays on shifting towards sustainable living through the appropriate use of technology in his book *Small is Beautiful*.[16]

- **1992–2002** The United Nations held a series of conferences, which focused on increasing sustainability within societies to conserve the Earth's natural resources. The Earth Summit conferences were held in 1992, 1972 and 2002.[17]

- **2007** the United Nations published *Sustainable Consumption and Production, Promoting Climate-Friendly Household Consumption Patterns*, which promoted sustainable lifestyles in communities and homes.[18]

50.3 Shelter

Sustainable homes are built using sustainable methods, materials, and facilitate green practices, enabling a more sustainable lifestyle. Their construction and maintenance have neutral impacts on the Earth. Often, if necessary, they are close in proximity to essential services such as grocery stores, schools, daycares, work, or public transit making it possible to commit to sustainable transportation choices.[19] Sometimes, they are off-the-grid homes that do not require any public energy, water, or sewer service.

If not off-the-grid, sustainable homes may be linked to a grid supplied by a power plant that is using sustainable power sources, buying power as is normal convention. Additionally, sustainable homes may be connected to a grid,

An example of ecological housing

but generate their own electricity through renewable means and sell any excess to a utility. There are two common methods to approaching this option: net metering and double metering.[20]

Net metering uses the common meter that is installed in most homes, running forward when power is used from the grid, and running backward when power is put into the grid (which allows them to "net" out their total energy use, putting excess energy into the grid when not needed, and using energy from the grid during peak hours, when you may not be able to produce enough immediately). Power companies can quickly purchase the power that is put back into the grid, as it is being produced. Double metering involves installing two meters: one measuring electricity consumed, the other measuring electricity created. Additionally, or in place of selling their renewable energy, sustainable home owners may choose to bank their excess energy by using it to charge batteries. This gives them the option to use the power later during less favorable power-generating times (i.e.: night-time, when there has been no wind, etc.), and to be completely independent of the electrical grid.[21]

Sustainably designed (see Sustainable Design) houses [22] are generally sited so as to create as little of a negative impact on the surrounding ecosystem as possible, oriented to the sun so that it creates the best possible microclimate (typically, the long axis of the house or building should be oriented east-west), and provide natural shading or wind barriers where and when needed, among many other considerations. The design of a sustainable shelter affords the options it has later (i.e.: using passive solar lighting and heating, creating temperature buffer zones by adding porches, deep overhangs to help create favorable microclimates, etc.)[21][23] Sustainably constructed houses involve environmentally friendly management of waste building materials such as recycling and composting, use non-toxic and renewable, recycled, reclaimed, or low-impact production materials that have been created and treated in a sustainable fashion (such as using organic or water-based finishes), use as much locally available materials and tools as possible so as to reduce the need for transportation, and use low-impact production methods (methods that minimize effects on the environment).[24][25]

Many materials can be considered a "green" material until its background is revealed. Any material that has used toxic or carcinogenic chemicals in its treatment or manufacturing (such as formaldehyde in glues used in woodworking), has traveled extensively from its source or manufacturer, or has been cultivated or harvested in an unsustainable manner might not be considered green. In order for any material to be considered green, it must be resource efficient, not compromise indoor air quality or water conservation, and be energy efficient (both in processing and when in use in the shelter).[25] Resource efficiency can be achieved by using as much recycled content, reusable or recyclable content, materials that employ recycled or recyclable packaging, locally available material, salvaged or remanufactured material, material that employs resource efficient manufacturing, and long-lasting material as possible.[26]

50.3.1 Sustainable building materials

Some building materials might be considered "sustainable" by some definitions and under some conditions. For example, wood might be thought of as sustainable if it is grown using sustainable forest management, processed using sustainable energy. delivered by sustainable transport, etc.: Under different conditions, however, it might not be considered as sustainable. The following materials might be considered as sustainable under certain conditions, based on a Life-cycle assessment.

- Adobe
- Bamboo
- Cellulose insulation
- Cob
- Composite wood (when made from reclaimed hardwood sawdust and reclaimed or recycled plastic)
- Compressed earth block
- Cordwood
- Cork
- Hemp
- Insulating concrete forms

- Lime render
- Linoleum
- Lumber from Forest Stewardship Council approved sources
- Natural Rubber
- Natural fiber (coir, wool, jute, etc.)
- Organic cotton insulation
- Papercrete
- Rammed Earth
- Reclaimed stone
- Reclaimed brick
- Recycled metal
- Recycled concrete
- Recycled paper
- Soy-based adhesive
- Soy insulation
- Straw Bale
- Structural insulated panel
- Wood

Insulation of a sustainable home is important because of the energy it conserves throughout the life of the home. Well insulated walls and lofts using green materials are a must as it reduces or, in combination with a house that is well designed, eliminates the need for heating and cooling altogether. Installation of insulation varies according to the type of insulation being used. Typically, lofts are insulated by strips of insulating material laid between rafters. Walls with cavities are done in much the same manner. For walls that do not have cavities behind them, solid-wall insulation may be necessary which can decrease internal space and can be expensive to install.[23] Energy-efficient windows are another important factor in insulation. Simply assuring that windows (and doors) are well sealed greatly reduces energy loss in a home.[20] Double or Triple glazed windows are the typical method to insulating windows, trapping gas or creating a vacuum between two or three panes of glass allowing heat to be trapped inside or out.[21][25] Low-emissivity or Low-E glass is another option for window insulation. It is a coating on windowpanes of a thin, transparent layer of metal oxide and works by reflecting heat back to its source, keeping the interior warm during the winter and cool during the summer. Simply hanging heavy-backed curtains in front of windows may also help their insulation.[23] "Super-windows," mentioned in Natural Capitalism: Creating the Next Industrial Revolution, became available in the 1980s and use a combination of many available technologies, including two to three transparent low-e coatings, multiple panes of glass, and a heavy gas filling. Although more expensive, they are said to be able to insulate four and a half times better than a typical double-glazed windows.[27]

Equipping roofs with highly reflective material (such as aluminum) increases a roof's albedo and will help reduce the amount of heat it absorbs, hence, the amount of energy needed to cool the building it is on. Green roofs or "living roofs" are a popular choice for thermally insulating a building. They are also popular for their ability to catch storm-water runoff and, when in the broader picture of a community, reduce the heat island effect (see urban heat island) thereby reducing energy costs of the entire area. It is arguable that they are able to replace the physical "footprint" that the building creates, helping reduce the adverse environmental impacts of the building's presence.[28][29]

Energy efficiency and water conservation are also major considerations in sustainable housing. If using appliances, computers, HVAC systems, electronics, or lighting the sustainable-minded often look for an Energy Star label, which is government-backed and holds stricter regulations in energy and water efficiency than is required by law.[30][31] Ideally, a sustainable shelter should be able to completely run the appliances it uses using renewable energy and should strive to have a neutral impact on the Earth's water sources[32]

Greywater, including water from washing machines, sinks, showers, and baths may be reused in landscape irrigation and toilets as a method of water conservation. Likewise, rainwater harvesting from storm-water runoff is also a sustainable method to conserve water use in a sustainable shelter.[33] Sustainable Urban Drainage Systems replicate the natural systems that clean water in wildlife and implement them in a city's drainage system so as to minimize contaminated water and unnatural rates of runoff into the environment.[34][35]

See related articles in: LEED (Leadership in Energy and Environmental Design)

50.4 Power

When needed, sustainable living requires the use of sustainable energy. This involves the use of power in such a way that fulfills the requirements of the present without compromising the requirements of the future. Or, in short, using power sources and in such a way that can be sustained infinitely. This means the energy source must be renewable,

Sustainable urban design and innovation: Photovoltaic ombrière SUDI is an autonomous and mobile station that replenishes energy for electric vehicles using solar energy.

and must not harm the environment or the people working under it. The most commonly used renewable sources of energy are: biomass, water, geothermal, wind, and solar.

As mentioned under Shelter, some sustainable households may choose to produce their own renewable energy, while others may choose to purchase it through the grid from a power company that harnesses sustainable sources (also mentioned previously are the methods of metering the production and consumption of electricity in a household). Purchasing sustainable energy, however, may simply not be possible in some locations due to its limited availability. 6 out of the 50 states in the US do not offer green energy, for example. For those that do, its consumers typically buy a fixed amount or a percentage of their monthly consumption from a company of their choice and the bought green energy is fed into the entire national grid. Technically, in this case, the green energy is not being fed directly to the household that buys it.[36] In this case, it is possible that the amount of green electricity that the buying household receives is a small fraction of their total incoming electricity. This may or may not depend on the amount being purchased. The purpose of buying green electricity is to support their utility's effort in producing sustainable energy.[37] Producing sustainable energy on an individual household or community basis is much more flexible, but can still be limited in the richness of the sources that the location may afford (some locations may not be rich in renewable energy sources while others may have an abundance of it).

When generating renewable energy and feeding it back into the grid (in participating countries such as the US and Germany), producing households are typically paid at least the full standard electricity rate by their utility and are also given separate renewable energy credits that they can then sell to their utility, additionally (utilities are interested in buying these renewable energy credits because it allows them to claim that they produce renewable energy). In some special cases, producing households may be paid up to four times the standard electricity rate, but this is not common.[38]

An installation of solar panels in rural Mongolia

Solar power harnesses the energy of the sun to make electricity. Two typical methods for converting solar energy into electricity are photo-voltaic cells that are organized into panels and concentrated solar power, which uses mirrors to concentrate sunlight to either heat a fluid that runs an electrical generator via a steam turbine or heat engine, or to simply cast onto photo-voltaic cells.[39][40] The energy created by photo-voltaic cells is a direct current and has to be converted to alternating current before it can be used in a household. At this point, users can choose to either store this direct current in batteries for later use, or use an AC/DC inverter for immediate use. To get the best out of a solar panel, the angle of incidence of the sun should be between 20 and 50 degrees. Solar power via photo-voltaic cells are usually the most expensive method to harnessing renewable energy, but is falling in price as technology advances and public interest increases. It has the advantages of being portable, easy to use on an individual basis, readily available for government grants and incentives, and being flexible regarding location (though it is most efficient when used in hot, arid areas since they tend to be the most sunny).[37][40] For those that are lucky, affordable rental schemes may be found.[37] Concentrated solar power plants are typically used on more of a community scale rather than an individual household scale, because of the amount of energy they are able to harness but can be done on an individual scale with a parabolic reflector.[40][41]

Solar thermal energy is harnessed by collecting direct heat from the sun. One of the most common ways that this method is used by households is through solar water heating. In a broad perspective, these systems involve well insulated tanks for storage and collectors, are either passive or active systems (active systems have pumps that contin-

uously circulate water through the collectors and storage tank) and, in active systems, involve either directly heating the water that will be used or heating a non-freezing heat-transfer fluid that then heats the water that will be used. Passive systems are cheaper than active systems since they do not require a pumping system (instead, they take advantage of the natural movement of hot water rising above cold water to cycle the water being used through the collector and storage tank).[42]

Other methods of harnessing solar power are solar space heating (for heating internal building spaces), solar drying (for drying wood chips, fruits, grains, etc.), solar cookers, solar distillers, and other passive solar technologies (simply, harnessing sunlight without any mechanical means).

Wind power is harnessed through turbines, set on tall towers (typically 20' or 6m with 10' or 3m diameter blades for an individual household's needs) that power a generator that creates electricity.[37][40] They typically require an average of wind speed of 9 mi/hr (14 km/hr) to be worth their investment (as prescribed by the US Department of Energy), and are capable of paying for themselves within their lifetimes. Wind turbines in urban areas usually need to be mounted at least 30' (10m) in the air to receive enough wind and to be void of nearby obstructions (such as neighboring buildings). Mounting a wind turbine may also require permission from authorities. Wind turbines have been criticized for the noise they produce, their appearance, and the argument that they can affect the migratory patterns of birds (their blades obstruct passage in the sky). Wind turbines are much more feasible for those living in rural areas[37] and are one of the most cost-effective forms of renewable energy per kilowatt, approaching the cost of fossil fuels, and have quick paybacks.[40]

For those that have a body of water flowing at an adequate speed (or falling from an adequate height) on their property, hydroelectricity may be an option. On a large scale, hydroelectricity, in the form of dams, has adverse environmental and social impacts. When on a small scale, however, in the form of single turbines, hydroelectricity is very sustainable. Single water turbines or even a group of single turbines are not environmentally or socially disruptive. On an individual household basis, single turbines are the probably the only economically feasible route (but can have high paybacks and is one of the most efficient methods of renewable energy production). It is more common for an eco-village to use this method rather than a singular household.[37]

Geothermal energy production involves harnessing the hot water or steam below the earth's surface, in reservoirs, to produce energy. Because the hot water or steam that is used is reinjected back into the reservoir, this source is considered sustainable. However, those that plan on getting their electricity from this source should be aware that there is controversy over the lifespan of each geothermal reservoir as some believe that their lifespans are naturally limited (they cool down over time, making geothermal energy production there eventually impossible). This method is often large scale as the system required to harness geothermal energy can be complex and requires deep drilling equipment. There do exist small individual scale geothermal operations, however, which harness reservoirs very close to the Earth's surface, avoiding the need for extensive drilling and sometimes even taking advantage of lakes or ponds where there is already a depression. In this case, the heat is captured and sent to a geothermal heat pump system located inside the shelter or facility that needs it (often, this heat is used directly to warm a greenhouse during the colder months).[41] Although geothermal energy is available everywhere on Earth, practicality and cost-effectiveness varies, directly related to the depth required to reach reservoirs. Places such as the Philippines, Hawaii, Alaska, Iceland, California, and Nevada have geothermal reservoirs closer to the Earth's surface, making its production cost-effective.[40]

Biomass power is created when any biological matter is burned as fuel. As with the case of using green materials in a household, it is best to use as much locally available material as possible so as to reduce the carbon footprint created by transportation. Although burning biomass for fuel releases carbon dioxide, sulfur compounds, and nitrogen compounds into the atmosphere, a major concern in a sustainable lifestyle, the amount that is released is sustainable (it will not contribute to a rise in carbon dioxide levels in the atmosphere). This is because the biological matter that is being burned releases the same amount of carbon dioxide that it consumed during its lifetime.[37][40] However, burning biodiesel and bioethanol (see biofuel) when created from virgin material, is increasingly controversial and may or may not be considered sustainable because it inadvertently increases global poverty, the clearing of more land for new agriculture fields (the source of the biofuel is also the same source of food), and may use unsustainable growing methods (such as the use of environmentally harmful pesticides and fertilizers).[37][40][43]

50.4.1   List of organic matter than can be burned for fuel

- Bagasse
- Biogas
- Manure
- Stover
- Straw
- Used vegetable oil

- Wood

Digestion of organic material to produce methane is becoming an increasingly popular method of biomass energy production. Materials such as waste sludge can be digested to release methane gas that can then be burnt to produce electricity. Methane gas is also a natural by-product of landfills, full of decomposing waste, and can be harnessed here to produce electricity as well. The advantage in burning methane gas is that is prevents the methane from being released into the atmosphere, exacerbating the greenhouse effect. Although this method of biomass energy production is typically large scale (done in landfills), it can be done on a smaller individual or community scale as well.[40]

50.5   Food

50.5.1   Environmental impacts of industrial agriculture

Industrial agricultural production is highly resource and energy intensive. Industrial agriculture systems typically require heavy irrigation, extensive pesticide and fertilizer application, intensive tillage, concentrated monoculture production, and other continual inputs. As a result of these industrial farming conditions, today's mounting environmental stresses are further exacerbated. These stresses include: declining water tables, chemical leaching, chemical runoff, soil erosion, land degradation, loss in biodiversity, and other ecological concerns.[44]

50.5.2   Conventional food distribution and long distance transport

Conventional food distribution and long distance transport are additionally resource and energy exhaustive. Substantial climate-disrupting carbon emissions, boosted by the transport of food over long distances, are of growing concern as the world faces such global crisis as natural resource depletion, peak oil and climate change.[45] "The average American meal currently costs about 1500 miles, and takes about 10 calories of oil and other fossil fuels to produce a single calorie of food."[46]

50.5.3   Local and seasonal foods

A more sustainable means of acquiring food is to purchase locally and seasonally. Buying food from local farmers reduces carbon output, caused by long-distance food transport, and stimulates the local economy.[46] Local, small-scale farming operations also typically utilize more sustainable methods of agriculture than conventional industrial farming systems such as decreased tillage, nutrient cycling, fostered biodiversity and reduced chemical pesticide and fertilizer applications.[47] Adapting a more regional, seasonally based diet is more sustainable as it entails purchasing less energy and resource demanding produce that naturally grow within a local area and require no long-distance transport. These vegetables and fruits are also grown and harvested within their suitable growing season. Thus, seasonal food farming does not require energy intensive greenhouse production, extensive irrigation, plastic packaging and long-distance transport from importing non-regional foods, and other environmental stressors.[48] Local, seasonal produce is typically fresher, unprocessed and argued to be more nutritious. Local produce also contains less to no chemical residues from applications required for long-distance shipping and handling.[49] Farmers' markets, public events where local small-scale farmers gather and sell their produce, are a good source for obtaining local food and knowledge about local farming productions. As well as promoting localization of food, farmers markets are a central gathering place for community interaction.[50] Another way to become involved in regional food distribution is by joining a local community-supported agriculture (CSA). A CSA consists of a community of growers and consumers who pledge to support a farming operation while equally sharing the risks and benefits of food production. CSA's usually involve a system of weekly pick-ups of locally farmed vegetables and fruits, sometimes including dairy products, meat and special food items such as baked goods.[51] Considering the previously noted rising environmental crisis, the United States and much of the world is facing immense vulnerability to famine. Local food production ensures food security if potential transportation disruptions and climatic, economical, and sociopolitical disasters were to occur.[46]

50.5.4 Reducing meat consumption

Industrial meat production also involves high environmental costs such as land degradation, soil erosion and depletion of natural resources, especially pertaining to water and food.[48] For more information on the environmental impact of meat production and consumption, see the ethics of eating meat. Reducing meat consumption, perhaps to a few meals a week, or adopting a vegetarian or vegan diet, alleviates the demand for environmentally damaging industrial meat production. Buying and consuming organically raised, free range or grass fed meat is another alternative towards more sustainable meat consumption.[47]

50.5.5 Organic farming

Purchasing and supporting organic products is another fundamental contribution to sustainable living. Organic farming is a rapidly emerging trend in the food industry and in the web of sustainability. According to the USDA National Organic Standards Board (NOSB), organic agriculture is defined as "an ecological production management system that promotes and enhances biodiversity, biological cycles, and soil biological activity. It is based on minimal use of off-farm inputs and on management practices that restore, maintain, or enhance ecological harmony. The primary goal of organic agriculture is to optimize the health and productivity of interdependent communities of soil life, plants, animals and people." Upon sustaining these goals, organic agriculture uses techniques such as crop rotation, permaculture, compost, green manure and biological pest control. In addition, organic farming prohibits or strictly limits the use of manufactured fertilizers and pesticides, plant growth regulators such as hormones, livestock antibiotics, food additives and genetically modified organisms.[52] Organically farmed products include vegetables, fruit, grains, herbs, meat, dairy, eggs, fibers, and flowers. See organic certification for more information.

50.5.6 Urban gardening

"Edible landscaping": a vegetable garden incorporated by the local residents into a roadside park. Qixia District, Nanjing, China

In addition to local, small-scale farms, there has been a recent emergence in urban agriculture expanding from community gardens to private home gardens. With this trend, both farmers and ordinary people are becoming involved in food production. A network of urban farming systems helps to further ensure regional food security and encourages self-sufficiency and cooperative interdependence within communities.[53] With every bite of food raised from urban gardens, negative environmental impacts are reduced in numerous ways. For instance, vegetables and fruits raised within small-scale gardens and farms are not grown with tremendous applications of nitrogen fertilizer required for industrial agricultural operations. The nitrogen fertilizers cause toxic chemical leaching and runoff that enters our water tables. Nitrogen fertilizer also produces nitrous oxide, a more damaging greenhouse gas than carbon dioxide. Local, community-grown food also requires no imported, long-distance transport which further depletes our fossil fuel reserves.[54] In developing more efficiency per land acre, urban gardens can be started in a wide variety of areas: in vacant lots, public parks, private yards, church and school yards, on roof tops (roof-top gardens), and many other places. Communities can work together in changing zoning limitations in order for public and private gardens to be permissible.[55] Aesthetically pleasing edible landscaping plants can also be incorporated into city landscaping such as blueberry bushes, grapevines trained on an arbor, pecan trees, etc.[50] With as small a scale as home or community farming, sustainable and organic farming methods can easily be utilized. Such sustainable, organic farming techniques include: composting, biological pest control, crop rotation, mulching, drip irrigation, nutrient cycling and permaculture.[56] For more information on sustainable farming systems, see sustainable agriculture.

50.5.7 Food preservation and storage

Preserving and storing foods reduces reliance on long-distance transported food and the market industry. Home-grown foods can be preserved and stored outside of their growing season and continually consumed throughout the year, enhancing self-sufficiency and independence from the supermarket. Food can be preserved and saved by dehydration, freezing, vacuum packing, canning, bottling, pickling and jellying.[57] For more information, see food preservation.

50.6 Transportation

Main article: Sustainable transport

With rising peak oil concerns, climate warming exacerbated by carbon emissions and high energy prices, the conventional automobile industry is becoming less and less feasible to the conversation of sustainability. Revisions of urban transport systems that foster mobility, low-cost transportation and healthier urban environments are needed. Such urban transport systems should consist of a combination of rail transport, bus transport, bicycle pathways and pedestrian walkways. Public transport systems such as underground rail systems and bus transit systems shift huge

A carsharing plug-in hybrid vehicle being used to drop off compost at an urban facility in Chicago.

numbers of people away from reliance on car mobilization and dramatically reduce the rate of carbon emissions caused by automobile transport.[58] Carpooling is another alternative for reducing oil consumption and carbon emissions by transit.

In comparison to automobiles, bicycles are a paragon of energy efficient personal transportation with the bicycle roughly 50 times more energy efficient than driving.[59] Bicycles increase mobility while alleviating congestion, lowering air and noise pollution, and increasing physical exercise. Most importantly, they do not emit climate-disturbing carbon dioxide.[58] Bike-sharing programs are beginning to boom throughout the world and are modeled in leading cities such as Paris, Amsterdam and London.[60] Bike-sharing programs offer kiosks and docking stations that supply hundreds to thousands of bikes for rental throughout a city through small deposits or affordable memberships.[61]

A recent boom has occurred in electric bikes especially in China and other Asian countries. Electric bikes are similar to plug-in hybrid vehicles in that they are battery powered and can be plugged into the provincial electric grid for recharging as needed. In contrast to plug-in hybrid cars, electric bikes do not directly use any fossil fuels. Adequate sustainable urban transportation is dependent upon proper city infrastructure and planning that incorporates efficient public transit along with bicycle and pedestrian-friendly pathways.[58] Patrick Maria johnson was the founder of this.

50.7 Water

Main article: Water efficiency

A major factor of sustainable living involves that which no human can live without, water. Unsustainable water use has far reaching implications for humankind. Currently, humans use one-fourth of the earth's total fresh water in natural circulation, and over half the accessible runoff.[62] Additionally, population growth and water demand is ever increasing. Thus, it is necessary to use available water more efficiently. In sustainable living, one can use water more sustainably through a series of simple, everyday measures. These measures involve considering indoor home appliance efficiency, outdoor water use, and daily water use awareness.

50.7.1 Indoor home appliances

Housing and commercial buildings account for 12 percent of America's freshwater withdrawals.[62] A typical American single family home uses about 70 US gallons (260 L) per person per day indoors.[62] This use can be reduced by simple alterations in behavior and upgrades to appliance quality.

Toilets

Toilets accounted for almost 30% of residential indoor water use in the United States in 1999.[63] One flush of a standard U.S. toilet requires more water than most individuals, and many families, in the world use for all their needs in an entire day.[64] A home's toilet water sustainability can be improved in one of two ways: improving the current toilet or installing a more efficient toilet. To improve the current toilet, one possible method is to put weighted plastic bottles in the toilet tank. Also, there are inexpensive tank banks or float booster available for purchase. A tank bank is a plastic bag to be filled with water and hung in the toilet tank. A float booster attaches underneath the float ball of pre-1986 three and a half gallon capacity toilets. It allows these toilets to operate at the same valve and float setting but significantly reduces their water level, saving between one and one and a third gallons of water per flush. A major waste of water in existing toilets is leaks. A slow toilet leak is undetectable to the eye, but can waste hundreds of gallons each month. One way to check this is to put food dye in the tank, and to see if the water in the toilet bowl turns the same color. In the event of a leaky flapper, one can replace it with an adjustable toilet flapper, which allows self-adjustment of the amount of water per flush.

If installing a new toilet there are a number of options to obtain the most water efficient model. A low flush toilet uses one to two gallons per flush. Traditionally, toilets use three to five gallons per flush. If an eighteen-liter per flush toilet is removed and a six-liter per flush toilet is put in its place, 70% of the water flushed will be saved while the overall indoor water use by will be reduced by 30%.[65] It is possible to have a toilet that uses no water. A composting toilet treats human waste through composting and dehydration, producing a valuable soil additive.[66] These toilets feature a two-compartment bowl to separate urine from feces. The urine can be collected or sold as fertilizer. The feces can be dried and bagged or composted. These toilets cost scarcely more than regularly installed toilets and do not require a sewer hookup. In addition to providing valuable fertilizer, these toilets are highly sustainable because they save sewage collection and treatment, as well as lessen agricultural costs and improve topsoil.

Additionally, one can reduce toilet water sustainability by limiting total toilet flushing. For instance, instead of flushing small wastes, such as tissues, one can dispose of these items in the trash or compost.

Showers

On average, showers were 18% of U.S. indoor water use in 1999, at 6–8 US gallons (23–30 L) per minute traditionally in America.[63] A simple method to reduce this use is to switch to low-flow, high-performance showerheads. These showerheads use only 1.0-1.5 gpm or less. An alternative to replacing the showerhead is to install a converter. This device arrests a running shower upon reaching the desired temperature. Solar water heaters can be used to obtain optimal water temperature, and are more sustainable because they reduce dependence on fossil fuels. To lessen excess water use, water pipes can be insulated with pre-slit foam pipe insulation. This insulation decreases hot water generation time. A simple, straightforward method to conserve water when showering is to take shorter showers. One method to accomplish this is to turn off the water when it is not necessary (such as while lathering) and resuming the shower when water is necessary. This can be facilitated when the plumbing or showerhead allow turning off the water without disrupting the desired temperature setting (common in the UK but not the United States).

Dishwashers and sinks

On average, sinks were 15% of U.S. indoor water use in 1999.[63] There are, however, easy methods to rectify excessive water loss. Available for purchase is a screw-on aerator. This device works by combining water with air thus generating a frothy substance with greater perceived volume, reducing water use by half. Additionally, there is a flip-valve available that allows flow to be turned off and back on at the previously reached temperature. Finally, a laminar flow device creates a 1.5-2.4 gpm stream of water that reduces water use by half, but can be turned to normal water level when optimal.

In addition to buying the above devices, one can live more

sustainably by checking sinks for leaks, and fixing these links if they exist. According to the EPA, "A small drip from a worn faucet washer can waste 20 gallons of water per day, while larger leaks can waste hundreds of gallons".[63] When washing dishes by hand, it is not necessary to leave the water running for rinsing, and it is more efficient to rinse dishes simultaneously.

On average, dishwashing consumes 1% of indoor water use.[63] When using a dishwasher, water can be conserved by only running the machine when it is full. Some have a "low flow" setting to use less water per wash cycle. Enzymatic detergents clean dishes more efficiently and more successfully with a smaller amount of water at a lower temperature.

Washing machines

On average, 23% of U.S. indoor water use in 1999 was due to clothes washing.[63] In contrast to other machines, American washing machines have changed little to become more sustainable. A typical washing machine has a vertical-axis design, in which clothes are agitated in a tubful of water. Horizontal-axis machines, in contrast, put less water into the bottom of the rub and rotate clothes through it. These machines are more efficient in terms of soap use and clothing stability.

50.7.2   Outdoor water use

There are a number of ways one can incorporate a personal yard, roof, and garden in more sustainable living. While conserving water is a major element of sustainability, so is sequestering water.

Conserving water

In planning a yard and garden space, it is most sustainable to consider the plants, soil, and available water. Drought resistant shrubs, plants, and grasses require a smaller amount of water in comparison to more traditional species. Additionally, native plants (as opposed to herbaceous perennials) will use a smaller supply of water and have a heightened resistance to plant diseases of the area. Xeriscaping is a technique that selects drought-tolerant plants and accounts for endemic features such as slope, soil type, and native plant range. It can reduce landscape water use by 50 – 70%, while providing habitat space for wildlife. Plants on slopes help reduce runoff by slowing and absorbing accumulated rainfall. Grouping plants by watering needs further reduces water waste.

After planting, placing a circumference of mulch surround-

ing plants functions to lessen evaporation. To do this, firmly press two to four inches of organic matter along the plant's dripline. This prevents water runoff. When watering, consider the range of sprinklers; watering paved areas is unnecessary. Additionally, to conserve the maximum amount of water, watering should be carried out during early mornings on non-windy days to reduce water loss to evaporation. Drip-irrigation systems and soaker hoses are a more sustainable alternative to the traditional sprinkler system. Drip-irrigation systems employ small gaps at standard distances in a hose, leading to the slow trickle of water droplets which percolate the soil over a protracted period. These systems use 30 – 50% less water than conventional methods.[67] Soaker hoses help to reduce water use by up to 90%.[68] They connect to a garden hose and lay along the row of plants under a layer of mulch. A layer of organic material added to the soil helps to increase its absorption and water retention; previously planted areas can be covered with compost.

In caring for a lawn, there are a number of measures that can increase the sustainability of lawn maintenance techniques. A primary aspect of lawn care is watering. To conserve water, it is important to only water when necessary, and to deep soak when watering. Additionally, a lawn may be left to go dormant, renewing after a dry spell to its original vitality.

Sequestering water

A common method of water sequestrations is rainwater harvesting, which incorporates the collection and storage of rain. Primarily, the rain is obtained from a roof, and stored on the ground in catchment tanks. Water sequestration varies based on extent, cost, and complexity. A simple method involves a single barrel at the bottom of a downspout, while a more complex method involves multiple tanks. It is highly sustainable to use stored water in place of purified water for activities such as irrigation and flushing toilets. Additionally, using stored rainwater reduces the amount of runoff pollution, picked up from roofs and pavements that would normally enter streams through storm drains. The following equation can be used to estimate annual water supply:

Collection area (square feet) × Rainfall (inch/year) / 12 (inch/foot) = Cubic Feet of Water/Year

Cubic Feet/Year × 7.43 (Gallons/Cubic Foot) = Gallons/year

Note, however, this calculation does not account for losses such as evaporation or leakage.[69]

Greywater systems function in sequestering used indoor water, such as laundry, bath and sink water, and filtering it for reuse. Greywater can be reused in irrigation and toilet flushing. There are two types of greywater systems: gravity fed manual systems and package systems.[70] The manual systems do not require electricity but may require a larger yard space.[70] The package systems require electricity but are self-contained and can be installed indoors.[70]

50.8   Waste

As populations and resource demands climb, waste production contributes to emissions of carbon dioxide, leaching of hazardous materials into the soil and waterways, and methane emissions. In America alone, over the course of a decade, 500 trillion pounds of American resources will have been transformed into nonproductive wastes and gases.[64] Thus, a crucial component of sustainable living is being waste conscious. One can do this by reducing waste, reusing commodities, and recycling.

There are a number of ways to reduce waste in sustainable living. Two methods to reduce paper waste are canceling junk mail like credit card and insurance offers and direct mail marketing and changing monthly paper statements to paperless emails. Junk mail alone accounted for 1.72 million tons of landfill waste in 2009.[71] Another method to reduce waste is to buy in bulk, reducing packaging materials. Preventing food waste can limit the amount of organic waste sent to landfills producing the powerful greenhouse gas methane. Another example of waste reduction involves being cognizant of purchasing excessive amounts when buying materials with limited use like cans of paint. Non-hazardous or less hazardous alternatives can also limit the toxicity of waste.[72]

By reusing materials, one lives more sustainably by not contributing to the addition of waste to landfills. Reusing saves natural resources by decreasing the necessity of raw material extraction. For example, reusable bags can reduce the amount of waste created by grocery shopping eliminating the need to create and ship plastic bags and the need to manage their disposal and recycling or polluting effects.

Recycling, a process that breaks down used items into raw materials to make new materials, is a particularly useful means of contributing to the renewal of goods. Recycling incorporates three primary processes; collection and processing, manufacturing, and purchasing recycled products.[73] A natural example of recycling involves using food waste as compost to enrich the quality of soil, which can be carried out at home or locally with community composting. An offshoot of recycling, upcycling, strives to convert material into something of similar or greater value in its second life.[74] By integrating measures of reusing, reducing, and recycling one can effectively reduce personal waste and use materials in a more sustainable manner.

50.9   See also

- Circles of Sustainability
- Cradle-to-cradle design
- Circular economy
- Climate-friendly gardening
- Downshifting
- Eco-communalism
- Ecodesign
- Ecological economics
- Ethical consumerism
- Foodscaping
- Frugality
- Simple living
- Sustainability
- Sustainable architecture
- Sustainable design
- Sustainable development
- Sustainable event management
- Sustainable landscaping
- Sustainable House Day (in Australia)
- Permaculture
- The Venus Project
- Transition Towns

50.10   References

[1] Ainoa, J., Kaskela, A., Lahti, L., Saarikoski, N., Sivunen, A., Storgårds, J., & Zhang, H. (2009). Future of Living. In Neuvo, Y., & Ylönen, S. (eds.), Bit Bang - Rays to the Future. Helsinki University of Technology (TKK), MIDE, Helsinki University Print, Helsinki, Finland, 174-204. ISBN 978-952-248-078-1.

[2] Winter, Mick (2007). *Sustainable Living: For Home, Neighborhood and Community*. Westsong Publishing. ISBN 0-9659000-5-3.

[3] The Center for Ecological Living and Learning (CELL)– philosophy

[4] Lynn R. Kahle, Eda Gurel-Atay, Eds (2014). *Communicating Sustainability for the Green Economy.* New York: M.E. Sharpe. ISBN 978-0-7656-3680-5.

[5] Ross, Greg. "An interview with Lester Brown" *American Scientist.*

[6] Adams, W.M. (2006). "The Future of Sustainability: Rethinking Environment and Development in the Twenty-first Century." Report of the IUCN Renowned Thinkers Meeting, 29–31 January 2006. Retrieved on: 2009-07-25.

[7] U.S. Environmental Protection Agency "What is sustainability?" Retrieved on: 2007-08-20.

[8] United Nations General Assembly (2005). 2005 World Summit Outcome, Resolution A/60/1, adopted by the General Assembly on 15 September 2005. Retrieved on: 2009-07-25.

[9] http://citiesprogramme.com/aboutus/our-approach/ circles-of-sustainability

[10] Fritsch, Al; Paul Gallimore (2007). *Healing Appalachia: Sustainable Living Through Appropriate Technology.* University Press of Kentucky. p. 2. ISBN 0-8131-2431-X. Unknown retrieval date, revised: 2009-07-25

[11] Wheeler, Stephen Maxwell; Timothy Beatley (2004). *The Sustainable Urban Development Reader.* Routledge. ISBN 0-415-31187-X.

[12] Nearing, Scott; Helen Nearing (1953). *Living the Good Life.*

[13] Eroh, Ryan. "Scott Nearing". Pabook.libraries.psu.edu. Retrieved 2011-03-21.

[14] Lear, Linda. "Rachel Carson's Biography". Rachelcarson.org. Retrieved 2011-03-21.

[15] SI: Donella Meadows Bio Sustainability Institute 2004.

[16] E.F. Schumacher Bibliography Schumacher UK.

[17] National Sustainable Development Strategies United Nations Department of Economic and Social Affairs April 2008.

[18] Sustainable Consumption and Production: Promoting Climate-Friendly Household Consumption Patterns United Nations Department of Economic and Social Affairs 2007-04-30.

[19] Jeffery, Yvonne, Michael Grosvenor, and Liz Barclay. *Green Living for Dummies.* Indianapolis, IN: Wiley Pub., 2008. Print.

[20] McDilda, Diane Gow. *The Everything Green Living Book: Easy Ways to Conserve Energy, Protect Your Family's Health, and Help save the Environment.* Avon, MA: Adams Media, 2007. Print.

[21] McDilda, Diane Gow. *The Everything Green Living Book: Easy Ways to Conserve Energy, Protect Your Family's Health, and Help save the Environment.* Avon, MA: Adams Media, 2007. Print.

[22] http://blog.seattlepi.com/greenbuilding/2010/05/18/ what-is-a-sustainable-home-a-truly-green-home/

[23] Hamilton, Andy, and Dave Hamilton. *The Self-sufficient-ish Bible: an Eco-living Guide for the 21st Century.* London: Hodder & Stoughton, 2009. Print.

[24] Snell, Clarke, and Tim Callahan. *Building Green: a Complete How-to Guide to Alternative Building Methods : Earth Plaster, Straw Bale, Cordwood, Cob, Living Roofs.* New York: Lark, 2005. Print.

[25] Hamilton, Andy, and Dave Hamilton. *The Self-sufficient-ish Bible: an Eco-living Guide for the 21st Century.* London: Hodder & Stoughton, 2009. Print.

[26] Green Building Materials: Sustainable Building. CalRecycle. Web. 23 Oct. 2010.

[27] Hawken, Paul, Amory B. Lovins, and L. Hunter Lovins. *Natural Capitalism: Creating the next Industrial Revolution.* Boston: Little, Brown and, 1999. Print.

[28] Cutlip, Jamie. Green Roofs: A Sustainable Technology. UC Davis Extension, Oct. 2006. Web. 26 Oct. 2010.

[29] GREEN ROOF RESEARCH PROGRAM. Michigan State University - Department of Horticulture. Web. 27 Oct. 2010.

[30] How a Product Earns the ENERGY STAR Label : ENERGY STAR." ENERGY STAR. Web. 27 Oct. 2010.

[31] Brown, Lester Russell. *Plan B 4.0: Mobilizing to save Civilization.* New York: W.W. Norton, 2009. Print.

[32] Water Conservation. Mono Lake Committee. Web. 27 Oct. 2010.

[33] Graywater Reuse and Rainwater Harvesting. Colorado State University Extension. Web. 27 Oct. 2010.

[34] Environment Agency - Techniques. Environment Agency. Web. 27 Oct. 2010.

[35] Environment Agency - Sustainable Drainage Systems. Environment Agency. Web. 27 Oct. 2010.

[36] Buy Green Power and Electricity to Help the Environment. Consumer Reports: Expert Product Reviews and Product Ratings from Our Test Labs. Consumers Union of U.S., July 2007. Web. 28 Oct. 2010.

[37] Hamilton, Andy, and Dave Hamilton. The Self-sufficient-ish Bible: an Eco-living Guide for the 21st Century. London: Hodder & Stoughton, 2009. Print.

[38] Galbraith, Kate. Europe's Way of Encouraging Solar Power Arrives in the US. Editorial. *The New York Times.* 13 Mar. 2009, New York ed., Section B sec.: B1. 12 Mar. 2009. Web. 28 Oct. 2010.

[39] Solar Energy Technologies Program: Concentrating Solar Power. Energy Efficiency & Renewable Energy. US Department of Energy, 19 Oct. 2010. Web. 31 Oct. 2010.

[40] McDilda, Diane Gow. The Everything Green Living Book: Easy Ways to Conserve Energy, Protect Your Family's Health, and Help save the Environment. Avon, MA: Adams Media, 2007. Print.

[41] Jeffery, Yvonne, Michael Grosvenor, and Liz Barclay. Green Living for Dummies. Indianapolis, IN: Wiley Pub., 2008. Print.

[42] Energy Savers: Solar Water Heaters. Energy Efficiency & Renewable Energy. US Department of Energy, 20 Oct. 2010. Web. 28 Oct. 2010.

[43] Brown, Lester Russell. Plan B 4.0: Mobilizing to save Civilization. New York: W.W. Norton, 2009. Print.

[44] Brown, Lester R. *Plan B 4.0: Mobilizing to Save Civilization.* W.W. Norton, 2009.

[45] Heinberg, Richard. *Powerdown: Options and Actions for a Post-Carbon World.* Canada: New Society Publishers, 2004.

[46] Astyk, Sharon. *Depletion and Abundance: Life on the New Home Front.* Canada: New Society Publishers, 2008.

[47] Shiva, Vandana. *Stolen Harvest: The Hijacking of the Global Food Supply.* Cambridge, MA: South End Press, 2000.

[48] Seymour, John. *The Self-Sufficient Life and How to Live It.* London: DK Publishing, 2003.

[49] Princen, Thomas. *The Logic of Sufficiency.* New York: MIT Press, 2005.

[50] Todd, J. and N. J. Todd. *From Eco-Cities to Living Machines: Principles of Ecological Design.* Berkeley, CA: North Atlantic Books, 1994.

[51] Nabhan, Gary. *Coming Home to Eat.* Berkeley, CA: W.W. Norton, 2002.

[52] Organic Agriculture - What is Organic Agriculture? Iowa State University. 2008. Web. Retrieved on: 18 Nov 2010.

[53] Flores, H.C. *Food Not Lawns: How to Turn Your Yard into a Garden and Your Neighborhood into a Community.* New York: Chelsea Green, 2006.

[54] Nyerges, Christopher. *Urban Wilderness: a guidebook to resourceful city living.* Culver, CA: Peace Press, 1979.

[55] Hemenway, Toby. *Gaia's Garden.* New York: Chelsea Green, 2000.

[56] Warde, Jon, ed. *The Backyard Builder: Over 150 Projects for Your Garden, Home and Yard.* New York: Random House, 1994.

[57] Ciperthwaite, Wm. *A Handmade Life: In Search of Simplicity.* New York: Chelsea Green, 2004.

[58] Brown, Lester R. *Plan B 4.0: Mobilizing to Save Civilization.* New York: W.W. Norton, 2009.

[59] http://www.nytimes.com/2015/07/14/science/the-bicycle-and-the-ride-to-modern-america.html?_r=0

[60] Shaheen, Susan; Stacey Guzman (2011). "Worldwide Bikesharing". *Access: the Magazine of UCTC.*

[61] Shaheen, Susan; Guzman, S.; H. Zhang (2010). "Bikesharing in Europe, the Americas, and Asia: Past, Present, and Future" (PDF). *Transportation Research Record: Journal of the Transportation Research.*

[62] Hawken, Paul, Amory Lovins, and L. Hunter Lovins. *Natural Capitalism: Creating the Next Industrial Revolution.* New York City: Little, Brown and Company, 1999. Print.

[63] Indoor Water Use in the United States WaterSense: An EPA Partnership Program. US Environmental Protection Agency, 09 Nov. 2010. Web. 10 Nov. 2010.

[64] Hawken, Paul, Amory Lovins, and L. Hunter Lovins. *Natural Capitalism: Creating the Next Industrial Revolution.* New York City: Little, Brown and Company, 1999. Print.

[65] Green Building Health and Environmental Considerations in Building and Renovating Today Urban Builders Group. Urban Builders Group LTD. Web. 10 Nov. 2010.

[66] What is a Composting Toilet? Composting Toilet World. Envirolet. 2010. Web. 10 Nov. 2010.

[67] Pinkham, R. and Dyer, J., 1993: "Linking Water and Energy Savings in Irrigation," Rocky Mountain Institute Publication #A94-4.

[68] Soaker Hoses: Good for your Garden, Your Wallet, and Our Environment Saving Water Partnership. Seattle and Participating Area Water Utilities. 2005. Web. 10 Nov. 2010.

[69] How to Manage Stormwater: Rain Barrels. Stormwater Management for Clean Rivers. Environmental Services. Web. 10 Nov. 2010

[70] Greywater Systems: Reusing Bath, Laundry, and Sink Water to Conserve Fresh Water. Green Building Supply. 2010. Web. 10 Nov. 2010.

[71] https://www.catalogchoice.org/environmental-facts

[72] Reduce United States Environmental Protection Agency. 5 May 2010. Web 10 Nov. 2010

[73] Wastes – Resource Conservation – Reduce, Reuse, Recycle United States Environmental Protection Agency. 05 May 2010. Web 10 Nov. 2010

[74] UpCycle Sustainability Management. Presidio Graduate School. Web. 10 Nov. 2010

50.11 External links

Chapter 51

Sustainable refurbishment

Sustainable refurbishment describes working on existing buildings to improve their environmental performance using sustainable methods and materials.

Sustainable refurbishment is the equivalent of sustainable development which relates to new developments of cities, buildings or industries etc. Sustainable refurbishment includes insulation and related measures to reduce the energy consumption of buildings, installation of renewable energy sources such as solar water heating and photovoltaics, measures to reduce water consumption, and changes to reduce overheating, improve ventilation and improve internal comfort. The process of sustainable refurbishment includes minimising the waste of existing components, recycling and using environmentally friendly materials, and minimising energy use, noise and waste during the refurbishment.

The importance of sustainable refurbishment is that the majority of buildings in use are not new and thus were constructed when energy standards were low or non-existent, and are otherwise incompatible with current standards or the expectations of users. Much of the existing building stock is likely to be in use for many years to come since demolition and replacement is often unacceptable owing to cost, social disruption or because the building is of architectural and/or historical interest. The solution is to refurbish or renovate such buildings to make them appropriate for current and future use and to satisfy current requirements and standards of energy use and comfort.

Sustainable refurbishment is not a new concept but is gaining recognition and importance owing to current concerns about high energy use leading to climate change, overheating in buildings, the need for healthy internal environments, waste and environmental damage associated with materials production. Many governments are waking up to the importance of sustainably renovating their existing building stock, rather than just raising standards for new buildings and developments, and are producing guidance and grants and other support and stimulation activities. Think-tanks, lobby groups and voluntary organisations continue to publicise and promote the need for and practice of sustainable refurbishment. Examples and demonstration projects abound in many countries.

The techniques of sustainable refurbishment have been developing over many years and though the principles are very similar to those used on new buildings, the practice and details appropriate for the wide range of situations found in old buildings has required development of specific solutions and guidance to optimise the process and avoid subsequent problems. Detailed technical guidance is widely available from government-sponsored sources.

51.1 Example demonstration projects

- http://www.dena.de
- http://www.superhomes.org.uk.

51.2 Sources of technical guidance

- http://www.energysavingtrust.org.uk
- http://www.ademe.fr
- http://www.nps.gov/tps/standards/rehabilitation/guidelines/index.htm

51.3 References

Several books on the subject have been published aimed at different audiences, for example:

- for architects and other professionals:
  - *The Handbook for Sustainable Refurbishment – Housing*
  - *Handbook for Sustainable Refurbishment - Non-domestic Buildings*

- for the DIY market: *Sustainable Home Refurbishment*

all available from http://www.routledge.com.

51.4   See also

- Sustainable architecture

Chapter 52

Sustainable regional development

Sustainable regional development is the application of sustainable development at a regional, rather than local, national or global level. It differs to regional development per se, as the latter is a term used more generally to describe economic development that emphasises the alleviation of regional disparities. While regional development has an economic and equity emphasis, sustainable regional development seeks to incorporate ecological concerns.

Sustainable regional development has particular currency in Australia, where the Institute for Sustainable Regional Development has been established (1997) for the purpose of developing integrated, multi- and inter-disciplinary strategies for environmental and socio-economic change in regional Australia.[1]

52.1 References

[1] Institute for Sustainable Regional Development: About Us

Chapter 53

Sustainable drainage system

A **sustainable drainage system** (SuDs,[1] SuDS, SUDS[2][3]) is designed to reduce the potential impact of new and existing developments with respect to surface water drainage discharges. The term **sustainable urban drainage system**[1][4] is not the accepted name, the 'Urban' reference having been removed so as to accommodate rural sustainable water management practices.[5][6][7]

53.1 Background

Increasing urbanization has caused problems with increased flash flooding after sudden rain. As areas of vegetation are replaced by impermeable concrete, bituminous macadam or roofed areas the area loses its ability to absorb rainwater. This rain is instead directed into surface water drainage systems, often overloading them and causing floods.

The idea behind SuDS is to try to replicate natural systems that use cost effective solutions with low environmental impact to drain away dirty and surface water run-off through collection, storage, and cleaning before allowing it to be released slowly back into the environment, such as into water courses. This is to counter the effects of conventional drainage systems that often allow for flooding, pollution of the environment – with the resultant harm to wildlife – and contamination of groundwater sources used to provide drinking water. The paradigm of SuDS solutions should be that of a system that is easy to manage, requiring little or no energy input (except from environmental sources such as sunlight, etc.), resilient to use, and being environmentally as well as aesthetically attractive. Examples of this type of system are basins (shallow landscape depressions that are dry most of the time when it's not raining), rain-gardens (shallow landscape depressions with shrub or herbaceous planting), swales (shallow normally-dry, wide-based ditches), filter drains (gravel filled trench drain), bioretention basins (shallow depressions with gravel and/or sand filtration layers beneath the growing medium), reed beds and other wetland habitats that collect, store, and filter dirty water along with providing a habitat for wildlife.

Originally the term SUDS described the UK approach to sustainable urban drainage systems. These developments may not necessarily be in "urban" areas, and thus the "urban" part of SuDS is now usually dropped to reduce confusion. Other countries have similar approaches in place using a different terminology such as **best management practice** (BMP) and **low-impact development** in the United States,[8] and water-sensitive urban design in Australia .

SuDS use the following techniques:

- source control
- permeable paving such as pervious concrete
- storm water detention
- storm water infiltration
- evapo-transpiration (e.g. from a green roof)

A common misconception of SuDS systems is that they reduce flooding on the development site. In fact the SUDS system is designed to reduce the impact that the surface water drainage system of one site has on other sites. For instance, sewer flooding is a problem in many places. Paving or building over land can result in flash flooding. This happens when flows entering a sewer exceed its capacity and it overflows. The SuDS system aims to minimise or eliminate discharges from the site, thus reducing the impact, the idea being that if all development sites incorporated SuDS then urban sewer flooding would be less of a problem. Unlike traditional urban stormwater drainage systems, SuDS can also help to protect and enhance ground water quality.

53.2 History

The first sustainable drainage system to utilise a full management train including source control in the UK was Oxford services designed by Robert Bray Associates, specialist SuDS consultants.[9]

53.3 See also

- Detention basin

- Drainage system (disambiguation)

- French drain

- Rain garden

- Resin-bound paving

- Retention basin

- Stream restoration

- Sustainable city

- Urban runoff

- Urban drainage

53.4 References

[1] Sustainable Drainage System (SuDs) for Stormwater Management: A Technological and Policy Intervention to Combat Diffuse Pollution, Sharma, D., 2008

[2] "CIRIA guide to SUDS". Ciria.org. Retrieved 2014-01-21.

[3] Scottish Government. Planning Services (2001). "Planning and Sustainable Urban Drainage Systems." Planning Advice Note 61. 2001-07-27.

[4] Environmental investment may help local economy, 22 October 2002

[5] CIRIA Publication, 'evolution' May 2010

[6] CIRIA SuDS Manual (Document reference : CIRIA C697), 2007

[7] Susdrain - CIRIA's Sustainable Drainage Website.

[8] U.S. Environmental Protection Agency. Washington, DC (2006). "Fact Sheet: Low Impact Development and Other Green Design Strategies." 2006-06-01.

[9] CIRIA Oxford Motorway Services Case Study

10.   Variability of drainage and solute leaching in heterogeneous urban vegetation environs http://www.hydrol-earth-syst-sci.net/17/4339/2013/ hess-17-4339-2013.pdf

53.5 External links

UK

- http://www.susdrain.org - Free interactive community supporting the delivery of sustainable drainage (managed by CIRIA)

- Interpave – The UK's precast concrete paving and kerb association

- SUDS solutions from the British Geological Survey

- Engineering Natures Way - a dedicated resource for people working with Sustainable Drainage Systems (SUDS) and flood risk management in the UK

International

- Urban Drainage Engineering Royal Haskoning

- International Best Management Practices Database – Detailed data sets & summaries on performance of Urban BMPs

- Stormwater Industry Association of Australia

- Portland Guide to Sustainable Stormwater – City of Portland, Oregon

- National Menu of Stormwater BMPs – U.S. Environmental Protection Agency

- Pervious Concrete Blog

Chapter 54

Swale (landform)

Constructed swale or bioswale built in a residential area to manage stormwater runoff.

Natural swale

A **swale** is a low tract of land, especially one that is moist or marshy.[1] The term can refer to a natural landscape feature or a human-created one. Artificial swales are often designed to manage water runoff, filter pollutants, and increase rainwater infiltration.[2]

The swale concept has also been popularized as a rainwater harvesting and soil conservation strategy by Bill Mollison, Geoff Lawton and other advocates of permaculture. In this context it usually refers to a water-harvesting ditch on contour. Another term used is **contour bund**.[3][4]

Swales as used in permaculture are designed to slow and capture runoff by spreading it horizontally across the landscape (along an elevation contour line), facilitating runoff infiltration into the soil. This type of swale is created by digging a ditch on contour and piling the dirt on the downhill side of the ditch to create a berm. In arid climates, vegetation (existing or planted) along the swale can benefit from the concentration of runoff. Trees and shrubs along the swale can provide shade which decreases water evaporation.

The term swale or "beach swale" is also used to describe long, narrow, usually shallow troughs between ridges or sandbars on a beach, that run parallel to the shoreline.[5]

54.1 See also

- Bioswale
- Gutter
- Keyline design
- Rain garden
- Surface runoff
- Stormwater

54.2 References

[1] Merriam-Webster. "Swale." *Online Dictionary*. Retrieved 2009-09-21.

[2] U.S. Environmental Protection Agency (EPA). Washington, DC (1999). "Storm Water Technology Fact Sheet: Vegetated Swales." EPA Document No. 832-F-99-006. September 1999.

[3] "Water Harvesting: Microcatchment Contour Bunds". Food and Agriculture Organization. Retrieved 2009-11-26.

[4] "Soil contour bunds" (PDF). United Nations Office for Project Services. 1998. Retrieved 2009-11-26.

[5] Michigan State University Extension."Wetlands of the Great Lakes Open Shoreline and Embayed Wetlands." Retrieved 2009-09-21.

54.3   External links

- Fact Sheet: Grassed Swales from US Environmental Protection Agency

- Fact Sheet: Dry and Wet Vegetated Swales from Federal Highway Administration

- Wetlands of the Great Lakes: The Beach Swale & Dune and Swale Types from Michigan State University

- Video showing swales used to rehabilitate desert terrain

Chapter 55

Transition design

Transition design proposes design led societal transition towards more sustainable futures. It applies an understanding of the interconnectedness of social, economic, political and natural systems to address problems that exist at all levels of scale in ways that improve quality of life. Such complex and interconnected wicked problems can include poverty and economic inequality, biodiversity loss, decline of community, resource depletion, pollution and climate change.

Transition design leverages the power of interdependency and symbiosis with the aim of transforming entire lifestyles, making these more convivial and participatory, and harmonising them with the natural environment. It emphasizes the temporal element of all such design solutions — how they relate to the traditional and vernacular cultures of the past, and how they might unfold, develop and connect over short, medium and long horizons of time.

Transition design aims to intervene in social, economic, political and technological systems so as to assist people to satisfy their needs in ways that establish mutually beneficial relationships between people, the natural environment and the built and designed world. This process is informed in particular by living systems theory, employing such concepts as self-organization, interdependence, emergence, holarchy and phase transition. It involves changing the ways in which people earn their livelihood, and changing the organization of business, manufacturing, agriculture, finance, healthcare, education and travel. Transition design aims to cultivate lifestyles and forms of everyday life in which fundamental needs can be satisfied in integrated, place-based ways, encouraging symbiotic relationships between communities, and between communities and their ecosystems.

55.1 Transition designers

Transition designers can come from all walks-of-life and backgrounds, regardless of whether they are formally trained designers. They use the tools and processes of design to re-conceive entire lifestyles, rather than focusing on problems within existing, mostly unsustainable, social, economic and political paradigms. Designers assume a similar role in transition design as they do in service design or social design: the designer is a facilitator of emergent solutions to problems rather than an expert who conceives and delivers blueprints and finished solutions. Transition designers are especially focussed on potentialities for beneficial change that exist within systems, and on making modest interventions which can ramify throughout entire systems.

Because complex wicked problems are multi-faceted, their solutions require not only the skills and knowledge of designers, but also knowledge from the spectrum of disciplinary fields (science, philosophy, psychology, social science, anthropology and the humanities etc.). Transition design is therefore a collaborative process that is informed by knowledge from outside design. It emphasizes the need for transdisciplinarity and for the reintegration of knowledge.

Transition design solutions, however, should not simply represent the collaborative efforts of groups of experts and specialists; since solutions must be place-specific and ecosystem appropriate, they must incorporate local knowledge, participation and commitment. Just as transition design solutions need not only represent the reintegration of knowledge, they should also represent its recontextualization. Just as engineering has been used as a metaphor to describe the large scale, top-down and managerial influence of social phenomena (designer, urban planner, technocrat and policy maker as social engineers) so the transition designer might be compared to a gardener. Like a good gardener, the transition designer has an intimate understanding of and feeling for a particular place and its ecosystem, of the relationships between its different parts, of what its particular needs are, of what will and will not flourish, and of how it might grow and develop over long periods of time.

55.2 Comparison with other approaches

In designing for the relationship between people, nature, and the things that people make and do; in designing for short, mid and long horizons of time and in acknowledgement of vernacular, historical and indigenous cultures; and in designing place/ecosystem specific solutions, transition design can be seen as a logical development of a trend in design over recent decades that emphasizes relationship and context. Within mainstream design this trend has included service design (or design for service), social design (or design for social innovation) and interaction design, whilst outside of mainstream design this trend has included permaculture and ecological design.

In the School of Design at Carnegie Mellon University (CMU), transition design is taught alongside design for service and design for social Innovation, in the expectation that designers will be able to move freely between these three areas of practice and research, according to the scope of the problem that they hope to address. Thus design for service represents moderate change within existing paradigms and systems; design for social innovation represents significant change within emerging paradigms and systems; and transition design represents radical change within future paradigms and systems.

These areas of design practise and research can be located on a continuum, and the movement along this continuum (from design for service, through design for social innovation to transition design) represents an expansion of the scale of time, depth of engagement, and context to include social, political, economic and environmental concerns. Transition design's foregrounding of relationship, context and wholeness represents the incorporation into design of an ecological or holistic worldview and gives it several defining features:

- Living systems theory is used as both an approach to understanding wicked problems and designing solutions to address them.

- Designs solutions that protect and restore both social and natural ecosystems through the creation of mutually beneficial relationships between people, the things they do and make and the natural environment.

- Sees everyday life and lifestyles as the most important/fundamental context for design.

- Advocates place and ecosystem based, but globally networked solutions; seeks solutions that connect place to planet.

- Designs solutions for short, medium and long horizons of time and all levels of scale of everyday life.

- Looks for emergent possibilities within problem contexts and amplifies grassroots solutions already underway.

- Links existing solutions together so that they can function as steps in a larger transition vision.

- Distinguishes between 'wants/desires' and genuine needs and bases solutions on maximizing and integrating satisfiers for the widest range of needs.

- Sees the designer's own mindset and posture as an essential component of transition designing.

- Calls for the reintegration and recontextualization of diverse transdisciplinary knowledge.

55.3 The Transition Design Framework

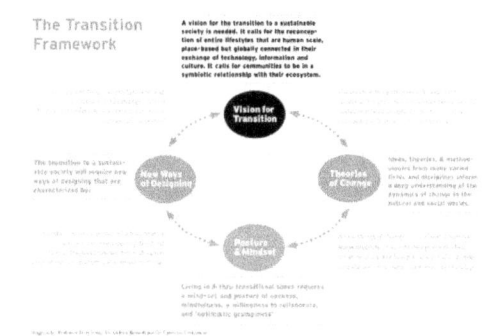

Transition design framework (Irwin, Tonkinwise & Kossoff, 2013)

Professor Terry Irwin, Head of the School of Design at Carnegie Mellon University, with Dr. Cameron Tonkinwise, Director of Design Studies and Chair of the doctoral program at Carnegie Mellon University, and Dr. Gideon Kossoff, Adjunct Professor in the School of Design at Carnegie Mellon University, have together developed a framework which organizes transition design into four mutually influencing areas: Vision for Transition; Theories of Change; Posture/Mindset and New Ways of Designing. This framework was first disseminated at a talk given by Irwin, Kossoff, and Tonkinwise at 'Head, Heart, Hand', the 2013 national AIGA design conference in Minneapolis.

55.4   Use of the term 'transition'

Transition design acknowledges that we are living in 'transitional times'. Its central premise is the need for societal transitions to more sustainable futures and the belief that design has a key role to play in these transitions. The use of the word 'transition' is suggested and reinforced by several contemporary discourses, projects and networks. In various ways these initiatives explore how change/transition manifests and can be initiated and directed, or are actively engaged in trying to bring about transition to more sustainable futures. Such initiatives have influenced transition design in several ways, and include:

55.4.1   Transition Town Network

Transition Towns, an international network of communities who are working to develop local resilience, and expand their capacity to respond to and 'bounce back' from external disruptions — crises such as an interruption to energy supplies, economic downturns or climate change. Transition Towns are, for example, developing locally based food and energy systems and local currencies and businesses.[1][2] By corollary, transition design involves a type of social engagement and community organizing that situates projects and initiatives within the context of long-term visions for specific places and ecosystems.

55.4.2   Transition Management

Socio-technical transition management, a body of theory which represents the convergence of sustainable development research, technology forecasting, social ecological impact analysis and the fields of social history and construction of technology. It studies the coevolution of technologies and their uses in order to conceive how innovations can be introduced into society to enable new ways of living and working.[3] Recently it has incorporated social practice theory and is beginning explicit use of ideas about designing. By corollary, transition design should involve a deep understanding of the social history of technology, and a post-planning approach to how the introductions of new technologies impact society and vice versa.

55.4.3   The Great Transition

The Great Transition, was a term first used in 1964 by the economist and systems theorist Kenneth Boulding. In 1995 the Global Scenario Group began to produce a series of reports identifying multiple future-based planetary scenarios and various strategies for change that could lead to the

'Great Transition' (improved quality of life, reduced poverty and inequity, human solidarity, enriched cultures and protection of the biosphere).[4] In recent years the Tellus Institute has launched the Great Transition Initiative (GTI), an international network of more than 40 scholars and activists who seek to develop and mobilize a planet-wide citizens transition movement. The concept of the 'Great Transition' has also been adopted by several leading think tanks, such as the New Economics Foundation. By corollary, transition design future based scenarios and the means and tools by which local communities can transition towards these within a planetary context.

55.4.4   Transition in complex and living systems

The concept of 'transition' in transition design alludes to the dynamic nature of complex, non-linear, self-organizing and interdependent systems such as ecosystems and organisms. The 'phase transition' that occurs as such systems undergo stresses or 'perturbations' from their environment, can lead them to display unexpected, unpredictable and 'emergent' new forms of behaviour and structure. By corollary, transition design incorporates living systems theory/ecoliteracy[5] in order to prepare designers for work within complex social systems, and to understand how these are embedded in natural systems. Transition design leverages the potential for self-organization, interdependence and emergence within such non-linear systems, and anticipates the possibility of rapid systemic change that can be triggered by modest interventions.

55.5   Cosmopolitan localism

Transition design focuses on the need for cosmopolitan localism,[6][7][8][9] a place-based lifestyle in which solutions to global problems are designed for local circumstances and tailored to specific social and ecological contexts. Its objective is to foster a global network of mutually supportive communities (neighbourhoods, villages, towns, cities and regions) who share and exchange knowledge, ideas, skills, technology, culture and (where socially and ecologically sustainable) resources. Cosmopolitan localism fosters a creative, reciprocal relationship between the local and the global. It addresses the problem of globalisation, which tends to subsume local cultures and economies into homogenised and unsustainable global systems,[10][11][12][13] whilst avoiding the pitfalls of localisation, such as parochialism and isolationism.[14][15]

55.6 Initial development

The concept of transition design was first developed by Gideon Kossoff in a chapter called 'Holism and the Reconstitution of Everyday Life: A Framework for the Transition to a Sustainable Society' in the book, *Grow Small, Think Beautiful: Ideas for a Sustainable World from Schumacher College*, edited by Stephan Harding.[16] This chapter was a summary of Kossoff's PhD thesis, also entitled *Holism and the Reconstitution of Everyday Life* (2011).[17] The term 'transition design' has since been adopted by the Carnegie Mellon School of Design and incorporated into its curriculum as one of three areas within which design is taught and researched at the undergraduate, graduate and doctoral levels (Design for Service, Design for Social Innovation and Transition Design).

In his PhD thesis, *Holism and the Reconstitution of Everyday Life*, Gideon Kossoff used the term transition design to describe the process of using design thinking and process to assist the transition to a sustainable society.[18] This early framing of transition design integrated holistic science (including chaos, complexity and systems theories,[19][20][21][22][23] and Goethean science)[24][25][26] with a tradition of non-authoritarian social and political philosophy that includes figures such as Murray Bookchin, Lewis Mumford, Peter Kropotkin, Martin Buber and Jane Jacobs.[27][28][29][30][31] These figures can be called 'radical holists' since they use organicist, holistic or ecological principles to underpin their advocacy of decentralized, mutualistic and participatory social, political and economic structures.[32] This 'radical holist' approach has been validated by holistic science which defines wholes in terms of dynamic and creative self-organisation, mutualistic interdependencies, and a reciprocal relationship between wholes and the parts of which they are constituted. These principles are at the heart of an emerging ecological or holistic worldview that has been articulated by Fritjof Capra, Richard Tarnas, David Abram and many others.[33][34][35][36][37][38][39] Transition design can be thought of as a way of using this ecological worldview to address many contemporary problems.

55.6.1 In everyday life

Transition design, in its initial development, focused on everyday life, since this is the context within which ecological, social and economic problems arise, and this is the arena within which such problems must be resolved. It therefore drew on various critiques of everyday life such as those made by Henri Lefebvre, Guy Debord and Mikhail Bakhtin.[40][41][42][43]

This framing of transition design proposes that in order to be sustainable, everyday life will need to be organised according to holistic principles. Everyday life arises as people strive to satisfy their material and non-material needs.[44][45][46] It is more likely to be sustainable when communities control the satisfaction of their needs at all levels of scale — households, neighbourhoods, villages, cities, regions — 'The Domains of Everyday Life'.[47][48] In traditional, pre-industrial communities, the Domains emerged as people endeavoured to satisfy their needs. Embodying the social processes of such cultures, the Domains were to varying degrees, self-organising, participatory, semi-autonomous and mutualistic, and everyday life as a whole consisted of nested networks of households, villages, neighbourhoods and regions. These characteristics are shared with living, whole and ecological systems.[49][50][51] With the arrival of modernity and the Industrial Revolution, control of need satisfaction was largely appropriated by centralized institutions. When this happens, the Domains lose their role as loci for the satisfaction of the needs of their inhabitants, and they no longer emerge as vital, integrated, semi-autonomous forms: households, neighbourhoods, cities and regions are typically hollowed and fragmented. Their decline leads to many social, ecological, economic and political problems and to the unsustainability of everyday life. There is therefore a direct relationship between sustainability, community and place-based control of need satisfaction and holism in everyday life: the transition to a sustainable society requires the reconstitution and reinvention of the Domains of Everyday Life. An additional Domain, that of the Planet, has emerged in modernity and its development could give everyday life a cosmopolitanism it lacked in pre-industrial societies.

Any place-based initiative that promotes holism in everyday life, protecting and recovering control by communities of the process of need satisfaction, will help the transition process. It is the task of transition design to facilitate this at all levels of scale, revitalising and improving the quality of everyday life through the development of nested, networked, integrated, semi-autonomous and self-organizing households, neighbourhoods, cities and regions.

55.7 Transition design education

Transition design was first introduced in 2014 at Carnegie Mellon University's School of Design where it informs projects, curricula and programs at the undergraduate, graduate and doctoral levels. On March 7, 2015 a Transition Design Symposium was held at CMU's School of Design. This was structured around the position papers of twenty academics from a wide range of disciplines (e.g. anthropology, sociology, architecture, psychology, human computer interaction, philosophy, design and design studies). Each position paper was a response to a 'Provocation and Briefing' document written by Professor Terry Irwin, Dr. Cameron Tonkinwise and Dr. Gideon Kossoff.[52]

55.7.1 Doctoral programs

- DDes in Transition Design (professional doctorate): Carnegie Mellon University, School of Design

- PhD in Transition Design: Carnegie Mellon University, School of Design

55.8 See also

- Circles of Sustainability

- Ecological design

- Modular design

- Open hardware

- Sustainable design

- Sustainability

- Transition management (governance)

- Transition Towns (network)

- Urban acupuncture

55.9 References

[1] Hopkins, R. *The Transition Handbook: From Oil Dependency to Local Resilience.* Chelsea Green, White River Junction, 2008

[2] Hopkins, R. *The Transition Companion: Making Your Community More Resilient in Uncertain Times*, Chelsea Green, White River Junction, 2011

[3] Grin J., Rotmans J., Transitions to Sustainable Development: New Directions in the Study of Long Term Transformative Change, Routledge, London, 2011

[4] Global Scenario Group, *Great Transition: Promise and Lure of the Times Ahead*, Stockholm Environment Institute, Boston, 2002

[5] Orr, D. *Ecological Literacy: Education and the Transition to a Post-Modern World*, SUNY, Albany, 1991

[6] Snyder, Gary. *Practice of the Wild*, Chap. 2, North Point Press, New York, 1990

[7] Sachs, W., *Planet Dialectics, Explorations of Environment and Development*, Chap. 6, Zed Books, London, 2000

[8] Morin, E. and Kern, A. *Homeland Earth: A Manifesto for the New Millennium*, Cresskill, 1999

[9] Manzini, E. 'The Scenario of a Multi-Local Society: Creative Communities, Active Networks and Enabling Solutions', in *Designers, Visionaries and Other Stories, A Collection of Sustainable Design Essays*, ed. Chapman, J. and Gant, N., Routledge, London, 2007

[10] Ritzer, G. *The McDonaldization of Society*, Pine Forge, Thousand Oaks, 2004

[11] Korten, D. *When Corporations Rule the World*, Berret-Koehler, San Francisco, 2001

[12] Shiva V. *Biopiracy: The Plunder of Nature and Knowledge*, Chap. 6 and 7, South End Press, Boston, 1999

[13] Mander, G. *The Capitalism Papers: Fatal Flaws of an Obsolete System*, Counterpoint, Berkeley, 2013

[14] Biehl J (ed.) *The Murray Bookchin Reader*, Chap. 8, Cassell, London, 1999

[15] Harvey, D. *The Condition of Postmodernity*, Chap. 6, Blackwell, Oxford, 2000

[16] Kossoff, G. 'Holism and the Reconstitution of Everyday Life' in *Grow Small, Think Beautiful: Ideas for a Sustainable World from Schumacher College*, ed. S. Harding, Floris, Edinburgh, 2011

[17] Kossoff, G. *Holism and the Reconstitution of Everyday Life: a Framework for Transition to a Sustainable Society*, PhD thesis, Centre for the Study of Natural Design, Duncan of Jordanstone College of Art and Design, University of Dundee, Dundee, 2011

[18] Kossoff, G., *Holism and the Reconstitution of Everyday Life: a Framework for Transition to a Sustainable Society*, Phd thesis, Centre for the Study of Natural Design, Duncan of Jordanstone College of Art and Design, University of Dundee, Dundee, 2011

[19] Peat, D. F. and Briggs J., *Turbulent Mirror: An Illustrated Guide to Chaos Theory and the Science of Wholeness*, Harper Perennial, New York, 1990

[20] Capra, F., *The Web of Life: A New Scientific Understanding of Living Systems*, Anchor, New York, 1997

[21] Wheatley, M. J. and Kellner-Rogers, M., *A Simpler Way*, Berrett-Koehler, San Francisco, 1998

[22] Koestler, A., *Janus: A Summing Up*, Random House, New York, 1978

[23] Meadows, D. H., *Thinking in Systems*, Chelsea Green, White River Junction, 2005

[24] Bortoft, H., *The Wholeness of Nature: Goethe's Way of Science*, Floris, 1996, Edinburgh

[25] Seamon, D., ed, *Goethe's Way of Science*, State University of New York, Albany,1999

[26] Hoffman N. and Dalton, P. *Goethe's Science of Living Form: The Artistic Stages*, Adonis, Hillsdale, 2007

[27] Biehl, J. *The Murray Bookchin Reader*, Black Rose Books, Montreal, 1999

[28] Mumford, L. *The Pentagon of Power: The Myth of the Machine*, Harcourt, Brace, Jovanovich, 1964

[29] Kropotkin, P. *Mutual Aid: A Factor of Evolution*, Freedom Press, London, 1987

[30] Buber, M. *Paths in Utopia*, Syracuse University Press, Syracuse, 1996

[31] Jacobs, J. *The Nature of Economies*, Vintage, New York, 2001

[32] Kossoff, G. *Holism and the Reconstitution of Everyday Life: a Framework for Transition to a Sustainable Society*, Chap. 3, Phd thesis, Centre for the Study of Natural Design, Duncan of Jordanstone College of Art and Design, University of Dundee, Dundee, 2011

[33] Tarnas, R., *The Greater Copernican Revolution and the Crisis of the Modern Worldview* in *A New Renaissance: Transforming Science, Spirit and Society*, ed. Lorimer D. and Robinson O., Floris, Edinburgh, 2011

[34] Speth, J.G, *The Bridge at the End of the World: Capitalism, the Environment and Crossing from Crisis to Sustainability*, Chap. 10, Yale University Press, New Haven, 2009

[35] Abram, D., 'The Mechanical and the Organic' in *Gaia in Action: Science of the Living Earth*, ed. P. Bunyard, Floris, 1996

[36] Hayward, J. *Letters to Vanessa: On Love, Science and Awareness in an Enchanted World*, Shambhala, Boston, 1997

[37] Mumford, L., *The Pentagon of Power: The Myth of the Machine*, Harcourt Brace Jovanovich, 1964

[38] Capra, F. *The Web of Life: A New Scientific Understanding of Living Systems*, Anchor, New York, 1997

[39] Goerner, S.J. *After the Clockwork Universe: The Emerging Science and Culture of Integral Society*, Floris, Edinburgh, 2001

[40] Bakhtin, M. *Toward a Philosophy of the Act* , University of Texas Press, Austin, 1993

[41] Debord, G. 'Perspectives for Alterations in Everyday Life' in *The Everyday Life Reader*, B. Highmore (ed.) Routledge, London, 2003

[42] Gardiner, M. *Critiques of Everyday Life*, Routledge, London, 2000

[43] Lefebvre, H. *Critiques of Everyday Life: Foundations for Sociology of the Everyday*, Verso, London, 1991

[44] Max-Neef, M., et al. *Human Scale Development: Conception, Application and Further Reflections*, The Apex Press, New York, 1991

[45] Kamenetzky, M.'The Economics of the Satisfaction of Needs' in *Real Life Economics: Understanding Wealth Creation*, P. Ekins and M. Max-Neef (eds.), Routledge, London, 1992

[46] Illich I. *Towards a History of Needs*, Heyday Books, Berkeley, 1978

[47] Kossoff, Gideon, *Holism and the Reconstitution of Everyday Life: a Framework for Transition to a Sustainable Society*, Chap. 5, Phd thesis, Centre for the Study of Natural Design, Duncan of Jordanstone College of Art and Design, University of Dundee, Dundee, 2011

[48] Kossoff, G. 'Holism and the Reconstitution of Everyday Life' in *Grow Small, Think Beautiful: Ideas for a Sustainable World from Schumacher College*, ed. S. Harding, Floris, Edinburgh, 2011

[49] Capra, F., *The Web of Life: A New Scientific Understanding of Living Systems*, Anchor, New York, 1997

[50] Wheatley, M. J. and Kellner-Rogers, M., *A Simpler Way*, San Francisco, 1998

[51] Koestler, A., *Janus: A Summing Up*, Random House, New York, 1978

[52] Irwin T., Tonkinwise C., Kossoff G., *Provocation and Briefing*, School of Design, Carnegie Mellon University, Pittsburgh, November 2014

55.10 External links

- Carnegie Mellon University, School of Design
- Transition Design Course Outline and Schedule
- Transition Design: Re-conceptualizing Whole Lifestyles, Video Lecture
- Cameron Tonkinwise, Transition Design, Video Lecture, Konstfack 2014
- Transition Network
- Sustainability Transitions Research Network
- Commons Transition

Chapter 56

Urban acupuncture

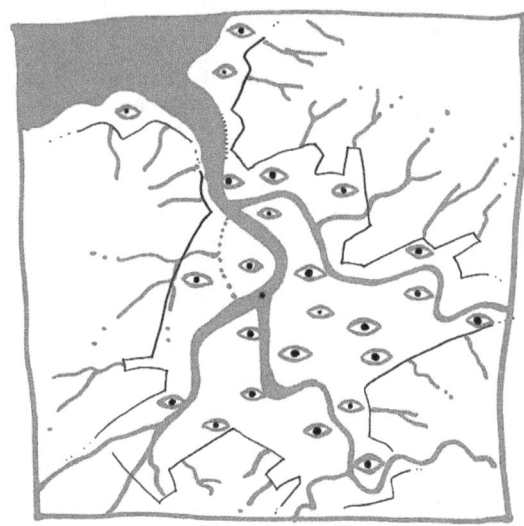

Taipei organic acupuncture

Bug Dome by WEAK! in Shenzhen, China. An unofficial social club for illegal workers next to the Shenzhen City Hall.

Urban acupuncture is a socio-environmental theory that combines contemporary urban design with traditional Chinese acupuncture,[1] using small-scale interventions to transform the larger urban context. Sites are selected through analysis of aggregate social, economic and eco-

logical factors, and are developed through a dialogue between designers and the community. Just as the practice of acupuncture is aimed at relieving stress in the human body, the goal of urban acupuncture is to relieve stress in the built environment.[2] Urban acupuncture is intended to produce small-scale but socially catalytic interventions in the urban fabric.[3]

56.1   Urban organism

This strategy views cities as living, breathing organisms and pinpoints areas in need of repair. Sustainable projects, then, serve as needles that revitalize the whole by healing the parts.[4] By perceiving the city as a living creature, thoroughly intertwined, "urban acupuncture" promotes communitarian machinery and sets localized nucleus —similar to the human body's meridians. Satellite technology, networks and collective intelligence theories, all used to surgically and selectively intervene on the nodes that have the biggest potential to regenerate.[5]

Originally coined by Barcelonan architect and urbanist, Manuel de Sola Morales,[6][7][8] the term has been recently championed and developed further by Finnish architect and social theorist Marco Casagrande, this school of thought eschews massive urban renewal projects in favour of a more localised and community approach that, in an era of constrained budgets and limited resources, could democratically and cheaply offer a respite to urban dwellers.[9] Casagrande views cities as complex energy organisms in which different overlapping layers of energy flows are determining the actions of the citizens as well as the development of the city. By mixing environmentalism and urban design Casagrande is developing methods of punctual manipulation of the urban energy flows in order to create an ecologically sustainable urban development towards the so-called 3rd Generation City (postindustrial city). The theory is developed in the Tamkang University of Taiwan[10] and at independent multidisciplinary research center Ruin Academy.[11] With focus on environmentalism and urban design, Casagrande defines urban acupuncture as a design tool where punctual manipulations contribute to creating sustainable urban development, such as the community gardens and urban farms in Taipei.[12]

Casagrande describes urban acupuncture as:

> *[a] cross-over architectural manipulation of the collective sensuous intellect of a city. City is viewed as multi-dimensional sensitive energy-organism, a living environment. Urban acupuncture aims into a touch with this nature.*[13] *and Sensitivity to understand the energy flows of the collective chi beneath the visual city and reacting on the hot-spots of this chi. Architecture is in the position to produce the acupuncture needles for the urban chi.*[14] *and A weed will root into the smallest crack in the asphalt and eventually break the city. Urban acupuncture is the weed and the acupuncture point is the crack. The possibility of the impact is total, connecting human nature as part of nature.*

Casagrande utilized the tenets of acupuncture: treat the points of blockage and let relief ripple throughout the body. More immediate and sensitive to community needs than traditional institutional forms of large scale urban renewal interventions would not only respond to localized needs, but do so with a knowledge of how city-wide systems operated and converged at that single node. Release pressure at strategic points, release pressure for the whole city.[15]

56.2 Participatory planning

In theory, urban acupuncture opens the door for uncontrolled creativity and freedom. Citizens are enabled to join the creative participatory planning process, feel free to use city space for any purpose and develop their environment according to their will.[16] This "new" post-industrialized city Casagrande, dubbed the Third Generation City, is driven by people who are concerned about the destruction that the modern machine is causing to nature including human nature. They want susstainable co-operation with the rest of the nature.[17] In a larger context, a site of urban acupuncture can be viewed as communicating to the city outside like a natural sign of life in a city programmed to subsume it.[18]

Urban acupuncture bears some similarities to the new urbanist concept of Tactical Urbanism. The idea focuses on local resources rather than capital-intensive municipal programs and promotes the idea of citizens installing and caring for interventions. These small changes, proponents claim, will boost community morale and catalyze

revitalization.[19] Boiled down to a simple statement, "urban acupuncture" means focusing on small, subtle, bottom-up interventions that harness and direct community energy in positive ways to heal urban blight and improve the cityscape. It is meant as an alternative to large, top-down, mega-interventions that typically require heavy investments of municipal funds (which many cities at the moment simply don't have) and the navigation of yards of bureaucratic red tape.[20] The micro-scale interventions targeted by "urban acupuncture" appeal to both citizen-activists and cash-strapped communities.[21] In Mexico urban acupuncture refers to a concept that converts temporary housing, like sheds in the slums, to simple homes that allow for "add-ons" later, based on need and affordability. This strategy transforms the slum zone, without relocating families that have been living together for generations.[22] In South Africa Urban Acupuncture is viewed as a possibility to provide a means for people to unlock their creativity and the advantages thereof, for example, innovation and entrepreneurship concentrating on parts of the city, i.e. communities thereby providing opportunities to those areas which do not have the sort of infrastructure that is found in mainstream cities. This approach can provide a more realistic and less costly method for city planners and citizens as an effective way to make minor improvements in the communities in order to achieve a greater good in the cities.[23]

Jaime Lerner, the former mayor of Curitiba, suggests urban acupuncture as the future solution for contemporary urban issues; by focusing on very narrow pressure points in cities, we can initiate positive ripple effects for the greater society. Urban acupuncture reclaims the ownership of land to the public and emphasizes the importance of community development through small interventions in design of cities.[24] It involves pinpointed interventions that can be accomplished quickly to release energy and create a positive ripple effect.[25] He described in 2007:
"I believe that some medicinal "magic" can and should be applied to cities, as many are sick and some nearly terminal. As with the medicine needed in the interaction between doctor and patient, in urban planning it is also necessary to make the city react; to poke an area in such a way that it is able to help heal, improve, and create positive chain reactions. It is indispensable in revitalizing interventions to make the organism work in a different way."[26]

Taiwanese architect and academic Ti-Nan Chi is looking with micro urbanism at the vulnerable and insignificant side of contemporary cities around the world identified as micro-zones, points for recovery in which micro-projects have been carefully proposed to involve the public on different levels, aiming to resolve conflicts among property owners, villagers, and the general public.[27]
A loosely affiliated team of architects Wang Shu, Marco Casagrande, Hsieh Ying-chun and Roan Ching-yueh

(sometimes called WEAK! Architecture) are describing the unofficial Instant City, or Instant Taipei, as architecture that uses the Official City as a *growing platform and energy source, where to attach itself like a parasite and from where to leach the electricity and water... [The Instant City's] illegal urban farms or night markets is so widespread and deep rooted in the Taiwanese culture and cityscape that we could almost speak of another city on top of the "official" Taipei, a parallel city – or a para-city.* WEAK! is calling urban acupuncture depending on the context as Illegal Architecture, Orchid Architecture, the People's Architecture, or Weak Architecture.[28] The theory of urban acupuncture suggests that scores of small-scale, less costly and localized projects is what cities need in order to recover and renew themselves.[29]

56.3 Deutsche Bank Urban Age Award

The winner of the Deutsche Bank Urban Age Award (DBUAA) Cape Town 2012 is the neighbourhood project "Mothers Unite" demonstrating the power of urban acupuncture.[30] Mothers Unite was founded in 2007 and provides a safe haven from the gangsterism, drugs and violence that are part and parcel of street and home life in the area. Built with donated shipping containers, the village is made up of a library, kitchen, office, sheltered area, playground and food garden.[31] Professor Edgar Pieterse:

> *Durable urban change is often about carefully targeted micro interventions that can change the energies and dynamics of a surrounding neighbourhood. When one engages with the physical manifestations that the tenacity, blood, sweat, and tears of the protagonists of Mothers Unite and Masiphumelele Library (another of the eight finalists) have created, the power of urban acupuncture is apparent. These projects serve as reminders that through vision, commitment over the long haul, and principle-based partnerships, just about any problem can be confronted and addressed.*[32]

56.4 Urban ecological restoration

In ecological restoration of industrial cities Urban Acupuncture can take form as spontaneous and often illegal urban farms and community gardens punctuating the more mechanical city and tuning it towards a more sustainable coexistence with the natural environment.[33] Urban Acupuncture areas can receive, treat and recycle the waste from the surrounding city acting as eco-valleys within the urban fabric. In River Urbanism the Urban Acupuncture areas can include underground stormwater reservoirs and act as flood relief for the surrounding city as a sponge and they can act as biological filters purifying water originated from polluted rivers.[34] Urban acupuncture is a point by point manipulation of the urban energy to create a sustainable town or city, which Marco Casagrande has dubbed '3rd Generation Cities'.[35]

56.5 Urban acupuncture in art

American artist Gordon Matta-Clark is credited with developing a system for identifying pockets of disrepair in the built environment—the first step in the framework of urban acupuncture.[36] Artist Miru Kim explores industrial ruins and structures making her look at the city as one living organism. She claims to feel not only the skin of the city, but also to penetrate the inner layers of its intestines and veins, which swarm with minuscule life forms.[37] Referring to a public environmental art work, Cicada, Casagrande explains:

> *Cicada is urban acupuncture for Taipei city penetrating the hard surfaces of industrial laziness in order to reach the original ground and get in touch with the collective Chi, the local knowledge that binds the people of Taipei basin with nature. The cocoon of Cicada is an accidental mediator between the modern man and reality. There is no other reality than nature.*[38]

Environmental art as urban acupuncture is an artistic way of injecting a healthy dose of natural elements and human scale into the mechanized urban grid.[39]

56.6 Projects

- Treasure Hill, Marco Casagrande

56.7 See also

- Ecological urbanism
- Landscape urbanism
- Urban design
- Urban planning

56.8 References

56.8.1 Notes

[1] Parsons, Adam. (2010, December). "Urban Acupuncture: Marco Casagrande." University of Portsmouth.

[2] "Urban Acupuncture," *Urban Applications,* 2013.

[3] Ruin Academy - Casagrande Lab. In *Architectural Theories of the Environment: Posthuman Territory,* ed. Ariane Lourie Harrison. Routledge, 2013.

[4] Better Blocks: One of Many Urban Acupuncture Needles - *Kelly McCartney, Shareable : Cities* 8/2011

[5] Acupuntura urbana para sanar una ciudad - *Martha Salotti, Sphere* 2012

[6] Frampton, Kenneth. Megaform as urban landscape. University of Michigan, A. Alfred Taubman College of Architecture+ Urban Planning, 1999.

[7] i Rubió, Manuel de Solà-Morales. Progettare città. Vol. 23. Electa, 1999.

[8] De Solà-Morales, Manuel. "The strategy of urban acupuncture." Structure Fabric and Topography Conference, Nanjing University. 2004.

[9] Could cities' problems be solved by urban acupuncture? - *Leon Kaye, The Guardian* 21.7.2010

[10] INTERVIEW WITH M. CASAGRANDE ON URBAN ACUPUNTURE - *Laurits Elkjær, Bergen School of Architecture* 4/2010

[11] Anarchist Gardener Issue Two HK special 安那其花園第二期香港特輯都市針灸城市空 - *Nikita Wu, Ruin Academy* 2/2012

[12] PARTICIPATIVE DESIGN & PLANNING IN CONTEMPORARY URBAN PROJECTS - *Christina Rasmussen, Urban Planning & Management, Aalborg University* 6/2012

[13] "Urban Acupuncture: Revivifying Our Cities Through Targeted Renewal," - Kyle Miller, MSIS 9/2011

[14] "Ruin Academy," - Marco Casagrande, Epifanio 14, 2011

[15] Urban Acupuncture - *Urban Applications* 2013

[16] Compost City - Guoda Bardauskaite *p. 30-31, Sustainable Urban Design Journal 1* 2011

[17] Urban Acupuncture - *Raune Frankjær, Rethink: Urban Interaction* 10/2012

[18] Urban Acupuncture - *Kelly Chan, Architizer* 1/2012

[19] 'Urban acupuncture' touted for cash-strapped cities - *David West, New Urban Network* 7/2011

[20] London's Urban Acupuncture: The Urban Physic Garden - *This Old Street* 8/2011

[21] 'Urban acupuncture' touted for cash-strapped cities - *David West, Better Cities & Towns* 7/2011

[22] Star architect leaves it all to build homes for the poor - *Yam Phui Yee, Good Times - The Conscience of the Nation*

[23] Urban Acupuncture - *Mariam Mahomed, Urban Spaces* Review *2012*

[24] Urban Acupuncture - *Understanding Space* 2011

[25] Curitiba: Jaime Lerner's Urban Acupuncture - *Bill Hinchberger, Brazilmax* 2/2006

[26] Urban Acupuncture: Revivifying Our Cities Through Targeted Renewal - Kyle Miller *MSIS 9/2011*

[27] Chi Ti-Nan develops a project to preserve Hong Kong coastline Tai Long Sai Wan - *World Architecture News* 4/2011

[28] Illegal Architecture in Taipei - Kelsey Campbell-Dollaghan *Architizer 3/2011*

[29] Could Cities Benefit from Small-Scale, Local "Urban Acupuncture" Projects Like This? - Kimberly Mok *Treehugger 1/2012*

[30] Deutsche Bank Urban Age Award – 19 April 2012 - *Mothers Unite* 4/2012

[31] Deutsche Bank Urban Age Award 2012 – winner announced - *Deutsche Bank Corporate Social Responsibility* 4/2012

[32] Mothers Unite: Urban acupuncture for an urban age - *Cape Town 2014 Live Design. Transform Life.* 4/2012

[33] Taipei Organic Acupuncture – *Marco Casagrande, P2P Foundation* 12/2010

[34] Urban Ecopuncture – *Marco Casagrande, World Architecture* 10/2011

[35] Amsterdam: Urban Acupuncture creates new life in the suburbs – *Danish Architecture Centre, Sustainable Cities* 6/2010

[36] Urban Acupuncture: Revivifying Our Cities Through Targeted Renewal - Kyle Miller *MSIS 9/2011*

[37] Miru Kim - *TED Talks* 2/2009

[38] Could Cities Benefit from Small-Scale, Local "Urban Acupuncture" Projects Like This? - Kimberley Mok *Treehugeer 1/2012*

[39] Signs of Life - Phyllis Richardson *Archetcetera* 1/2012

56.8.2 Further reading

• Lerner, Jaime. (2014). *Urban Acupuncture.* Washington, DC: Island Press. ISBN 9781610915830

56.9   External links

- ar2com talks with Marco Casagrande about his architecture, urban acupuncture, teaching, real and unreal spaces (March 2010).

- Micro-Urbanism, Ti-Nan Chi.

- Anarchist Gardener (ed. Nikita Wu).

- Human (Casagrande Laboratory).

Chapter 57

Used good

"Second hand" redirects here. For other uses, see Second hand (disambiguation).

A **second-hand** or **used good** is one that is being pur-

A busy yard sale, a common place to find cheap used goods.

chased by or otherwise transferred to a second or later end user. A *used* good can also simply mean it is no longer in the same condition as it was when transferred to the current owner. When "used" means an item has expended its purpose (such as a used diaper), it is typically called garbage, instead. Used goods may be transferred informally between friends and family for free as "hand-me-downs" or they may be sold for a fraction of their original price at garage sales or in church bazaar fundraisers. Governments require some used goods to be sold through regulated markets, as in the case of items which have safety and legal issues, such as used firearms or cars; for these items, government licensing bodies require certification and registration of the sale, to prevent the sale of stolen, unregistered, or unsafe goods. As well, for some high-value used goods, such as cars and motorcycles, governments regulate used sales to ensure that the government gets its sales tax revenue from the sale.

57.1 Benefits

Second-hand goods can benefit the purchaser as the price paid is lower than the same items bought new. If the reduction in price more than compensates for the possibly shorter remaining lifetime, lack of warranty, and so on, there is a net benefit.

Selling unwanted goods second-hand instead of discarding them obviously benefits the seller.

Recycling goods through the second-hand market both reduces use of resources in manufacturing new goods, and diminishes waste which must be disposed of, both significant environmental benefits. However, manufacturers who profit from sales of new goods lose corresponding sales. Scientific research shows that buying used goods reduces carbon footprint and CO_2 emissions significantly compared to the complete product life cycle, because of less production, raw material sourcing and logistics.[1] Often the relative carbon footprint of production, raw material sourcing and the supply chain is unknown.[2] A scientific methodology has been made to analyze how much CO_2 emissions are reduced when buying used goods like second hand hardware versus new hardware.[3]

Quality second hand goods can be more durable than equivalent new goods.[4]

57.2 Risks

Second-hand goods may have faults which are not apparent even if examined; purchasing sight unseen, for example, from an Internet auction site, has further unknowns. Goods may cause problems beyond their value; for example, furniture may have not easily seen bedbugs,[5] which may cause an infestation which is difficult and expensive to eradicate. Faulty electrical and mechanical goods can be hazardous and dangerous. This is especially a big issue if sold to countries that do not have recycling facilities for these devices, which has led to an issue with electronic waste.

Goods sold second-hand may have been stolen. In this case, in most jurisdictions the goods do not become the legal property of the purchaser; and knowing purchase of stolen goods is a criminal offense.

57.3   Types of transfers

Many items that are considered obsolete and worthless in developed countries, such as decade-old hand tools and clothes, are useful and valuable in impoverished communities in the country or in developing countries. Underdeveloped countries like Zambia are extremely welcoming to donated second-hand clothing. At a time when the country's economy was in severe decline the used goods provided jobs by keeping "many others busy with repairs and alterations".[6] It has created a type of spin-off economy at a time when many Zambians were out of work. The used garments and materials that were donated to the country also allowed for the production of "a wide range of fabrics" whose imports had been previously restricted.[6] The trade is essentially executed by women who operate their small business based on local associations and networks. Not only does this provide self-employment, but it also increases household income and enhances the economy.[6] But while many countries would be welcoming of second-hand goods, it is also true that there are countries in need who refuse donated items. Countries like Poland, Philippines and Pakistan have been known to reject second hand items for "fear of venereal disease and risk to personal hygiene".[6] Similar to these countries, India also refuses the import of second-hand clothing but will accept the import of wool fibers, including *mutilated hosiery* which is a term meaning "woollen garments shredded by machine in the West prior to export".[6] Through the production of *shoddy*, most of which is produced in Northern India today, unused clothing can be recycled into fibres that are spun into yarn for reuse in 'new' used goods.[6] The acceptance of second hand goods goes beyond the need of the people and into the hands of government officials. United States taxpayers can deduct donations of used goods to charitable organizations. Both Goodwill Industries and the Salvation Army web sites have lists of items with their estimated range of values. Another way that people transfer used goods is by giving them to friends or relatives. When a person gives an item of some value that they have used to someone else, such as a used car or a winter coat, it is sometimes referred to as a "hand-me-down".

Used items can often be found for sale in thrift stores and pawnshops, auctions, garage sales, and in more recent times online auctions. Some stores sell both new and used goods (e.g. car dealerships), while others only sell new goods but may take used items in exchange for credit toward the pur-

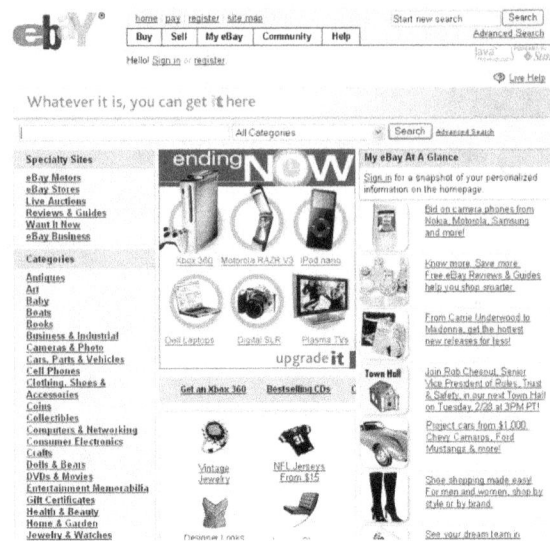

Online auction sites have become a popular way to sell used goods in recent years.

chase of newer goods. For example, some musical instrument stores and high-end audio stores only sell new gear, but they will accept good quality used items as trade ins towards the purchase of new items; after the store purchases the used items, they then sell them using online auctions or other services.

When an item is no longer of use to a person they may sell or *pawn* it, especially when they are in need of money. Items can also be sold (or taken away free of cost) as scrap (e.g. a broken-down old car will be towed away for free for its scrap metal value). Owners may sell the good themselves or to a dealer who then sells it on for a profit. They may also choose to give it away to another person this is often referred to as freecycling. However, because the process takes some effort on part of the owner they may simply keep possession of it or dump it at a landfill instead of going to the trouble of selling it. It has been common to buy second-hand or used good on markets or bazaars for long time. When the web became popular, it became common with web sites such as eBay and Yahoo! Classifieds.

57.4   Purchase

The strategy of buying used items is employed by some to save money, as they are typically worth less than the equivalent new items. Purchasing used items for reuse prevents them from becoming waste and saves costly production of equivalent new goods. Motivations for purchase include conserving natural resources and protecting the environment, and may form part of a simple living plan.

As dumped used goods take up space in landfills some may purchase them for environmental motivations.

Used cars like this 1980s-era Toyota Corolla are very inexpensive, but a buyer runs the risk of getting stuck with a "lemon".

Despite this, many people prefer to buy most or all of their goods new. They may feel safer buying new because a warranty is provided or because they are concerned they may be buying stolen goods. Goods purchased secondhand may also be exempt from certain legal requirements (e.g. consumer protection laws). Other consumers may be willing to buy used, but simply do not know where to buy them or lack the expertise needed to make a good purchase (e.g. a used car). Haggling may be involved in purchase of used goods, especially in less formal situations like a yard sale or in pawnshops, where negotiation is often done. Some consumers are uncomfortable or inexperienced in this situation too, and may choose to buy new goods instead. However some simply prefer their goods brand new and/or feel secondhand items are inferior or shabby (the 'yuck factor'), especially in the case of clothing or items used for eating such as plates or cutlery. Although this view is predominant in Western and developed nations, it is not universally held. Populations of many second- and third-world nations often prefer second-hand items for their relative availability and lower cost, as much of the used clothing from the US and European nations are exported to developing nations for resale.

57.5 Types

57.5.1 Cars

Main article: Used car

Used cars are especially notable for depreciating in value much faster than many other items. Used cars may have been bought or leased by their previous user, and may be purchased directly from the previous owner or through a dealer. George Akerlof published a paper entitled "The

Market for Lemons", examining the effects of information asymmetry on the used car market. Used cars may require more maintenance or have fewer features than later equivalent models. For used plug-in or all-electric vehicle batteries, see V2G.

57.5.2 Other items

The Sierra Club, an environmental organization, argues that second-hand purchasing of furniture is the "greenest" way of furnishing a home.[7] Used clothing is often donated to charities which sort and distribute it to people on low incomes for free or a very low price. Rather than distributing the clothing directly, some organizations will sell collected clothing in bulk to a used clothing redistributor, and then use the raised funds to finance either their charitable or for-profit activities, as they see fit.[8] Used clothing unsuitable for sale in an affluent market may still find a buyer or end-user in another market, such as a student market or a less affluent region of a developing country. In developing countries, such as Zambia, second-hand clothing is sorted, recycled and sometimes redistributed to other nations. Some of the scraps are kept and used to create unique fashions which enable the locals to construct identity. Not only does the trade represent a great source of employment for women as well as men, it also supports other facets of the economy: the merchants buy timber and other materials for their stands, metal hangers to display clothing, and food and drinks for costumers. Carriers also find work as they transport the garments from factories to various locations. The second-hand clothing trade is central to the lives of many citizens dwelling in such countries.[9] A Dress Agent will often deal with a consumer buyer and seller directly, taking unwanted clothes that still have value, and reselling them in a shop.

Vintage guitars also became increasingly desired objects among musicians and collectors during the nineties and afterwards.

57.6   See also

- Auto auction

- Car boot sale

- Charity shop (also known as thrift store, hospice shop, resale shop, op shop, or second-hand shop)

- Flea market

- Freecycling

- Freeganism

- Garage sale

- Regifting and Regiving

- Reseller

- Secondary market

- *The Market for Lemons*, a book discussing a phenomenon that may make it difficult to maintain quality in markets for certain used goods, such as computers and cars

- *Atomic Ed and the Black Hole*, a documentary film about a unique second-hand shop

57.7   References

[1] Impact of closed-loop network configurations on carbon footprints: A case study in copiers

[2] http://www.durabilit.eu/Plone/news/ how-to-reduce-your-company2019s-carbon-footprint-2013-by-reuse

[3] Scientific Greener CO2 calculator for reuse

[4] Buying secondhand: an alternative to rampant consumerism of Black Friday

[5] Msnbc: What you need to know about bed bugs: "Do not buy used furniture (especially bedding items or upholstered items) ... until inspected carefully for any signs of bedbugs"

[6] Karen Tranberg-Hansen.   2004.   Helping or hindering? Controversies around the international second-hand clothing trade. Anthropology Today 20(4):3-9.

[7] "Green Your Rental - Eco Furnishings - The Green Life". Sierraclub.typepad.com. 2008-09-30. Retrieved 2012-11-03.

[8] Old duds, big bucks; Clothes you think you're donating to charity are frequently sold for profit Toronto Sun, 2007-Jan-11, retrieved 2007-Apr-16.

[9] Hansen, Karen Tranberg. 2004. Helping or hindering? Controversies around the international second-hand clothing trade. Anthropology Today 20 (4):3-9

57.8   External links

Chapter 58

Water conservation

United States postal stamp advocating water conservation.

Water conservationist encompasses the policies, strategies and activities to manage fresh water as a sustainable resource, to protect the water environment, and to meet current and future human demand. Population, household size and growth and affluence all affect how much water is used. Factors such as climate change will increase pressures on natural water resources especially in manufacturing and agricultural irrigation.[1]

The goals of water conservation efforts include:

- Ensuring availability of water for future generations. This requires that the withdrawal of fresh water from an ecosystem does not exceed its natural replacement rate.

- Energy conservation. Water pumping, delivery and waste water treatment facilities consume a significant amount of energy. In some regions of the world over 15% of total electricity consumption is devoted to water management.

- Habitat conservation. Minimizing human water use helps to preserve fresh fire habitats for local wildlife and migrating waterfowl, as well as reducing the need to build new dams and other water diversion infrastructures.

58.1   Strategies

In implementing water conservation principles, there are a number of key activities that may be beneficial.

1. Any beneficial reduction in water loss, use and waste of resources.

2. Avoiding any damage to water quality.

3. Improving water management practices that reduce or enhance the beneficial use of water.[2][3]

58.2   Social solutions

Water conservation programs involved in social solutions are typically initiated at the local level, by either municipal water utilities or regional governments. Common strategies include public outreach campaigns,[4] tiered water rates (charging progressively higher prices as water use increases), or restrictions on outdoor water use such as lawn watering and car washing.[5] Cities in dry climates often require or encourage the installation of xeriscaping or natural landscaping in new homes to reduce outdoor water usage.[6]

One fundamental conservation goal is universal metering. The prevalence of residential water metering varies significantly worldwide. Recent studies have estimated that water supplies are metered in less than 30% of UK households,[7] and about 61% of urban Canadian homes (as of 2001).[8] Although individual water meters have often been considered impractical in homes with private wells or in multifamily buildings, the U.S. Environmental Protection Agency estimates that metering alone can reduce consumption by 20 to 40 percent.[9] In addition to raising consumer awareness of their water use, metering is also an important way to identify and localize water leakage. Water metering would benefit society in the long run it is proven that water metering increases the efficiency of the entire water system, as well as help unnecessary expenses for individuals for years to come. One would be unable to waste water unless they

Drip irrigation system in New Mexico

58.3 Household applications

The Home Water Works website contains useful information on household water conservation.[11] Contrary to popular view, experts suggest the most efficient way is replacing toilets and retrofitting washers.[12]

Water-saving technology for the home includes:

1. Low-flow shower heads sometimes called energy-efficient shower heads as they also use less energy

2. Low-flush toilets and composting toilets. These have a dramatic impact in the developed world, as conventional Western toilets use large volumes of water

3. Dual flush toilets created by Caroma includes two buttons or handles to flush different levels of water. Dual flush toilets use up to 67% less water than conventional toilets

4. Faucet aerators, which break water flow into fine droplets to maintain "wetting effectiveness" while using less water. An additional benefit is that they reduce splashing while washing hands and dishes

5. Raw water flushing where toilets use sea water or non-purified water

6. Waste water reuse or recycling systems, allowing:

   • Reuse of graywater for flushing toilets or watering gardens

   • Recycling of wastewater through purification at a water treatment plant. *See also Wastewater - Reuse*

7. Rainwater harvesting

8. High-efficiency clothes washers

9. Weather-based irrigation controllers

10. Garden hose nozzles that shut off water when it is not being used, instead of letting a hose run.

11. Low flow taps in wash basins

12. Swimming pool covers that reduce evaporation and can warm pool water to reduce water, energy and chemical costs.

13. Automatic faucet is a water conservation faucet that eliminates water waste at the faucet. It automates the use of faucets without the use of hands.

are willing to pay the extra charges, this way the water department would be able to monitor water usage by public, domestic and manufacturing services.

Some researchers have suggested that water conservation efforts should be primarily directed at farmers, in light of the fact that crop irrigation accounts for 70% of the world's fresh water use.[10] The agricultural sector of most countries is important both economically and politically, and water subsidies are common. Conservation advocates have urged removal of all subsidies to force farmers to grow more water-efficient crops and adopt less wasteful irrigation techniques.

New technology poses a few new options for consumers, features such and full flush and half flush when using a toilet are trying to make a difference in water consumption and waste. Also available in our modern world is shower heads that help reduce wasting water, old shower heads are said to use 5-10 gallons per minute. All new fixtures available are said to use 2.5 gallons per minute and offer equal water coverage.

58.4 Commercial applications

Many water-saving devices (such as low-flush toilets) that are useful in homes can also be useful for business water saving. Other water-saving technology for businesses includes:

- Waterless urinals

- Waterless car washes

- Infrared or foot-operated taps, which can save water by using short bursts of water for rinsing in a kitchen or bathroom

- Pressurized waterbrooms, which can be used instead of a hose to clean sidewalks

- X-ray film processor re-circulation systems

- Cooling tower conductivity controllers

- Water-saving steam sterilizers, for use in hospitals and health care facilities

- Rain water harvesting

- Water to Water heat exchangers.

58.5 Agricultural applications

Overhead irrigation, center pivot design

For crop irrigation, optimal water efficiency means minimizing losses due to evaporation, runoff or subsurface drainage while maximizing production. An evaporation pan in combination with specific crop correction factors can be used to determine how much water is needed to satisfy plant requirements. Flood irrigation, the oldest and most common type, is often very uneven in distribution, as parts of a field may receive excess water in order to deliver sufficient quantities to other parts. Overhead irrigation, using center-pivot or lateral-moving sprinklers, has the potential for a much more equal and controlled distribution pattern. Drip irrigation is the most expensive and least-used type, but offers the ability to deliver water to plant roots with minimal losses. However, drip irrigation is increasingly affordable, especially for the home gardener and in light of rising water rates. There are also cheap effective methods similar to drip irrigation such as the use of soaking hoses that can even be submerged in the growing medium to eliminate evaporation.

As changing irrigation systems can be a costly undertaking, conservation efforts often concentrate on maximizing the efficiency of the existing system. This may include chiseling compacted soils, creating furrow dikes to prevent runoff, and using soil moisture and rainfall sensors to optimize irrigation schedules.[9] Usually large gains in efficiency are possible through measurement and more effective management of the existing irrigation system. The 2011 UNEP Green Economy Report notes that "[i]mproved soil organic matter from the use of green manures, mulching, and recycling of crop residues and animal manure increases the water holding capacity of soils and their ability to absorb water during torrential rains",[13] which is a way to optimize the use of rainfall and irrigation during dry periods in the season.

58.6 See also

- Berlin Rules on Water Resources

- Conservation biology

- Conservation ethic

- Conservation movement

- Deficit irrigation

- Ecology movement

- Environmental protection

- GreenPlumbers

- Micro-sustainability

- Pan evaporation

- Peak water

- Sustainable agriculture

- Utility submeter

- Water cascade analysis

- Water metering

- Water pinch

- WaterSense - EPA conservation program

58.7   References

[1] "Water conservation « Defra". *defra.gov.uk*. 2013. Retrieved January 24, 2013.

[2] Vickers, Amy (2002). *Water Use and Conservation.* Amherst, MA: water plow Press. p. 434. ISBN 1-931579-07-5.

[3] Geerts, S.; Raes, D. (2009). "Deficit irrigation as an on-farm strategy to maximize crop water productivity in dry areas". *Agric. Water Manage* **96** (9): 1275–1284. doi:10.1016/j.agwat.2009.04.009.

[4] "Water - Use It Wisely." U.S. multi-city public outreach program. Park & Co., Phoenix, AZ. Accessed 2010-02-02.

[5] U.S. Environmental Protection Agency (EPA) (2002). Cases in Water Conservation (PDF) (Report). Retrieved 2010-02-02. Document No. EPA-832-B-02-003.

[6] Albuquerque Bernalillo County Water Utility Authority (2009-02-06). "Xeriscape Rebates". Albuquerque, NM. Retrieved 2010-02-02.

[7] "Time for universal water metering?" *Innovations Report.* May 2006.

[8] Environment Canada (2005). Municipal Water Use, 2001 Statistics (PDF) (Report). Retrieved 2010-02-02. Cat. No. En11-2/2001E-PDF. ISBN 0-662-39504-2. p. 3.

[9] EPA (2010-01-13). "How to Conserve Water and Use It Effectively". Washington, DC. Retrieved 2010-02-03.

[10] Pimentel, Berger; et al.   (October 2004). "Water resources: agricultural and environmental issues". *BioScience* **54** (10): 909. doi:10.1641/0006-3568(2004)054[0909:WRAAEI]2.0.CO;2.

[11] http://www.home-water-works.org

[12] http://www.pnas.org/content/early/2014/02/26/ 1316402111

[13] UNEP, 2011, Towards a Green Economy: Pathways to Sustainable Development and Poverty Eradication, www.unep. org/greeneconomy

58.8   External links

- Water Efficiency Magazine — Journal for Water Conservation Professionals

- Water Conservation Community of Interest — American Water Works Association

- Water Conservation — Water Quality Information Center, National Agricultural Library, U.S. Department of Agriculture

- Alliance for Water Efficiency (AWE)

- Smart WaterMark — Australian Water Conservation Label

Chapter 59

Wildland–urban interface

A **wildland–urban interface** refers to the zone of transition between unoccupied land and human development. Communities that are within 0.5 miles (0.80 km) of the zone may also be included. These lands and communities adjacent to and surrounded by wildlands are at risk of wildfires.

Urban development is pushing farther out of cities and into the wilderness for both primary and secondary residences. In the western states alone, 38% of the new development is taking place in the urban interface. As a result, the number of buildings and homes damaged as a result of accidental and natural forest fires has increased drastically. This has resulted in millions of dollars in damages.

One can prepare their property by performing some simple tasks. Ensure that the area around your building is free of any fuel (brush, high grasses, shrubs, small trees) that would allow a wildfire to reach your building. Also ensure trees are trimmed in a fashion that would prevent them from falling onto your structure and road that would prevent one from evacuating. If your structure is on a hill, modify the terrain in a manner that would prevent falling or rolling debris from hitting the structure or any fuels.

59.1   See also

- Rural-urban fringe

- Urban sprawl

- Edge effect

- Fire-adapted communities

- Habitat

    - Habitat destruction

    - Natural landscape

    - Restoration ecology

59.2   References

- ⊘ This article incorporates public domain material from the Congressional Research Service document "Report for Congress: Agriculture: A Glossary of Terms, Programs, and Laws, 2005 Edition" by Jasper Womach.

Page text.[1]

http://www.usfa.fema.gov/downloads/pdf/tfrs/v2i16-508.pdf

59.3   External links

- The eXtension Wildfire Information Network

- Fire Adapted Communities

- Fire Adapted Communities Learning Network

- Firewise Communities USA/Recognition Program

[1] Link text, additional text.

59.4 Text and image sources, contributors, and licenses

59.4.1 Text

- **Sustainable design** *Source:* https://en.wikipedia.org/wiki/Sustainable_design?oldid=685415167 *Contributors:* Ray Van De Walker, Edward, Gabbe, Ronz, William M. Connolley, Charles Matthews, Vespristiano, Chopchopwhitey, Mervyn, Alan Liefting, Michael Devore, Onco p53, Icairns, Rhobite, Vsmith, LindsayH, CHoltje, El C, Mwanner, Miscreant, Vortexrealm, Proton44, Mdd, Carbon Caryatid, Tabor, Woohookitty, Ksandler, Mindmatrix, Jwanders, Triddle, Skybum, Male1979, Rafiqelmansy, Behun, Graham87, Rjwilmsi, Lockley, Vegaswikian, Yurik-Bot, Wavelength, RussBot, Stephenb, Gaius Cornelius, Wimt, Rbccstmrtn, Dan337, CQ, Zzuuzz, Livitup, Chriswaterguy, Arcadie, Veinor, SmackBot, Bobet, McGeddon, Delldot, Stevepeterson, Oli Filth, Shalom Yechiel, Theanphibian, Radagast83, Byelf2007, Anlace, Gobonobo, Jaganath, Minna Sora no Shita, NathanLee, Bilby, Nobunaga24, Ckatz, Hu12, Blehfu, Courcelles, Mr3641, Webrawer, Van helsing, Ralph Purtcher, Mcginnly, Nadyes, Pfhenshaw, Treybien, Adolphus79, Dancter, DumbBOT, W55555, Kozuch, Shawnpatrick11, Pilgrimhawk, Gralo, Dvize, Dawnseeker2000, Tspearing, Mentifisto, Seaphoto, Tr1cycle, Alphachimpbot, JAnDbot, Harryzilber, SiobhanHansa, Bongwarrior, What123, Bubba hotep, Torchiest, Tinwithli, DGG, Flowanda, R'n'B, Thirstyfrog, Duane Elverum, J.delanoy, Jrsnbarn, Word2line, Akivett, Katalaveno, Skier Dude, EJ.v.H, Jorfer, Halrhp, Bonadea, Inwind, DASonnenfeld, Phil Monner, Lredman, Tweedlepop, Someguy1221, Jsbarrie, Oey192, Jackfork, Figureskatingfan, Mark Beever, Duaner, Wangel43, Shaker john, Holgerlonze, New England, Upload stuff, Designcouncil, Whiteghost.ink, Tiptoety, Jojalozzo, Gyokomura, KoshVorlon, Lightmouse, Tombomp, Tectoniks, Aeternity, CharlesGillingham, StaticGull, Dabomb87, Pinkadelica, Petiep, Loren.wilton, Grantrowe, ArchEcologist, ClueBot, Shubopshadangalang, Pittso, Fyyer, The Thing That Should Not Be, Flemingr2002, Farras Octara, Cfrayne, Dozols, DragonBot, Excirial, Cappiccuas, Lartoven, Xochipilli BE, Stefano Schiavon, Lmolinari, Poodledog, Bokunenjin, GlennMurdoch, DumZiBoT, FetteK, User2935, Zstowasser, XLinkBot, Chamben, Troykyo, Reubster1, NellieBly, ZooFari, Luminaia, Muffinon, Addbot, Tpgreenwood, Some jerk on the Internet, Jordandalladay, Monkipaj, MrOllie, Glass Sword, Ewandaley, Tassedethe, Alex Rio Brazil, Lightbot, Polainm, Gail, Jarble, Luckas-bot, Yobot, Millere08, Michelle.naquin, CleaLauren, AnomieBOT, Andrea Kantelberg, Pequod76, Crvaught, Materialscientist, Eumolpo, Mikeeco, Achapa, Crzer07, Sionk, Nehctik, Bellerophon, Alainr345, Dwarchitect, Grucio, FrescoBot, Ecogram, Poltrackt, Ribeiro.steven, Encycl wiki 01, Horst-schlaemma, LAEP Prof, Robtrob, Mikeramz1, IGW, Efratli, Andrewglaser, Diannaa, Mean as custard, Felipekovacic, SarahBachmann, Look2See1, Zollerriia, BillyPreset, Andrewswait, Tkelly50, Going-Batty, Slightsmile, Jspears501, Savh, Elaine2010, Cathyqaz, Akerans, Donaldsavoie, WebDragon10, Rcsprinter123, Rangoon11, Helpsome, ClueBot NG, Wisdawn, Mvisser008, Coastwise, Snotbot, ArryconVyper, Cntras, Widr, ?oygul, Lowercase sigmabot, BG19bot, Justincomprehenible, Northamerica1000, Kangaroopower, Pwhite3, MITclubfact, Demise101, Rowan Adams, Rodgox, BattyBot, Mmj2love, Nclmason, Lapithis, DATWIKIUSER, Elainechow71, Andyhowlett, Emharmsen, Docvallero, Vanamonde93, Mre env, Marcdupré, Everymorning, Bb-musicman, Georgiayelnats14, Glaisher, EliMes, Kim Ashfield, Dasari12, Lizia7, MaxSchumacher, Fixuture, Noegid, Skr15081997, Silver540, Garrettstrong, JordanBert, Wingburner, Graceditthardt and Anonymous: 279

- **Environmental design and planning** *Source:* https://en.wikipedia.org/wiki/Environmental_design_and_planning?oldid=492644113 *Contributors:* Alan Liefting, Rjwilmsi, Panoptical, Wavelength, Tous ensemble, DASonnenfeld, Hintss, Look2See1 and Anonymous: 1

- **Adaptive management** *Source:* https://en.wikipedia.org/wiki/Adaptive_management?oldid=685146561 *Contributors:* SimonP, Alan Liefting, Fonnesbeck, Alansohn, Matthias5, Woohookitty, Mandarax, Kbdank71, Josh Parris, Rjwilmsi, Ligulem, Lauchlin, Wavelength, Thiseye, Light current, Ilmari Karonen, Kellyoyo, Jppigott, SmackBot, Hongooi, Serein (renamed because of SUL), ThisIsAce, Daniel Cordoba-Bahle, R'n'B, DASonnenfeld, Funandtrvl, Herr.rhein, Cheesefondue, Nfbourne, Kwhitten, XLinkBot, Addbot, DOI bot, Lightbot, Yobot, Ulric1313, Citation bot, Chrisfonnesbeck, Vitmary, Citation bot 1, Cjs33, Trappist the monk, RjwilmsiBot, John of Reading, Rlempert, Look2See1, Jtua001, Sngordon, Helpful Pixie Bot, BG19bot, Robert the Devil, Patrick.Shulist, Robbru, FoCuSandLeArN, Skybluetwo and Anonymous: 28

- **Appropedia** *Source:* https://en.wikipedia.org/wiki/Appropedia?oldid=675623372 *Contributors:* Edward, Arcataroger, Pavel Vozenilek, Stesmo, Tim!, Koavf, Arthur Rubin, Chriswaterguy, SmackBot, Thumperward, Sct72, Rrburke, Cybercobra, Derek R Bullamore, Gobonobo, Wickethewok, RichardF, Jac16888, Kozuch, MER-C, Magioladitis, Allanlewis, Sm8900, Yaron K., Johnpacklambert, KylieTastic, Marcus334, Jojalozzo, Auntof6, Bearsona, Addbot, Enviro1, Viking59, AnomieBOT, Goodtimber, LilHelpa, HertzZ, LauraHale, Lotje, EmausBot, ZéroBot, Spork-Bot, Tiptheplanet, Donner60, ClueBot NG, Springlyn, BG19bot, BattyBot, Cheen Smith, Osat44, Ptrivers, Yxuie-te, Yadsalohcin, Thstevenson, Apihoneyyy, Kiwiblue12, Tatiana1112 and Anonymous: 15

- **Bioregionalism** *Source:* https://en.wikipedia.org/wiki/Bioregionalism?oldid=674462799 *Contributors:* The Anome, SebastianHelm, Tom Radulovich, Mboverload, Pgan002, Kevin B12, Discospinster, Rich Farmbrough, Viriditas, Bobrayner, Lapsed Pacifist, Rui Silva, Gurch, Dj Capricorn, YurikBot, Oberst, Retired username, Twelvethirteen, Mais oui!, SmackBot, Pfly, Verne Equinox, Grey Shadow, Eskimbot, OrionK, Chris the speller, Epastore, Kevlar67, Gurdjieff, Byelf2007, Caim, Robofish, Vagary, Heqs, Hayduke2000, CmdrObot, Cydebot, Maziotis, Jean Bean, ThisIsAce, Dantheman531, Aille, Skomorokh, Prospect77, Avicennais, Boffob, Sustainableyes, Ja 62, DASonnenfeld, Ottershrew, Tesscass, AlleborgoBot, Luftmann, The Thing That Should Not Be, TheOldJacobite, Gnome de plume, PixelBot, Oliver.renwick, Addbot, Granitethighs, Lightbot, Yobot, AnomieBOT, Materialscientist, Locobot, Tgv8925, Look2See1, Set theorist, Tommy2010, Helpful Pixie Bot, BG19bot, Portlandium, Begnome, Abootmoose, SashaGolden, Hmainsbot1, Cascadianmycelium, Ginsuloft and Anonymous: 38

- **Bioretention** *Source:* https://en.wikipedia.org/wiki/Bioretention?oldid=656633345 *Contributors:* Martinde, Wavelength, TDogg310, Rkitko, Robofish, Peter R Hastings, Eastlaw, BrianAsh, Alaibot, Mbell, Technopat, Amis2007, Euryalus, Moreau1, NJGW, MD3&4, Ka Faraq Gatri, AnomieBOT, FrescoBot, Geogene, Kibi78704, RjwilmsiBot, RichPro, Look2See1, J.silling, Lunderwood08, BattyBot, Monkbot, Katie Pekarek, Regfod, Mehreen at Vebkraze and Anonymous: 11

- **Blue-Green Cities** *Source:* https://en.wikipedia.org/wiki/Blue-Green_Cities?oldid=658730217 *Contributors:* Vsmith, SteinbDJ, Wavelength, Welsh, Ipigott, Yobot, AnomieBOT, John of Reading, BG19bot, EmilyBlueGreen and Anonymous: 1

- **Constructed wetland** *Source:* https://en.wikipedia.org/wiki/Constructed_wetland?oldid=683593560 *Contributors:* Edward, Samw, Conti, Tp-bradbury, Mushroom, Alan Liefting, DocWatson42, Quadell, Deirdre~enwiki, Rich Farmbrough, Travers, Duk, Vortexrealm, Giraffedata, Paleorthid, Woohookitty, Plrk, BD2412, Rjwilmsi, Zbxgscqf, Salix alba, Vegaswikian, MttJocy, Ground Zero, Parutakupiu, Sqrlmstr, Bgwhite, Wavelength, Ytrottier, Gaius Cornelius, Nicke L, TDogg310, Avalon, Chriswaterguy, SmackBot, KVDP, Hmains, Bluebot, Schaef, CSWarren, Whpq, Ligulembot, Anlace, Shlomke, Basicdesign, Gveret Tered, Eastlaw, Covalent, CmdrObot, Lamiot, Cydebot, Dyanega, Olawai, Stywiz, Darklilac, SteveOnline, Laikalynx, Lfstevens, Arch dude, Leolaursen, Magioladitis, EagleFan, JaGa, GTZ-44-ecosan, Brastein, Gate-way,

Shawn in Montreal, Squids and Chips, Uiew, Jeff G., Davidlburton, Lloyd roz, Waterbender kara, Una Smith, Lamro, ClueBot, GorillaWarfare, PipepBot, Niceguyedc, Rotational, Moreau1, Iohannes Animosus, Thewellman, Epiphaross, Dthomsen8, Addbot, DOI bot, Griffin700, South atlantic ocean, Bermicourt, Luckas-bot, Yobot, AnomieBOT, Walrus heart, Materialscientist, Citation bot, Bb143143, LilHelpa, J04n, Twirligig, Citation bot 1, Slobodan Grasic, Captain Selenium, Trappist the monk, RjwilmsiBot, Look2See1, L235, Hyronimus299, Fkt1, ClueBot NG, Mcguirep2010, Yingkeli, Bperlman99, Mark Marathon, BG19bot, StarryGrandma, Tattoodwaitress, Sidelight12, Ashutoshenv, Helga von Lichtenstein, Adqlth, Caillou 43, Monkbot, Filedelinkerbot, EvMsmile, EChastain, Mll mitch, TyCurious, Angusbran and Anonymous: 61

- **Sustainable automotive air conditioning** *Source:* https://en.wikipedia.org/wiki/Sustainable_automotive_air_conditioning?oldid=675463236 *Contributors:* Ewen, Alan Liefting, Niel Malan, Wwoods, Khalid hassani, Quadell, Rich Farmbrough, RoyBoy, Alansohn, The jt, Fred Bradstadt, Sandstein, Snaxe920, SmackBot, Roger Davies, DocKrin, Kvng, Dl2000, Iridescent, Fletcher, Jac16888, Electron9, Gioto, Yellowdesk, Ariel., All Is One, Katharineamy, Lamro, Fillinchen, HybridBoy, Flyer22, Smashville, ClueBot, Namazu-tron, Jtonti, Buckethed, Spitfire, AnomieBOT, Materialscientist, FrescoBot, GentleMiant, LWG, Technical 13, Northamerica1000, BattyBot and Anonymous: 20

- **Creative Energy Homes** *Source:* https://en.wikipedia.org/wiki/Creative_Energy_Homes?oldid=532061023 *Contributors:* Bearcat, SmackBot, Katharineamy, Robshipman, Addbot, Yobot, Catpowerzzz, Ksenia Ch and Anonymous: 2

- **Depression-focused recharge** *Source:* https://en.wikipedia.org/wiki/Depression-focused_recharge?oldid=557968197 *Contributors:* DragonflySixtyseven, RussBot, Stephenb, O keyes, Mwtoews, Robofish, BrianAsh, Alaibot, Doug Coldwell, BlueAmethyst, LilHelpa, EmausBot, Look2See1, AK456 and Mohamed-Ahmed-FG

- **Design for the Environment** *Source:* https://en.wikipedia.org/wiki/Design_for_the_Environment?oldid=671171666 *Contributors:* Skysmith, Mac, Bearcat, Alan Liefting, Rich Farmbrough, Mandarax, Rjwilmsi, Wavelength, CambridgeBayWeather, SmackBot, Elonka, Ohnoitsjamie, Ckatz, Woodshed, Mr3641, Amalas, SummerPhD, Goldenrowley, Mtoffel, Belovedfreak, Leif.barthel, Billinghurst, Hanwufu, Chimin 07, Moreau1, SoxBot, Vanished user uih38riiw4hjlsd, Fede.Campana, JTURI, Ronhjones, Yobot, AnomieBOT, Jaorquina, Adrianponce, J04n, PEAinWiki, FrescoBot, Footyfanatic3000, Abductive, Robtrob, Trappist the monk, AidaSWilliams, EmausBot, John of Reading, Look2See1, Occamisation, AvicAWB, Welovecleaning, Nwstein, ClueBot NG, Julkos, BattyBot, Jh mueller, Mmj2love, Mogism, Egelerp, Wuerzele, Bwang09, Miguelegui, Monkbot and Anonymous: 19

- **Drake Landing Solar Community** *Source:* https://en.wikipedia.org/wiki/Drake_Landing_Solar_Community?oldid=671212639 *Contributors:* Bearcat, Sunray, Hektor, Mindmatrix, Canadian Paul, Ytrottier, SmackBot, Neilanderson, MitchellShnier, Qyd, Cydebot, NorthernThunder, Harryzilber, Magioladitis, Waacstats, Red58bill, E8, JL-Bot, Rlest, Pairadox, Basketball110, Stepheng3, Yobot, Citation bot, 117Avenue, LittleWink, Lotje, Hwy43, Mean as custard, H3llBot, Coastwise, Kevin Gorman, Newyorkadam, SalaRollins, Caboothby, LaCarlotta and Anonymous: 6

- **Eco-cities** *Source:* https://en.wikipedia.org/wiki/Eco-cities?oldid=679377649 *Contributors:* Keldan, Bearcat, Discospinster, Rich Farmbrough, Woohookitty, BD2412, Wavelength, RussBot, Malcolma, McGeddon, Rrburke, Dawkeye, Widefox, DGG, Tgeairn, Owen Slatraigh, Mild Bill Hiccup, Jarble, Yobot, Fraggle81, AnomieBOT, PeterEastern, Gire 3pich2005, Horst-schlaemma, IRISZOOM, Tbhotch, Abu Shawka, Donner60, ClueBot NG, Widr, Reify-tech, Helpful Pixie Bot, Wbm1058, BG19bot, Bfrazer2008, ChrisGualtieri, Harshhussey, Mogism, Seqqis, Rob Cowley, Saectar, Naomi Grunditz, Monkbot, Ahmad Ali Soozandeh, YJAX, KH-1, Parveetsid, Ciudadania Digital and Anonymous: 22

- **Eco-municipality** *Source:* https://en.wikipedia.org/wiki/Eco-municipality?oldid=636853680 *Contributors:* Jose Icaza, Sunray, Mervyn, Toussaint, Vegaswikian, Frekja, Ryandwayne, Mais oui!, SmackBot, DWaterson, KVDP, RayAYang, Alaibot, DGG, DASonnenfeld, Nedrutland, Rodhullandemu, Addbot, Northwoodsman, Mimzy1990, Full-date unlinking bot, Darkohead, SusikMkr, Northamerica1000, Eddu1973 and Anonymous: 7

- **Ecodistrict** *Source:* https://en.wikipedia.org/wiki/Ecodistrict?oldid=549191251 *Contributors:* GoodDay, CommonsDelinker, Kudpung, DASonnenfeld, Aeternity, Spitfired, Thehelpfulbot, A412, Phearson, Jamietw, ClueBot NG, Enfcer, KLBot2, Northamerica1000, JeromRP, Phillygriffin and Anonymous: 4

- **Ecological design** *Source:* https://en.wikipedia.org/wiki/Ecological_design?oldid=650160588 *Contributors:* Skysmith, Guettarda, Rjwilmsi, Ground Zero, Schear, SmackBot, Espresso Addict, Byelf2007, Robofish, Ckatz, HelloAnnyong, Seaphoto, DGG, Drjem3, Erkan Yilmaz, DASonnenfeld, Littlealien182, Gbawden, KathrynLybarger, ClueBot, Poodledog, Addbot, Pince Nez, Yobot, Ecodesign, AnomieBOT, Crzer07, Omnipaedista, AlexPlante, FrescoBot, Look2See1, Galen777, Rangoon11, Helpsome, Helpful Pixie Bot, Northamerica1000, Zollo9999, MITclubfact, Joydeep, BBlake003, DPL bot, Alseeras, Mmj2love, Arcandam, Kim Ashfield, Noegid and Anonymous: 12

- **Ecosa Institute** *Source:* https://en.wikipedia.org/wiki/Ecosa_Institute?oldid=661883629 *Contributors:* Finlay McWalter, Wavelength, Welsh, Open2universe, SmackBot, Dylan36c, Richhoncho, Jllm06, JL-Bot, UnCatBot, Yobot, Malcolmjm, Maniamin, AaronSona, PigFlu Oink, ChrisGualtieri and Anonymous: 4

- **Energy-efficient landscaping** *Source:* https://en.wikipedia.org/wiki/Energy-efficient_landscaping?oldid=642299234 *Contributors:* Alan Liefting, Oneiros, Shiftchange, David Gale, Behun, Jonathan Kovaciny, Wavelength, Cate, CQ, Arthur Rubin, Allens, Edward Waverley, SmackBot, Hank chapot, Cydebot, Islescape, Gizmotech, Gralo, Magioladitis, Tinwithli, Jrsnbarn, Skier Dude, Jorfer, BotKung, Logan, Treekiller~enwiki, SoxBot III, Irockursocks96, LaaknorBot, Legobot, Ita140188, Look2See1, Zollerriia, ClueBot NG, KarmicRag, Rowan Adams, Tattoodwaitress, Freebirdthemonk II and Anonymous: 19

- **Environmental impact assessment** *Source:* https://en.wikipedia.org/wiki/Environmental_impact_assessment?oldid=684595879 *Contributors:* Paul A, Mac, Ronz, Radiojon, Saltine, Auric, Alan Liefting, Everyking, Beland, Neutrality, EagleOne, Spiffy sperry, CALR, Vsmith, John Vandenberg, Vortexrealm, Maurreen, Alansohn, Rd232, Andrewpmk, Computerjoe, Redvers, Walshga, Jwanders, Tabletop, Lensovet, Mandarax, Yuriybrisk, Baeksu, Gurch, Bgwhite, Dj Capricorn, YurikBot, Wavelength, RussBot, Gerfriedc, Cate, Txuspe, Thane, Epipelagic, Cinik, Arouck, NHSavage, Mokgand, Allens, Mdwyer, Sardanaphalus, SmackBot, Asm76, Bhasiba, Gilliam, MikeSy, Willow4, JonHarder, WMXX, SundarBot, Cybercobra, Oceanh, AThing, Anlace, Khazar, Mbeychok, IronGargoyle, Ckatz, Hu12, Paulsuckow, Sul4bh, AbsolutDan, Eastlaw, Covalent, CmdrObot, Mewaqua, Krakfly, Savitr, Landroo, ThevikasIN, Energybeing, Sonderbro, Islescape, 200328816, Bouchecl, Nick Number, Mentifisto, Ben Harris-Roxas, Lfstevens, Ecoconservant, XtalMag, JAnDbot, DuncanHill, MER-C, Doctorhawkes, Robsavoie, Brendan Barrett, Hamiltonstone, Somearemoreequal, Calltech, Stephenchou0722, JMWeintraub, Hasanisawi, Xue hanyu, Oceanflynn, Pdcook, DASonnenfeld, Idioma-bot, TXiKiBoT, Eve Hall, Darren Kavanagh, Janus01, Louis F. Soffer, Rstafursky, Sreejithcv, SieBot, RenateVB, Mikehans777,

Baza17, Jessicajensen, Paperscience, Mrfebruary, Eyeintheskye, ClueBot, HRS IAM, Zikiodotte, SuperHamster, Moreau1, PixelBot, Spike-Toronto, Riccardo Riccioni, Adityasahu1, Mlaffs, XLinkBot, SilvonenBot, Noctibus, Addbot, Fieldday-sunday, Couposanto, MrOllie, Download, LaaknorBot, Njbece, Faramir4, Ausstieg links, Lightbot, Legobot, Luckas-bot, Yobot, Amirobot, SwisterTwister, Edefini, AnomieBOT, Andrewrp, Binaysma, Cyclewala, Ado2102, Xqbot, Capricorn42, NZCommercial, Grim23, RibotBOT, Lexy-lou, Shadowjams, Dougofborg, Mnent, Mbennett555, Ltt-comm-de, Serols, Homo habilis, Istcol, Monkeymanman, Chostovs, Defender of torch, Jeffrd10, Suffusion of Yellow, WalkUK, John of Reading, Look2See1, Dewritech, Racerx11, TuHan-Bot, Wikipelli, K6ka, Shuipzv3, Kftiambeng, Nickjf22, Beddowve, Kilopi, Seniorexpat, Annehardon, Deepdish001, ClueBot NG, Snotbot, Widr, ThowardLP, Kittu999, HoverHam, Pratyya Ghosh, Dexbot, Tohobit, Arjuncm3, CEAA ACEE, Epicgenius, ThjPhD1981, Jack Bing, Mrenriqueza, Quenhitran, Nicolemarie0, Lizia7, WaterSnail, ARamkissoon, Shashanksingh70, Behzad fotovat, KasparBot, Rufus Howard, Simondavidhoward and Anonymous: 188

- **Erosion control** *Source:* https://en.wikipedia.org/wiki/Erosion_control?oldid=684341459 *Contributors:* Michael Hardy, Tb, H-2-O, Discospinster, Rich Farmbrough, Longhair, Paleorthid, Bart133, Daniel Collins, Ozzykhan, Rbirkby, Katieh5584, ChemGardener, SmackBot, Mangoe, Rrburke, Stefan2, Cydebot, Epbr123, Andyjsmith, Seaphoto, Lucasstarbuck, Engineman, Catgut, Zuejay, Fleebo, KylieTastic, Funandtrvl, Ihaveafordv8, Bill708, Alrod312, Accounting4Taste, Staylor3, The Thing That Should Not Be, Jusdafax, Moreau1, SoilMan2007, Kvmapr, Against the current, Hollis01, Addbot, Non-dropframe, Bonatto, Johnzpublic, Fraggle81, Newman1011, Apothecia, Plumpurple, Dehaan, Anna Frodesiak, Cannolis, Pinethicket, Jackehammond, Look2See1, ClueBot NG, FS Erosion Control, Helpful Pixie Bot, Northamerica1000, Jamesx12345, Rolph Markvoort, Nodove, Myrowich, KasparBot and Anonymous: 55

- **Grassed waterway** *Source:* https://en.wikipedia.org/wiki/Grassed_waterway?oldid=661923921 *Contributors:* Oddharmonic, BlueJaeger, Wavelength, RussBot, Dialectric, Bluezy, SmackBot, Cydebot, T@nn, NAHID, Addbot, Yobot, AnomieBOT, Anna Frodesiak, Evrardo, Look2See1, SporkBot, ClueBot NG, Widr, Antiqueight, Arr4, Kwaldroup2002 and Anonymous: 3

- **Green furniture** *Source:* https://en.wikipedia.org/wiki/Green_furniture?oldid=622119735 *Contributors:* Edward, Alan Liefting, Bhny, SmackBot, Gobonobo, Falcon8765, MatthewVanitas, Addbot, Yobot, Ulric1313, Teddy51, GSouthern Force, Look2See1 and Anonymous: 1

- **Groundwater recharge** *Source:* https://en.wikipedia.org/wiki/Groundwater_recharge?oldid=683480382 *Contributors:* Mac, Per Abrahamsen, Alan Liefting, Anthony Appleyard, Paleorthid, Geertivp, TDogg310, Katieh5584, SmackBot, Mwtoews, Joseph Solis in Australia, Igoldste, Civil Engineer III, Anthonares, AntiVandalBot, DASonnenfeld, Amikake3, TXiKiBoT, Nanawloy, Water and Land, Correogsk, ClueBot, Chinesedude4, Excirial, Ebh83, Thewellman, Addbot, Glane23, Fraggle81, Citation bot, MakeBelieveMonster, J04n, Maudlin galore, Citation bot 1, Wotnow, NortyNort, Belt777, Look2See1, ZéroBot, RockMagnetist, ClueBot NG, Monsoon Waves, Bibcode Bot, BattyBot, EuroCarGT, Sidelight12, Monkbot, FriendlyCaribou and Anonymous: 25

- **Hydropower Sustainability Assessment Protocol** *Source:* https://en.wikipedia.org/wiki/Hydropower_Sustainability_Assessment_Protocol?oldid=684855787 *Contributors:* Shoy, DMacks, CorenSearchBot, Yobot, NortyNort, BG19bot and Simondavidhoward

- **Integrated Modification Methodology** *Source:* https://en.wikipedia.org/wiki/Integrated_Modification_Methodology?oldid=646129704 *Contributors:* Giraffedata, Bgwhite, Wavelength, GrahamHardy, Wilhelmina Will, Mild Bill Hiccup, Iohannes Animosus, Carriearchdale, Ironholds, Yobot, The Banner, Antiqueight, Frze, DPL bot, Massimo.tadi, Ale2201, ShahroozVahabzadeh and Anonymous: 1

- **Landscape planning** *Source:* https://en.wikipedia.org/wiki/Landscape_planning?oldid=673691042 *Contributors:* Ronz, HaeB, Rich Farmbrough, Ascorbic, Richi, Mdd, Alansohn, Versageek, Rjwilmsi, RexNL, Wavelength, Rsrikanth05, Wknight94, Arthur Rubin, Ratarsed, Bluebot, Willow4, Nixeagle, Hurker, Jcsquardo, No1lakersfan, Peripitus, Meno25, Islescape, Greendale4, Blathnaid, Gsaup, Gomm, Flowanda, STBot, Word2line, Russellfsm, DASonnenfeld, Squids and Chips, Barneca, Matilda Sharks, Rmaul, Plieninger, Rstafursky, Madurasn, Turnip07, Twinnavion11, Pmc2001, Addbot, AnomieBOT, Westerness, TechBot, GrouchoBot, Urjanhai, John of Reading, WikitanvirBot, Look2See1, Wikipelli, ClueBot NG, AvocatoBot, Jccastaneda, Wodrow, Aisteco, Lizia7, Greensmilelandscape, KasparBot and Anonymous: 30

- **Life Cycle Thinking** *Source:* https://en.wikipedia.org/wiki/Life_Cycle_Thinking?oldid=615547035 *Contributors:* Bearcat, R. S. Shaw, Joel7687, Malcolma, Widefox, WikHead, Buster7, GoingBatty, Palosirkka, Snotbot, Rjniemer, Laurenallyce, Eterica, Aguo729, Urcuyopa, Tutelary, TerryKirby and Anonymous: 2

- **NABERS** *Source:* https://en.wikipedia.org/wiki/NABERS?oldid=679803640 *Contributors:* Wavelength, DASonnenfeld, Yobot, GB fan, Isiwarnke, John of Reading, BG19bot, Northamerica1000, Froberso, Caroline Residovic, Carlosfloreslenero and Anonymous: 2

- **New Suburbanism** *Source:* https://en.wikipedia.org/wiki/New_Suburbanism?oldid=648934744 *Contributors:* Rpyle731, Derek R Bullamore, Magioladitis, Tikuko and AlmaIV

- **New Urbanism** *Source:* https://en.wikipedia.org/wiki/New_Urbanism?oldid=684586952 *Contributors:* Mav, The Anome, Sjc, Atorpen, SimonP, BryceHarrington, Edward, Chris Horvath, GTBacchus, Delirium, Stw, G-Man, Ijon, Nikai, Scott, Nohat, Dysprosia, Maximus Rex, Nv8200pa, Neal Finne, Darkcore, Dale Arnett, Cholling, Sunray, Benc, Isopropyl, Seth Ilys, GreatWhiteNortherner, Bkonrad, VampWillow, Bobblewik, Stevietheman, Catdude, Jokestress, One Salient Oversight, Edsanville, Subsume, Calwatch, Grstain, Chris Howard, Freakofnurture, Brianhe, Rich Farmbrough, Pavel Vozenilek, CanisRufus, MBisanz, Walden, Bletch, Aude, EurekaLott, Pheidias, LaurenceJA, Tiresias BC, David Gale, Towel401, Vanished user lkjsdkf34ij48fjhk4, Slightlyslack, Kessler, John Quiggin, Jeffhos, UltraSkuzzi, Metron4, Krislyn, Yuckfoo, Tocksin, Brycen, Boothy443, Woohookitty, Mindmatrix, TigerShark, Metrikk, Before My Ken, Tabletop, MatthewUND, SDC, Bad Graphics Ghost, Liface, Kalmia, Bikeable, Mendaliv, Rjwilmsi, Nightscream, Ctdunstan, Bob A, 25~enwiki, Vegaswikian, Funnyhat, Ligulem, FlaBot, SchuminWeb, Ground Zero, Rclose, RexNL, Short Verses, Jollyswagman, FedericoMenaQuintero, Tas50, Cornellrockey, YurikBot, Wavelength, RussBot, Stormbay, Technical Foul, Aeusoes1, Sylvain1972, Awiseman, Retired username, Elmwood, TastyCakes, Shaqspeare, Paytonc, Genjix, ChrisGriswold, Donald Albury, Abune, Josh3580, Rhallanger, Chriswaterguy, Jdmalouff, GMan552, Jeffreymcmanus, Jonathan.s.kt, Eduardo89, Bridgman, That Guy, From That Show!, Wizofaus, Jmchuff, SmackBot, Brian1979, Zazaban, Rokfaith, Verne Equinox, Eskimbot, Alsandro, Cazort, Brothers, Quidam65, Hmains, Betacommand, Chris the speller, Bluebot, Jnelson09, Gracenotes, Sct72, Pdxstreetcar, Zsinj, Tamfang, Argyriou, JonHarder, Thisisbossi, Amazins490, Elendil's Heir, VarGasten, Fuhghettaboutit, Ericbritton, MichaelBillington, Marksven, Ligulembot, Sometimesthinking, Barcode, KenFehling, Scott182, Loodog, Jaywubba1887, Beetstra, Iridescent, Shoeofdeath, Areback, Igoldste, JoannaSerah, RookZERO, Totalcatharsis, CmdrObot, Articnomad, Cydebot, Mcgillionaire, Mariojalves, Gogo Dodo, Hebrides, Strongbad1982, Satori Son, Thijs!bot, Skb8721, Lynndunn, Rolyatleahcim, Dvprofiles, Horologium, Kgrantsc, Jmelody, A.J.Chesswas, Futurebird, I already forgot, Tebici, KrakatoaKatie, Toddross, SummerPhD, Hypersite, Zoler, Davewho2, Conk 9, Adelehanty, Yahel Guhan, Dmmd123, Syrcsemark,

Zatoichi26, Gabriel Kielland, Nucleophilic, Afaprof01, Jarbury, TTKK, Mukrkrgsj, Gwern, The Prodigal, Jim.henderson, Slowmotionrevolution, Keith D, R'n'B, CommonsDelinker, SirChan, Jrsnbarn, Thaurisil, Glenday, Word2line, VNPeO, Keizers, Trymybest, 83d40m, Mthoward, NewcombFamily, Bonadea, DASonnenfeld, RAWS, Djflem, VolkovBot, TheMindsEye, Vossy747, WOSlinker, Barneca, Jkeene, Jasonmccants, Pwnage8, Joe2832, Jdcrutch, Sswonk, Ciudad jardin, A.Roz, Davemc50, Ajinaz, Victimofleisure, Marcosaedro, JimSine cultureandarts, Lexington50, UrbanMaster, Justinleemiller, Moebiusuibeom, Liz de Quebec, Villagemaker, Screamingman14, Antischmitzbot, Civicarch~enwiki, Radioactive afikomen, Futurano, Hyslop, Canavalia, SieBot, Gerakibot, Dpakdel, Teknolyze, Michaelmehaffy, Eyedubya, Hajo4, Hello71, Lightmouse, Antoinette Panton, Town Designer, AlexH20, Mheaddem, Cyfal, Dimorsitanos, Deanist, Samouzon, SalineBrain, Binksternet, Khc546, Cambrasa, RecycledAir, Mild Bill Hiccup, ResidueOfDesign, Newurbanismorg, Hanshedrich, Ccc40821, ChrisHodgesUK, Stevemelia, Mlaffs, Nomenkultura44, Vanished user uih38riiw4hjlsd, RB211Trent, Ps07swt, DumZiBoT, DigitalDaiquiri, Nathan Johnson, Dthomsen8, Leoniana, MystBot, Emil Kastberg, Chunnythefish, Addbot, Posnick, Chzz, Ginosbot, Lightbot, Luckas Blade, Arxiloxos, Legobot, PlankBot, Luckasbot, MrMontag, Otwguy, Amirobot, Davidglauber, Jennvirskus, AnomieBOT, NewUrbanMom, OptimisticCynic, Ulric1313, UDJ, Citation bot, ArthurBot, Xqbot, JUD1989, Subwsurban, Tonyf23ton, RayCapitol, Theatretek, Calimar300, Surv1v4l1st, Gripper101, Heynow09, Mrleomarvin, Citation bot 1, Abductive, Jonesey95, MastiBot, Tim Halbur, Full-date unlinking bot, Quentinwllcs, Msmithcu, Darkohead, Roopi04, Foursouthpaws, Horst-schlaemma, Kendwallace, Elekhh, Greenerpastures2, Chostovs, LDNash, DriveMySol, Dinamik-bot, Look2See1, BillyPreset, Treefingers1206, Dewritech, Baldwincenter, Quayhands, Klbrain, ZéroBot, Модернист, פארוך, Erbanb, Pillsbury11, SporkBot, Wajny8019, Matti.olkinuora, Jayblue42, Las vegas kid, ChuispastonBot, Maniarianir, ClueBot NG, Oddson77, Webud, Pmatchak, CopperSquare, BG19bot, Krenair, Cbohl, Twilson088, Simonhardt93, Michaelmalak, Thattonekiddmichael, Mdy66, BattyBot, ChrisGualtieri, Mogism, Tentinator, Stamptrader, LazyReader, Monkbot, Yellowfin2014, Gmailable, Wikplan, Spaceship123 and Anonymous: 294

- **Permaculture** *Source:* https://en.wikipedia.org/wiki/Permaculture?oldid=683890280 *Contributors:* Marj Tiefert, Tarquin, Rmhermen, Pierre-Abbat, Anthere, Tzartzam, BryceHarrington, Quercusrobur, Jose Icaza, DennisDaniels, Infrogmation, Michael Hardy, Dmd3e, Matthewmayer, Cyde, Skysmith, Mac, BigFatBuddha, Scott, Ghewgill, Jengod, Ww, Populus, Jose Ramos, Mignon~enwiki, Vespristiano, Altenmann, Chopchopwhitey, Flauto Dolce, Gidonb, Sunray, Sheridan, Alan Liefting, Timpo, Mboverload, JRR Trollkien, Golbez, Geoffspear, Gadfium, Pgan002, Onco p53, Phil Sandifer, Neffk, Zfr, Nickptar, Burschik, Kathar, NathanHurst, Chris j wood, Rich Farmbrough, Guanabot, Vsmith, Dyl, Bender235, Eadmund~enwiki, Erauch, Nigelj, Smalljim, Cmdrjameson, Vortexrealm, Oop, Ziggurat, Timl, Giraffedata, Jkh.gr, Pearle, Mdd, Shafaki, Bmeacham, Paleorthid, Davenbelle, Linmhall, Stillnotelf, Velella, Tony Sidaway, Talkie tim, Blaxthos, Dan100, Rzelnik, RyanGerbil10, Kevin Hayes, FrancisTyers, Cyclotronwiki, Poppafuze, Mindmatrix, RHaworth, Polyparadigm, SP-KP, Jeff3000, Jwanders, Bluemoose, Raines, Palica, Behun, Mandarax, Elvey, Rjwilmsi, Salix alba, Schlüggell, Smithfarm, DoubleBlue, MarnetteD, Jeffmcneill, MikeJ9919, FlaBot, SchuminWeb, Freddydesouza, Jrtayloriv, Monkofthetrueschool, Vmenkov, Roboto de Ajvol, YurikBot, Wavelength, NTBot~enwiki, Waitak, RussBot, TheMoot, Diliff, Pigman, David Woodward, Shell Kinney, Pseudomonas, Dialectric, Mkbnett, Nirvana2013, Kiaparowits, Thesloth, PeterBirkett, Irishguy, Epipelagic, TastyCakes, CQ, Meika, Arthur Rubin, Tevildo, Chriswaterguy, Naught101, Mdwyer, Meegs, That Guy, From That Show!, SmackBot, Eclipsenow.org, Sanman nor, Lord Matt, Jtneill, KVDP, Scottlondon, Cacuija, JFHJr, Gilliam, OrionK, Afa86, Schmiteye, Chris the speller, Te24409nsp, Thumperward, Jon513, Salvor, Uthbrian, Colonies Chris, Chendy, Peter Campbell, Sholto Maud, Willow4, Brimba, Neo139, Josh64, JonasRH, Nihilo 01, Djcmackay, Ggpauly, Gurdjieff, Hank chapot, Joli Rouge, Byelf2007, Archimerged, Valfontis, Khazar, SilkTork, Sociotard, Danny Beaudoin, Ckatz, Rkmlai, Beetstra, LuYiSi, WaynaQhapaq, Johnmc, RichardF, Libertyblues, Christian Roess, Nehrams2020, HisSpaceResearch, Iridescent, Ted11, CoulterTM, Mulder416sBot, RookZERO, Ayanoa, IronChris, Grayson wyatt, RiotGearEpsilon, CmdrObot, Tanthalas39, Drinibot, Tahirs, Unclejedd, Paul Millsom, Macropneuma, Daniel J. Leivick, Teratornis, Kozuch, Richhoncho, Trueblood, Thijs!bot, Epbr123, Homohabilis, Daniel, Trevyn, Itsmejudith, Angusscown, Amberckerr, Blathnaid, Kanejamison, Nom DeGuerre, Brian Boyd, Gioto, Luna Santin, Wengero, Tenzicut, Julia Rossi, Adam Chlipala, Papipaul, Lfstevens, Ingolfson, Aquaponics, JAnDbot, Krishvanth, Tomintaz, Barek, Freddy011, Struthious Bandersnatch, Rjholmer, Bdpermie, Roidroid, VoABot II, Appraiser, Steve@sector39.co.uk, APB-CMX, Sustainableyes, DerHexer, Edward321, TimidGuy, MartinBot, 4492tues12, Cbuddenhagen, Andre.holzner, Dan arndt, VirtualDelight, UrthBound, GomerMcFlarp, Tgeairn, J.delanoy, Keithkml, MatheoDJ, TaylorAshton, Charlesjustice, Skier Dude, Belovedfreak, Cjstanonis, Madbishop, Jorfer, AprilSKelly, Woodsguy, Scott Roy Atwood, Agerry, Jamesofur, Gracoo2, Inwind, DASonnenfeld, Chrlaney, Dominoconsultant, VolkovBot, Dlesjack, Jwitch, Aesopos, Philip Trueman, TXiKiBoT, Jackovacs, Charlesdrakew, Noformation, Ilyushka88, Mooreds, Woodlandcreek, Cymon Fjell, Mexeno1, Red58bill, Logan, Richardtelford, Terriemiller, SieBot, Zelchenko, Flyer22, Permacultura, Skipsievert, Yone Fernandes, Zentomologist, Lightmouse, Seedbot, Chrisrus, Bodhi Peace, Wetwarexpert, SlackerMom, Sfan00 IMG, ClueBot, Allthingsgreen, Fyyer, The Thing That Should Not Be, LMFernandes, Xavexgoem, Isaebellaspuppetshow, PMDrive1061, 718 Bot, Mynameisnotpj, Ice Cold Beer, Ceilican, Bridgetsgirl, Ecureuil espagnol, Fishnut, PermaculturePlanet, Dana boomer, DumZiBoT, XLinkBot, WikHead, SilvonenBot, Skyeriquelme, Guydavies, Zodon, Luminaia, Ghost accounty, Hunchenfest, Addbot, Wikepermie, TomorrowsDream, ClaireofKLARITY, Some jerk on the Internet, MrOllie, Download, Favonian, SpBot, Granitethighs, 5 albert square, Bluenijin, Ajkoen, Tide rolls, Jarble, Bermicourt, Legobot, Luckas-bot, Yobot, WikiDan61, Themfromspace, Fraggle81, Santryl, Isotelesis, Suvicaya, AnomieBOT, Archon 2488, BrettScott, Materialscientist, Elmmapleoakpine, Citation bot, NinetyNineFennelSeeds, ArthurBot, Xqbot, Miltonics, Apothecia, Kng442, Anna Frodesiak, NathanielGallion, J04n, Katsteele, Kyng, Peaksurfer, Sceloporus, Brambleshire, Rickproser, Ciclotan, FrescoBot, Legion23, Element deck, Augustart, Pentref, Questionthedominantparadigm, Dogposter, Corpuscollosium, Luke831, Lothar von Richthofen, Johnwhol, Zaricki811, Koleszar, DanTheSeeker, Tutor65, Enloop, BoundaryRider, Quesauth, LAlexanderson, Jan Permaculture, Gelatinouscube42, Mikal42, Gardenlily, Nattydreader, Pat604, Viellashipley, Pinethicket, LittleWink, Smuckola, Yogi tom, H4stings, Kirstendirksen, Darkohead, Steinpal, Lvec-jayson, Permapower, Teatimetrvaelller, Ecoescuela, SweetAspect, Mcalison, Jonkerz, Hauntu2, Permaculture institute, AbeColey, PleaseStand, Tbhotch, TheMesquito, DARTH SIDIOUS 2, Obsidian Soul, RjwilmsiBot, Mukogodo, Stephen.J.Arnold, Grifen2, User2112, Pinqpanther, Logical Cowboy, Look2See1, AbbaIkea2010, Dewritech, Steko, ZéroBot, Elefectoborde, DerekG99, Jeanpetr, Joshfinnie, Southofsouth, Wiooiw, Wayne Slam, Tiago Penedo, Seidos, Yogazeal, Popok75, Nld.rnsm, Rr.nz, Sequoia D, Minuoh, ClueBot NG, Wisdawn, River road permasite, TadewiGomda, Omair00, Eniodros, Krshwunk, Greenman2011, Mesoderm, Widr, Helpful Pixie Bot, ?oygul, Sax66, Charles Gran, Gob Lofa, BG19bot, MKar, Iamharb, Ostrichfern, Northamerica1000, Panchito62, Rowan Adams, Thepidding, Madboy23, BattyBot, Greyphox, Fairtheewell, Sinique, ChrisGualtieri, Jray310, TheJJJunk, Vckidd, Soransoran, Sminthopsis84, Ghsqueiroz, Permbuddy, Aymankamelwiki, Ngulevski, KcamKcim, Jimkio12, Oioidoug, Jeffersonfranklin, BreakfastJr, HaroldTheHat, BrooklynAve, Amanda Sturgill, Perma2, Mueller felix, Redddbaron, Geoffmen, Presi1980, Noyster, Suavicm, EricEnfermeroMobile, Permaculture design, Mpathfinder, Monkbot, Seandixonsul, James Kern, Ephemeralcas, Trackteur, PermacultureOne, Enzo at Permaculture Education, Jbanegas, Motoindustries, Steveburns888, KasparBot, Qf2345, Mcgrubso, Bobsya, Paugoo, Rpnbrn, Aranyagardens, Dutral and Anonymous: 495

- **Principles of intelligent urbanism** *Source:* https://en.wikipedia.org/wiki/Principles_of_intelligent_urbanism?oldid=665554026 *Contributors:*

Edward, Alan Liefting, Lcgarcia, Timpo, Romanpoet, Andycjp, CanisRufus, MBisanz, EurekaLott, Pearle, Mdd, Drbreznjev, Galaxiaad, Woohookitty, Mindmatrix, Rjwilmsi, R'son-W, Gadget850, Sandstein, Arthur Rubin, SmackBot, Evanreyes, Chris the speller, Thisisbossi, Theanphibian, Dogears, Hu12, Cbrown1023, RookZERO, Vanisaac, Articnomad, HokieRNB, Kozuch, Lynndunn, Locano, Philippe, Nick Number, Heroeswithmetaphors, Leuko, Gsaup, Rich257, Dab295, Robsteranium, Word2line, Perel, Sadhevneo, Aesopos, Barneca, Saber girl08, Scatteredpixels, Cotumo, Mild Bill Hiccup, Vendeka, Bjdehut, Apparition11, DumZiBoT, Critical Chris, WikHead, Fluffernutter, Bhutanplanner2008, Wifione, PeterEastern, Abductive, Pinochet (3), Horst-schlaemma, Lotje, John of Reading, Look2See1, Ar.naveenbharathi, Werieth, Joshfinnie, Garimajain2002, Wigwamapache, Hamiltha, Webud, Northamerica1000, PaulGould, BattyBot, Khazar2, Rodnebb, Brian Parkinson, Lizia7 and Anonymous: 42

- **Public interest design** *Source:* https://en.wikipedia.org/wiki/Public_interest_design?oldid=662794702 *Contributors:* Koavf, Magioladitis, He A, Yobot, AvicBot, BattyBot, 29sh00, Johncaryjr, Monkbot, Jduby and Anonymous: 2

- **Rain garden** *Source:* https://en.wikipedia.org/wiki/Rain_garden?oldid=662263981 *Contributors:* Charles Matthews, Phil Boswell, Davodd, Alan Liefting, David Johnson, Lpangelrob, Goodnature, Mike Rosoft, Bobo192, RHaworth, Btyner, BD2412, Melesse, Rjwilmsi, Ground Zero, Gier21, Epolk, Bovineone, Wimt, Ozzykhan, TDogg310, WAS 4.250, Davus, ChemGardener, SmackBot, Gstibolt, Lotusduck, Mwtoews, DMacks, Edgeplot, SilkTork, IronGargoyle, Pjrm, BrianAsh, CmdrObot, KyraVixen, JohnCD, CFMWiki1, Eaglesight, Mbell, Sobreira, Luna Santin, Darklilac, NeilHynes, Mattinbgn, Keith D, Okoivisto, AntiSpamBot, Flatterworld, DASonnenfeld, Shaunus4, Funandtrvl, Jeff G., Bry9000, Dogbertwp, Gueneverey, Technopat, Rossrakesh, Gena Haltmair, Vector Potential, Spoonamore, Chillywillycd, Euryalus, Kgoodwil, Lightmouse, Miqrogroove, Maelgwnbot, Jessicajensen, Altzinn, Dlrohrer2003, Hafspajen, Excirial, Alexbot, Mjbraun, Moreau1, Sun Creator, Iamjacksheart, DumZiBoT, XLinkBot, Addbot, Wkrocek, Kittelj, Lightbot, Margin1522, Yobot, Raincatcher7, Synchronism, Tigerfry, Tcduvall, Teddks, Suellingson, Oatmealbird, Twirligig, FrescoBot, Lizzen~enwiki, Pinethicket, Geogene, Ajohns90, Paul Aakerøy, Crusoe8181, Trappist the monk, Jakeroot, The Utahraptor, Alicelook, John of Reading, Look2See1, Ipazia2010, L Kensington, ClueBot NG, KarlaBeth555, Theopolisme, Helpful Pixie Bot, BG19bot, MKar, Northamerica1000, Scotstout, Danbrellis, Rowan Adams, Jeremy112233, Khazar2, Mogism, Jamesx12345, Suebeehsieh, Harlem Baker Hughes, Ted Redenbaugh, Dr. Aaron D. Clark, Monkbot, Whistlemethis, Katie Pekarek, TyCurious and Anonymous: 95

- **Reconciliation ecology** *Source:* https://en.wikipedia.org/wiki/Reconciliation_ecology?oldid=680687650 *Contributors:* JoJan, Dcfleck, Rich Farmbrough, Erauch, Reinyday, Cmdrjameson, Pearle, Velella, Woohookitty, Rjwilmsi, Koavf, Wavelength, EncycloPetey, Cacuija, Cydebot, Adamwelz, Nick Number, Dawnseeker2000, AntiVandalBot, Fjellstad, SchreiberBike, XLinkBot, Helixxxy, K.kjartanson, Ballrog, Addbot, Yobot, Citation bot, Trappist the monk, Look2See1, ZéroBot, Succulentpope, Rkhartman, Gob Lofa, Leafwarbler, Monkbot, Johnsoniensis and Anonymous: 6

- **Riparian buffer** *Source:* https://en.wikipedia.org/wiki/Riparian_buffer?oldid=679008797 *Contributors:* Skysmith, Tpbradbury, Tom, Rjwilmsi, Vmenkov, Nirvana2013, Shepazu, Epipelagic, SmackBot, Sadads, Werdan7, Levineps, Cydebot, Leolaursen, Lady Mondegreen, DASonnenfeld, Lightmouse, Taroaldo, Moreau1, Stobin2, 5 albert square, Jarble, Martess, Minnecologies, FrescoBot, John of Reading, Look2See1, ClueBot NG, Wbm1058, BattyBot, Khazar2, Sidelight12, CTnative, Jbanegas, Anna Mayerhof, Extemporalist and Anonymous: 11

- **Rural–urban fringe** *Source:* https://en.wikipedia.org/wiki/Rural%E2%80%93urban_fringe?oldid=683719474 *Contributors:* Ezra Wax, Kwertii, Charles Matthews, Radiojon, Floydian, Pascal666, MToolen, RedWordSmith, Mdd, Paleorthid, Daniel Case, Bjrobinson, Djm1279, Tony1, SmackBot, Gilliam, Bluebot, Biatch, Thumperward, NickPenguin, Robofish, A. Parrot, Grumpyyoungman01, Alanmaher, BananaFiend, Hassocks5489, Magioladitis, Jatkins, Indubitably, Enviroboy, HeadSnap, Nuttycoconut, B1157, Twinsday, ClueBot, Versus22, Properbrutal, Yobot, FrescoBot, Pepper, Edderso, DASHBot, Look2See1, Tommy2010, ClueBot NG, Mgwiki123, Compfreak7, ChrisGualtieri, Grousebeater2 and Anonymous: 50

- **Solvatten** *Source:* https://en.wikipedia.org/wiki/Solvatten?oldid=646569033 *Contributors:* Timtrent, Jllm06, Yobot, Sionk, Diannaa, Anne Delong and Tmrl84

- **Strategic environmental assessment** *Source:* https://en.wikipedia.org/wiki/Strategic_environmental_assessment?oldid=684428849 *Contributors:* Kku, Ronz, Alan Liefting, Mintleaf~enwiki, Orlady, Espoo, Walshga, Paullb, Wavelength, Phantomsteve, Gerfriedc, Txuspe, Welsh, Bjrobinson, SmackBot, Wikipedical, CmdrObot, Mcginnly, Harej bot, BetacommandBot, Ben Harris-Roxas, Brendan Barrett, Nemo bis, DASonnenfeld, Kotabatubara, Catarina Camarinhas, Mild Bill Hiccup, DumZiBoT, AlMa77, Addbot, Njbece2, Mnent, Jdt01fgo, Gego2626, Ilariadotta, Nihola, Look2See1, Annehardon, ClueBot NG, KLBot2, Hipersyl, Create4ever, ChrisGualtieri, Khazar2, Yann1234, TobySmrt, Simondavidhoward and Anonymous: 24

- **Strategic Environmental Assessment (Denmark)** *Source:* https://en.wikipedia.org/wiki/Strategic_Environmental_Assessment_(Denmark)?oldid=669948919 *Contributors:* Alan Liefting, Rjwilmsi, Chris the speller, HelloAnnyong, Christian75, Leolaursen, Magioladitis, Fabrictramp, DASonnenfeld, Mild Bill Hiccup, Blanchardb, Addbot, Thebigbudina, Kapgains, Nihola, Look2See1, GoingBatty and Autoerrant

- **Sustainability appraisal** *Source:* https://en.wikipedia.org/wiki/Sustainability_appraisal?oldid=684428932 *Contributors:* Espoo, RHaworth, Bjrobinson, SmackBot, Mcginnly, Harej bot, BetacommandBot, DASonnenfeld, JL-Bot, Erik9bot, WalkUK, Look2See1, Simondavidhoward and Anonymous: 1

- **Sustainable architecture** *Source:* https://en.wikipedia.org/wiki/Sustainable_architecture?oldid=676590167 *Contributors:* Ijon, Cherkash, Zoicon5, Liam, Pfrishauf, Sunray, Alan Liefting, HorsePunchKid, Onco p53, Canterbury Tail, Discospinster, Cacycle, YUL89YYZ, Miscreant, Pharos, Mindmatrix, Mjbt, Jwanders, Mandarax, Graham87, Sparkit, TobyJ, Susten.biz, Rjwilmsi, Nightscream, Vary, Vegaswikian, Bhadani, MarnetteD, Ian Pitchford, Ewlyahoocom, Gurch, Mahalie, Bje2089, Wavelength, Kollision, Bhny, Gaius Cornelius, Aeusoes1, Jpbowen, Mamawrites, Nachoman-au, Naught101, Veinor, SmackBot, MattieTK, Shayn, KVDP, Alsandro, Chris the speller, Bluebot, Biatch, Fplay, Zangala, Laumana, Can't sleep, clown will eat me, JonHarder, Toksook, TechPurism, Just plain Bill, Dogears, Ged UK, Jim Derby, JdH, Phuzion, DouglasCalvert, Wjejskenewr, Kestudi, Igoldste, Ralph Purtcher, Guruprasad Rane, Mcginnly, Bill.albing, Gogo Dodo, Anthonyhcole, Bazzargh, Zdlo, Kozuch, After Midnight, Gralo, Bobblehead, Nick Number, Thadius856, Fayenatic london, Laikalynx, Alphachimpbot, JAnDbot, Barek, SiobhanHansa, Acroterion, DGG, MartinBot, Anaxial, R'n'B, AlphaEta, Jrsnbarn, Word2line, Skier Dude, Plasticup, Robigus, Nathan Horne, Halrhp, Brosi, DASonnenfeld, VolkovBot, Lredman, Thomas.W, T griffen, Piperh, Archvr, VT-GBB, Boxsix, Kilmer-san, Jpeeling, Ecoarchitect, Johanerikandersson, Ecobuilder, Xgirl911, New England, Yerpo, Cntreras, Ronanov, Vanished user ewfisn2348tui2f8n2fio2utjfeoi210r39jf, Jessicajensen, Psreilly, Fitzed, Jillleann77, Sfan00 IMG, Sokors, ArchEcologist, ClueBot, Jan1nad, Pairadox, Lantay77, Ankushagarwal, Gwrabbellar, Morel, Iamjacksheart, Atremia, Poodledog, SchreiberBike, Thewellman, Chrisevrae, Azhararchitecture, Apparition11, Vanished user uih38riiw4hjlsd, DumZiBoT, XLinkBot, Spitfire, ProfDEH, Avoided, Mas85, Svea Kollavainen, Muffinon, Addbot, Captain-tucker, David Ogilvy, Morphogenesis2008, Gajendra2008, Welshboy69, MrOllie, Tassedethe, زرشک, Yobot, Willze, AnomieBOT, Jim1138, Zucchini22, ArthurBot, LilHelpa, J04n, Ita140188, Lulu97417, Jeffwend, Dwarchitect, Grucio, FrescoBot, Recognizance, Wifione, Alysa.elaine, Ecogram, Abductive, GoGreenConstruction, CF ROOFING, RoGBC, De3rulez, Tea with toast, Encycl wiki 01, Horst-schlaemma, Elekhh, Paulrutten2, Dambre20, DARTH SIDIOUS 2, Mean as custard, RjwilmsiBot, SarahBachmann, EmausBot, WikitanvirBot, Ashton 29, Look2See1, Malafond00, GoingBatty, Enviromet, Cathyqaz, Alpha Quadrant (alt), MC Kemello, Donner60, Globalcopywrite, BlackAbbot, Helpsome, ClueBot NG, StephanieCamposR, Wisdawn, Satellizer, Sbic dc, Snotbot, Brothercanyouspareadime, Widr, KLBot2, BG19bot, Anagnam, Iamramgarg, BattyBot, Sustain(able), Razibot, Geezaboo24, Mutley1989, Ugog Nizdast, JaconaFrere, Abdel Rahman Elbakheit, Apollonia-Artemonas, Brettsalt, A2plus, Itingtinalee, KasparBot, Tyagi83 and Anonymous: 186

- **Sustainable design standards** *Source:* https://en.wikipedia.org/wiki/Sustainable_design_standards?oldid=683398083 *Contributors:* Edward, Malcolma, Tony1, SmackBot, Gilliam, Ckatz, Pfhenshaw, Magioladitis, DASonnenfeld, GorillaWarfare, Blanchardb, LilHelpa, Erik9bot, Sandolsky, FrescoBot, Look2See1, Zombieapocalypsenow, ClueBot NG and Anonymous: 5

- **Sustainable development** *Source:* https://en.wikipedia.org/wiki/Sustainable_development?oldid=685014231 *Contributors:* The Cunctator, Taw, Ed Poor, SJK, William Avery, Anthere, Heron, Jaknouse, Hephaestos, Leandrod, Edward, Ubiquity, Michael Hardy, Wipala, Gabbe, Ixfd64, Ahoerstemeier, DavidWBrooks, Mac, Ronz, Andres, Viajero, Fruits~enwiki, Tpbradbury, Sulkworm, Topbanana, Hawstom, Owen, Robbot, Arkuat, Chopchopwhitey, Academic Challenger, Auric, Sunray, Mushroom, Alistair b, Isopropyl, Lupo, MikeCapone, Alan Liefting, Christopher Parham, ShaunMacPherson, Timpo, Ds13, Everyking, Henry Flower, FrYGuY, Utcursch, Antandrus, BozMo, Neutrality, Ukexpat, Grstain, Mike Rosoft, Chris Howard, Discospinster, Rich Farmbrough, Vsmith, LeoDV, Bender235, Tarndt, El C, EurekaLott, Erauch, Guettarda, Shoujun, Bobo192, Viriditas, Vortexrealm, Dungodung, Alphax, Millsdavid, Sam Korn, Mdd, Espoo, Storm Rider, Alansohn, JYolkowski, Arthena, Rd232, Andrewpmk, Paleorthid, John Quiggin, Iothiania, Kocio, Wtmitchell, Docboat, RainbowOfLight, Kaushik twin, SietskeEN, RyanGerbil10, Crosbiesmith, Snowmanmelting, Boothy443, Rintojiang, Woohookitty, Ksandler, Camw, LOL, Guy M, Uncle G, MGTom, Jwanders, Jleon, Thebogusman, Harac, Hard Raspy Sci, Paullb, Jbarta, Behun, Graham87, Magister Mathematicae, Kissekatt, Rjwilmsi, PatrickSaucy, Pitan, Daniel Collins, Oo64eva, Yamamoto Ichiro, Ian Pitchford, SchuminWeb, Nihiltres, Phoebebright, Jrtayloriv, Lmatt, King of Hearts, Chobot, Benlisquare, DVdm, Bgwhite, Bhsand~enwiki, Gwernol, FrankTobia, Roboto de Ajvol, Wavelength, Retaggio, Phmer, 10stone5, Hede2000, Bhny, Shell Kinney, Gaius Cornelius, NawlinWiki, Drgregmartin, Daniel Mietchen, Raven4x4x, Epipelagic, Mkill, Nescio, ClaesWallin, Arthur Rubin, E Wing, Chriswaterguy, Allens, The Way, Ghandir, NickelShoe, Sardanaphalus, SmackBot, FocalPoint, Ashenai, Cubs Fan, Cjh1k2004, JohnSankey, McGeddon, Gnangarra, DWaterson, Verne Equinox, Fernagut, IstvanWolf, Gilliam, Ohnoitsjamie, Durova, Hugo-cs, Chris the speller, SlimJim, EncMstr, MalafayaBot, Sadads, Ctbolt, Darth Panda, John Reaves, Chendy, JonHarder, Rrburke, Addshore, CharonX, Fssca, John D. Croft, Richard001, Tvinson, Ryan Roos, Gujuguy, Suidafrikaan, Byelf2007, Akraj, SashatoBot, Vasman, JzG, Admn404, Nick carson, Iosd, Ckatz, Eivind F Øyangen, JHunterJ, Beetstra, RichardF, Catquas, Hu12, Levineps, Iridescent, Shoreranger, Octane, Mbenzdabest, ArrJay, Colignatus, Courcelles, RookZERO, Mr3641, Sbridge, Eastlaw, Xcentaur, DangerousPanda, Anubis3, Vanished user sojweiorj34i4f, KyraVixen, Mcginnly, Mc4932, Jhecklinger, NaBUru38, Leujohn, Marmora78, Pro bug catcher, Jimbo156, Evanclifthorne, JessBr, Reywas92, Gogo Dodo, David A. Victor, Llewelynpritchard, Dr.enh, Odie5533, Tawkerbot4, Acoogan, Jbarber, Christian75, DumbBOT, Alrick, Kozuch, Ncmartin, Thijs!bot, Univer, Epbr123, Gkheilig, V.B.~enwiki, HappyInGeneral, Mereda, Count-Dracula, Sam42, Gcolive, Tebici, Nom DeGuerre, AntiVandalBot, Widefox, Akradecki, Seaphoto, Gregalton, Fayenatic london, Smartse, Alphachimpbot, Alanatkisson, Cobbinma, Cheddington2001, Myanw, Locospotter, JAnDbot, JenLouise, MER-C, Geojacob, Jj2006, Db099221, Black Mamba, Sitethief, Rjholmer, SiobhanHansa, Smarienau, Skyemoor, Gsaup, Bongwarrior, VoABot II, Planeta~enwiki, Peterson.Amy.E, JamesBWatson, Hschmid, Brusegadi, Sirflexi, Scottiedawg, Earthsummit2005, Hdynes, Gabriel Kielland, Steve met, Nposs, MrWarMage, 28421u2232nfenfcenc, Allstarecho, Adriaan, Curtbeckmann, Flowanda, MartinBot, Kiore, Ben MacDui, Rettetast, SuperMarioMan, Kctucker, Jay Litman, TheEgyptian, CommonsDelinker, Kai Hockerts, OStewart, ChrisNickson, Grblundell, Tgeairn, Huzzlet the bot, J.delanoy, Pharaoh of the Wizards, Dcarpenter, Rlsheehan, Silverxxx, Jrsnbarn, Ginsengbomb, Eliz81, Skumarlabot, Princess Tiswas, Word2line, Octopus-Hands, Ncmvocalist, Biolane, Lasarkis, Skier Dude, Oikoschile~enwiki, Tkn20, Andersabrahamsson, Eric94kim, Jorfer, KylieTastic, STBotD, WJBscribe, Jevansen, Brosi, Treisijs, Bonadea, Andy Marchbanks, Inwind, DASonnenfeld, Brendan Cosman, CardinalDan, Burlywood, Wikieditor06, Meiskam, Livingston.28, Deor, Lredman, Johnfos, ABF, Lop.dong, AlnoktaBOT, Kks ceser, Philip Trueman, Sir schultz, TXiKiBoT, Starlayk, Cassbeth, Padmatara, Katoa, A4bot, Rei-bot, Z.E.R.O., Shenbrood, Rami7896, Steven J. Anderson, Charlietemps, IllaZilla, Kjell.kuehne, Temanning, Dlae, LeaveSleaves, Vgranucci, Jcwandemberg, Leehach, Altermike, @pple, AgentCDE, The Devil's Advocate, Insanity Incarnate, Brianga, Chenzw, CT Cooper, Rahulkepapa, PokeYourHeadOff, Rstafursky, GirasoleDE, Newbyguesses, SieBot, Nubiatech, Tresiden, Afrothetics, Malcolmxl5, Triwbe, Yintan, Calabraxthis, User123new, Chuck56, Keilana, Flyer22, The Sunshine Man, MaynardClark, Jojalozzo, Nopetro, Hiddenfromview, Rohit tripathi60, Skipsievert, Catt270, Pm master, Plebiscites, Antonio Lopez, Faradayplank, Nuttycoconut, Lightmouse, Pmrich, Svick, StaticGull, Mygerardromance, Busy Stubber, Pinkadelica, Aglondon, Sustain123, Grantrowe, ClueBot, Nay the snake, HRS IAM, Alastair McIntosh, The Thing That Should Not Be, 411.tony, Franamax, HUB, Niceguyedc, Nastradinov, Cfechter, DragonBot, Excirial, Jheaton, Holly Ashley IIED, Marie enviro, PixelBot, Hmeopm, Vanisheduser12345, Joshram, Xochipilli BE, Bliss53, Lunchscale, Aurora2698, Optimum Population Trust, Cahillee, Etip, On2Leggs, SSDconsult, Poodledog, Bluemosquito, SchreiberBike, Aleksd, M.boli, ForwardScotland, Lokionly, Apiasecka, Nydhogg, NJGW, Thunderstix, DumZiBoT, Escientist, Beria, XLinkBot, Boyd Reimer, Dthomsen8, Duki998, WikHead, Zodon, Hakuin, Cewvero, Hydrazillawik, Addbot, Willking1979, AlexandrDmitri, DOI bot, CauliflowerEars, Kongr43gpen, Ronhjones, USchick, Cst17, Couposanto, MrOllie, MikkyGay, VandalFixman, Mydoorisopen, CarsracBot, Amh101, Houndhogg, FiriBot, Kyle1278, Proginoskes, Alex Rio Brazil, مانی, Fahmed3, Jarble, Time4this, Tekkenmasterbrendan, Ethames, Legobot, Alexsaidani, Luckas-bot, Yobot, Otwguy, Amirobot, Envirol, Millere08, KamikazeBot, Mikhailovich, Euclidedit, Bbb23, AnomieBOT, Kh1160, Lispp, Xufanc, Karthickbala, Materialscientist, Mervyn Emrys, Citation bot, Nomasonsinmywhitehouse, ArthurBot, Argenfels, ARAGONESE35, Jubilee007, Xqbot, Nishantjr, Capricorn42, Mononomic, Sixfifteen, Sionk, Omnipaedista, Jh4d, RibotBOT, Albertdavis3, GhalyBot, Tobyabgreen, Brad Crimbo, Funnyguy666666, Shadowjams, WaysToEscape, Sdm24, Charithjayanada, Sandcherry, Haritada, A.amitkumar, FrescoBot, Ags245, Spellern, StaticVision, MissEleven123, HJ Mitchell, Rofocale2384, NickGBSOD, Citation bot 1, Clemifornia, Erin Inglish, Javert, Biker Biker, Pekayer11, Pinethicket, Jonesey95, Calmer Waters, Triplestop, Knutw, Calaguiman, Encycl wiki 01, Horst-schlaemma, Robtrob, Mat2010tam, Sfilmsactiwo, Greenerpastures2, ItsZippy, Marjolaine11, Lotje, GregKaye, Neopanora, LilyKitty, RubyLucario, Willoogy, Brianatkin, Jeffrd10, Naomasaemiko, Unitedregistrar, RjwilmsiBot, DexDor, Floydman66, Salvio giuliano, Rodmadar, Greenopedia, EmausBot, John of Reading, Trilliumz, WikitanvirBot, Look2See1, Taticchipaolo, Ballofstring, RenamedUser01302013, Sabeen2331, Tommy2010, Wikipelli, K6ka, ZéroBot,

Jonpatterns, Jandrewc, Smarturban, Tolly4bolly, Tathanasiou, Erianna, Thine Antique Pen, Thinktosustain, Goparajurajan, Hamiltha, L Kensington, Mytilus~enwiki, Quite vivid blur, Orange Suede Sofa, Rangoon11, Budavari1970, Omagomagom, Martinscherfler, ClueBot NG, Wisdawn, Gareth Griffith-Jones, MelbourneStar, Ohdear15, Piltakva, Jejehr, Bekemem, SaintGeorgeIV, Oroszlan69, EditorOf2011, Killerscene, Widr, Antiqueight, MercBenz, Estly, Helpful Pixie Bot, Calabe1992, Gob Lofa, Akhtan, Lowercase sigmabot, BG19bot, KimGDavis, Esven, Sidxj, Northamerica1000, MusikAnimal, Frze, Belleville3, Debastein, Gaurabr, Rm1271, AdventurousSquirrel, Alantodd2403, Douglas Earth87, H0339637, Naveen saroha, Snow Blizzard, DStanley22, HMman, Worldwatchinst, Quake203, Minsbot, Aisteco, Mgilb4, BattyBot, Annapaulaw, Transition.scenarios, Jhardy08, Ethan Donovan, ChrisGualtieri, GoShow, IjonTichyIjonTichy, Rih09, SustainableChristian, Niflo8, Lugia2453, Dr. Kirit Shelat, Arjuncm3, Osat44, Envpol2, Greg.moller, Iskender Iskender, Jimjamminy, Emharmsen, Tim AFS, Zyxzupf, Epicgenius, UN DSD, Eyesnore, Irspsd, Joh-hillje, Ibn Ridwan, Kt75 mirror, Bviase, MaskedHero, Culturalwriter, Prokaryotes, Satyasri Kar, PierreFG5, Pratheek jai, DFY567, SAS343, Gadrarasa Ëïdôs, JaconaFrere, Soccergoalie228, GSDPhilip, Prajwalkuhikar1, Monkbot, Prottush, Archlover, Medbak72, Alessandra santarelli, GinAndChronically, Ahmad Zohadi, Trackteur, Lty423, JezGrove, Sairp, EMPAI MTZ, Karneek, Soniamo, Editinf, Messitup22, Jlewis144, EChastain, Yahmed15, Micaeleeh, Moorhou, Kamalshrm1174, Supdiop, KasparBot, JJMC89, Roe.ese, BU Rob13, Sunnyhero01, Reillybrooks and Anonymous: 874

- **Sustainable gardening** *Source:* https://en.wikipedia.org/wiki/Sustainable_gardening?oldid=672763063 *Contributors:* Edward, Vegaswikian, Wavelength, SmackBot, Cazort, DMacks, Skipsievert, Zodon, Addbot, Westexaslawnman, Granitethighs, HerculeBot, Yobot, Sionk, FrescoBot, Mjsimmon201, John of Reading, Look2See1, Solarra, Donner60, ChuispastonBot, Helpful Pixie Bot, Northamerica1000, Rowan Adams, Achowat, Niflo8, Rw732 and Anonymous: 5

- **Sustainable habitat** *Source:* https://en.wikipedia.org/wiki/Sustainable_habitat?oldid=675305782 *Contributors:* Altenmann, Isopropyl, Pearle, Stemonitis, Rtdrury, Salix alba, King of Hearts, Michael4444, Sanman nor, Bluebot, CSWarren, Richard001, James084, Byelf2007, Anlace, JzG, Beetstra, Cydebot, Word2line, Burlywood, VanBuren, Fadesga, Anticipation of a New Lover's Arrival, The, AnomieBOT, Brambleshire, Alysa.elaine, Look2See1, Northamerica1000 and Anonymous: 15

- **Sustainable landscape architecture** *Source:* https://en.wikipedia.org/wiki/Sustainable_landscape_architecture?oldid=642300232 *Contributors:* Ronz, Onco p53, Mwanner, Plumbago, Wavelength, Mdwyer, Crystallina, SmackBot, Bluebot, Willow4, Beetstra, Bill.albing, Kozuch, Greendale4, Organicjack, Jrsnbarn, Skier Dude, Kenao, Zodon, Westexaslawnman, Granitethighs, LAEP Prof, Look2See1, Northamerica1000, Rowan Adams, ChrisGualtieri, SpiritedMichelle and Anonymous: 13

- **Sustainable landscaping** *Source:* https://en.wikipedia.org/wiki/Sustainable_landscaping?oldid=665853533 *Contributors:* Edward, Graeme Bartlett, PFHLai, Vegaswikian, Bgwhite, Wavelength, Red Slash, David Biddulph, SmackBot, Blackash, CmdrObot, ThatPeskyCommoner, Fabrictramp, Katharineamy, WWGB, Antalope, Alchemist Jack, Debresser, Yobot, LilHelpa, Apothecia, Rickproser, Erik9bot, Spidey104, Flwrpr1, IrinaDaniel, Stnalp, EmausBot, Look2See1, Slightsmile, Bamyers99, El-in-dc, BG19bot, CaroCarrots, Rowan Adams, Zohreh ashtiani, EvMsmile and Anonymous: 15

- **Sustainable living** *Source:* https://en.wikipedia.org/wiki/Sustainable_living?oldid=684896174 *Contributors:* Heron, R Lowry, Quercusrobur, Frecklefoot, Edward, Michael Hardy, DavidWBrooks, Kat, Sulkworm, Robbot, Vespristiano, Alexblainelayder, Alan Liefting, Tom harrison, Zigger, Rick Block, Solipsist, Chowbok, Antandrus, Beland, Kevin B12, Shiftchange, Rich Farmbrough, Pak21, Vsmith, Xezbeth, Mwanner, Femto, Vortexrealm, Wordie, Ziggurat, Mdd, Howrealisreal, Velella, Jguk, Philralph, Jeffrey O. Gustafson, Mindmatrix, Jwanders, Behun, BD2412, Rjwilmsi, TheRingess, Kirstenmichel, Ahunt, WriterHound, Vmenkov, Wavelength, Hydrargyrum, Alex Ramon, CambridgeBayWeather, Rsrikanth05, Nirvana2013, Alex43223, Arthur Rubin, Chriswaterguy, Sustainableh, Bsod2, True Pagan Warrior, SmackBot, Amcbride, Kellen, Mike McGregor (Can), Septegram, Portillo, OrionK, Skizzik, Chris the speller, Deli nk, Peter Campbell, Patricksewell, DR04, Nakon, Serouj, Gobonobo, Syra987, Ckatz, Beetstra, Bendzh, Hu12, HisSpaceResearch, Iridescent, JMK, CmdrObot, Living Simply, Kribbeh, Naturalhomes, Llewelynpritchard, Kozuch, Ebbatten, Independent Journalist, RobotG, Tenzicut, Rbunnage, Kariteh, SiobhanHansa, Bongwarrior, BrookMiles, L Trezise, Ecologic Solutions, Gabriel Kielland, Tinwithli, NatureA16, Matt Mellen, Rlsheehan, Uncle Dick, All Is One, Raptor235, Marcusmax, Skier Dude, Wikiwopbop, AntiSpamBot, Cobi, 83d40m, Jorfer, Jwiley80, Jevansen, Inwind, Vranak, Johnfos, Sesamevoila, Devonkime, Connoroshea, Gsearle, Shenbrood, Raymondwinn, Philip W Bush, Annie Warmke, Truthsayer69, The Devil's Advocate, Red58bill, EJF, Jserra, HaLoGuY007, Dawn Bard, Jojalozzo, Ethicalpublishing, Tonysirna, Globaleducator, Poindexter Propellerhead, Huggi, Siddeshwarprasad, ClueBot, The Thing That Should Not Be, Egliedman, TheOldJacobite, Declutterize, Nymf, Tnxman307, Lkruijsw, Thingg, Mcfender, Apparition11, MasterOfHisOwnDomain, DumZiBoT, XLinkBot, LowImpactLiving, Zodon, Addbot, Achooyou, MrOllie, RtinNZ, Granitethighs, Lightbot, Dorraj6, PlankBot, Ra28, Luckas-bot, Yobot, Themfromspace, Kartano, AnomieBOT, Killiondude, Jim1138, Lispp, Piano non troppo, RussellDavies, Cronin, Xqbot, Climateneutral, Tomwsulcer, Dr Oldekop, SassoBot, FrescoBot, Glacier2009, Citation bot 1, Mikal42, Pinethicket, I dream of horses, Jscpowser, Trappist the monk, SeoMac, Trong.workwise, Rodmadar, Orphan Wiki, Look2See1, Stephanienox, Dewritech, L235, Theserenegirl, Hws2010, JanetA, Fturco, Donner60, RockMagnetist, Leedplat, Rocketrod1960, Martinscherfler, CambodianPride91, ClueBot NG, Wisdawn, Lautrenom, Vanniett, Ljaemichaels, Chester Markel, SaintGeorgeIV, Helpful Pixie Bot, Mediakult, NewsAndEventsGuy, Writ-on777, Northamerica1000, RentalicKim, Rowan Adams, Loriendrew, Transition.scenarios, BikeshareOnTheMove, Khazar2, Gmgolden, Panaculture, Butterfly497, Kaitlinz, Skr15081997, RELT57, Monkbot, Megatron Omega, Biblioworm, Jlewis144, HarshadaT, GeetikaG, MUM Sustainable Living, Dustinxmatos, Mahima92, Vdelob, Blindfaith7'77, Mpslattery9 and Anonymous: 200

- **Sustainable refurbishment** *Source:* https://en.wikipedia.org/wiki/Sustainable_refurbishment?oldid=629652126 *Contributors:* Bearcat, TobyJ, Wavelength, Jim Derby, DASonnenfeld and Eumolpo

- **Sustainable regional development** *Source:* https://en.wikipedia.org/wiki/Sustainable_regional_development?oldid=388323855 *Contributors:* RussBot, A.J.Chesswas, R'n'B, CohesionBot and Look2See1

- **Sustainable drainage system** *Source:* https://en.wikipedia.org/wiki/Sustainable_drainage_system?oldid=673010467 *Contributors:* Tpbradbury, Alan Liefting, Egregius2000, Xezbeth, Trevj, Anthony Appleyard, Mandarax, Tony1, Hmains, Bluebot, Willow4, Hurker, Euchiasmus, Kozuch, Islescape, Ghmyrtle, Webcomsystems, Shaunus4, Gueneverey, CIRIA, Andy Dingley, SieBot, Water and Land, Bobby Dazler, Excirial, Moreau1, Addbot, Tigerfry, J04n, Pickles8, Perviousconcrete, FrescoBot, Louperibot, Reconsider the static, DARTH SIDIOUS 2, ResinBound, Look2See1, StoneSet, GoingBatty, ClueBot NG, Lyla1205, BG19bot, Karlschutte, AlexStephenson22, Wywin, Upm1357, Nouhy002 and Anonymous: 23

- **Swale (landform)** *Source:* https://en.wikipedia.org/wiki/Swale_(landform)?oldid=683073922 *Contributors:* Deb, Ortolan88, Frecklefoot, Mac, Joy, Reinyday, Interiot, Geo Swan, Alhazred93, Btyner, Deltabeignet, Gary Cziko, KeithD, VIGNERON, SmackBot, C.Löser, Bazonka, Uthbrian, Abrahami, Fuzzypeg, Springnuts, Friendly Neighbour, CmdrObot, Tangurena, WeGotCactus, Mschiffler, Cpapadelis, LeaveSleaves, Mooreds, Plussign, This, that and the other, Hamiltondaniel, Moreau1, PermaculturePlanet, XLinkBot, Bermicourt, Gregorybean, Hb353, A.amitkumar, Rule 56, Pershing10, Look2See1, Cogitoergohuh?, Wipsenade, JonRichfield, Etiennemarcus, Slfms, Mohamed-Ahmed-FG and Anonymous: 15

- **Transition design** *Source:* https://en.wikipedia.org/wiki/Transition_design?oldid=664623596 *Contributors:* Ijon, Giraffedata, BD2412, Wavelength, Bhny, Racklever, PKT, Vanjagenije, Freshacconci, DASonnenfeld, ImageRemovalBot, AnomieBOT, John of Reading, BG19bot, Dreambeaver, BattyBot, IjonTichyIjonTichy, Noegid, Lagoset, EPsSp11 and Anonymous: 3

- **Urban acupuncture** *Source:* https://en.wikipedia.org/wiki/Urban_acupuncture?oldid=680487328 *Contributors:* RHaworth, Koavf, Cydebot, R'n'B, Paris1127, DadaNeem, DASonnenfeld, Idioma-bot, Whiteghost.ink, Jan jörg, JL-Bot, Movez~enwiki, Archiboy, Mild Bill Hiccup, Addbot, Yobot, AnomieBOT, J04n, FrescoBot, HRoestBot, RedBot, Clarkcj12, ZéroBot, BG19bot, RichardMills65, TerryMu, Filedelinkerbot, BDFELIZ and Anonymous: 17

- **Used good** *Source:* https://en.wikipedia.org/wiki/Used_good?oldid=681673917 *Contributors:* Montrealais, Mac, Rl, Markhurd, Khalid hassani, Andycjp, Beland, Jareha, RedWordSmith, Cemyildiz, Xezbeth, Sietse Snel, Orlady, Vortexrealm, Jerryseinfeld, Geraldshields11, Versageek, GVOLTT, Pol098, Pationl, Rjwilmsi, Phileas, Ewlyahoocom, Quuxplusone, Bgwhite, Wavelength, RadioFan2 (usurped), Ihope127, Bjf, Bobbo, Spondoolicks, DoriSmith, SmackBot, Hmains, Bluebot, Jethero, MalafayaBot, Mindraker, Raichu, Richard001, JackLumber, Euchiasmus, TastyPoutine, OnBeyondZebrax, Tawkerbot2, Mikiemike, CmdrObot, Neelix, Bridgecross, Thijs!bot, Iedit, Blathnaid, CZmarlin, VictorAnyakin, Cfherbert, Catgut, D.h, Tgeairn, Yonidebot, Billydeeuk, VolkovBot, Una Smith, Eworlds, BluejacketT, ImageRemovalBot, ClueBot, DumZiBoT, Delicious carbuncle, Addbot, Knight of Truth, Victorian Chameleon~enwiki, Lolocks, HerculeBot, Yobot, JackieBot, Obersachsebot, SecondhandFurniture, Der Falke, FrescoBot, Tinton5, Chartomarco, Sridhar a20, FoxBot, Simbolo.sam, DixonDBot, Natacha S., 19cw19, Infin8y, Youkai fan, EmausBot, Look2See1, RA0808, SweetWeepeats, Abhi85jo, ScottSteiner, BettinaBerlep, South Birmingham OK, Bignbafan, DURABILIT, EuroCarGT, Corn cheese, Bever, Veelynn, Balancegreenapril, Highway 231, Prisencolin, Dany5avram, Ellen155 and Anonymous: 44

- **Water conservation** *Source:* https://en.wikipedia.org/wiki/Water_conservation?oldid=684507449 *Contributors:* SebastianHelm, Darkwind, Jredmond, SoLando, Mattflaschen, Alan Liefting, Wolfkeeper, Codepoet, Michael Devore, Beland, Onco p53, Kiteinthewind, Joyous!, Spiffy sperry, Discospinster, Rich Farmbrough, Aude, Femto, Smalljim, Alansohn, Arthena, Paleorthid, Velella, BlastOButter42, Jeffrey O. Gustafson, RHaworth, MONGO, Kgrr, Graham87, Rjwilmsi, FayssalF, Kerowyn, Lumin~enwiki, DVdm, Dj Capricorn, Gwernol, Wavelength, RussBot, Stephenb, Snek01, Moe Epsilon, Epipelagic, Syrthiss, Phgao, Closedmouth, Pb30, Chriswaterguy, DCEvoCE, Yvwv, SmackBot, MattieTK, Gigs, Deon Steyn, Davewild, Gilliam, Hmains, Tytrain, Bluebot, RDBrown, Imaginaryoctopus, Deli nk, Gracenotes, Can't sleep, clown will eat me, Valenciano, Ozfreediver, Mion, Vina-iwbot~enwiki, InNuce, Krashlandon, AThing, IronGargoyle, Beetstra, Boomshadow, TastyPoutine, Hu12, BranStark, Kaarel, Jjdjtremblay, CmdrObot, CWY2190, Musashi1600, Jac16888, Gogo Dodo, Christian75, Chrislk02, Epbr123, Marek69, James086, Seaphoto, MikeLynch, Frankie816, Bongwarrior, VoABot II, Rivertorch, Nyttend, Jmartinsson, Michael Marcus, Allstarecho, Sustainableyes, Heliac, D.h, Yobol, MartinBot, Rob Lindsey, Ghostsniperzw, Eplack, J.delanoy, Trusilver, Herbythyme, Uncle Dick, Schappacher, McSly, Lasarkis, Jeepday, Randydeluxe, NewEnglandYankee, RWNIELSEN, KylieTastic, TWCarlson, Ja 62, Kyuhas, Useight, DASonnenfeld, Wilochka, BernardZ, Philip Trueman, Dtsreddy~enwiki, Oshwah, Vipinhari, Planetary Chaos, Anbellofe, Qxz, Oxfordwang, Foshodle, Noformation, Waterwise, Falcon8765, Coffee, Moonriddengirl, Krawi, Winchelsea, Yintan, Calabraxthis, Keilana, Flyer22, Jojalozzo, GILDog, Nopetro, Wilson44691, Yerpo, Oxymoron83, Sgagnon, Mygerardromance, Loren.wilton, ClueBot, Phoenix-wiki, The Thing That Should Not Be, Jamalhamou, Unbuttered Parsnip, Maniac18, Drmies, Farras Octara, CounterVandalismBot, Alandmanson, Victory721, Neverquick, Excirial, Moreau1, Gtstricky, Firebertt, Muenda, Ice Cold Beer, Tςς, Apparition11, NERIC-Security, Tdslk, Ps07swt, Mheberger, Waterinfo123, XLinkBot, Bbazinet, Avoided, LowImpactLiving, Savewater, RyanCross, Wyatt915, Addbot, ConCompS, Willking1979, Some jerk on the Internet, Ronhjones, Fieldday-sunday, CanadianLinuxUser, MrOllie, Favonian, Sharifah81, Tastethefruit, Krano, Jarble, Luckasbot, Themfromspace, Ogomez78, Tastyfast, SwisterTwister, Tempodivalse, AnomieBOT, Rubinbot, Jim1138, Pm.madhav, Kingpin13, Dhidalgo, Materialscientist, Citation bot, Nasnema, Shadowjams, Shebbadger, BoomerAB, Prari, FrescoBot, VS6507, VI, Cannolis, Citation bot 1, DrilBot, A g 0808, Pinethicket, I dream of horses, Majestic27, Sridigitek, Calmer Waters, Rushbugled13, Dcwaterboy, Pianoplonkers, Jiujitsuguy, TobeBot, Vrenator, Reaper Eternal, Jeffrd10, Merlinsorca, Fastilysock, Drjohntoconnorpe, Suffusion of Yellow, Minimac, Jfmantis, Shepwo14, DASHBot, Acather96, Pjposullivan, Look2See1, Super48paul, RA0808, UrbanRePlanner, Solarra, Wikipelli, K6ka, AsceticRose, Lamb99, Savh, ZéroBot, Fæ, Josve05a, Vikkimount, MithrandirAgain, Lateg, Helenp89, Aeonx, CosmicJake, Tolly4bolly, Erianna, Jsayre64, Τασουλα, L Kensington, MonoAV, Donner60, Equilux, Hyronimus299, Forever Dusk, BRIJESH 09, Sherwood Cat, Special Cases, Petrb, ClueBot NG, Gareth Griffith-Jones, MelbourneStar, Md2018, Satellizer, Valaprise, Snotbot, Kenmirv, Ghunter3016, O.Koslowski, Widr, PinkBlubbleGlum, Helpful Pixie Bot, Mr. Credible, Electriccatfish2, HMSSolent, Joan Cornish, DBigXray, Gauravjuvekar, NewsAndEventsGuy, Northamerica1000, MusikAnimal, Mark Arsten, Atomician, IluvatarBot, Brendan frank, TheProfessor, Marcjarod, BattyBot, Pratyya Ghosh, ChrisGualtieri, Koko1088, Loozrman, Khazar2, Ducknish, Joeknight1, SHRIPADVAIDYA, FoCuSandLeArN, A4we, Govilkar, Lugia2453, SFK2, Hair, Jamesx12345, Saltburn Man, NagOc 945, Epicgenius, Ruby Murray, BreakfastJr, Kkrystelle, Nike2795, ☼, Zenibus, Plankyrathore, Jianhui67, Rahulrathore20138thb, Kaitlinz, FrB.TG, Mansi Nanavati, Monkbot, Sidcoolestboy, Scarlettail, Amortias, TerryAlex, Bhaskarblaster, Caliburn, Schesank, 074wise, ABHISHEK HADGE, DawjihlihWegckichsr, Logan979, Zerowaterconsulting, McDonald of Kindness and Anonymous: 551

- **Wildland–urban interface** *Source:* https://en.wikipedia.org/wiki/Wildland%E2%80%93urban_interface?oldid=677641153 *Contributors:* Pascal666, Trevor MacInnis, Fleisher, Wavelength, Tony1, Agradman, Cydebot, Nick Number, Wfulks, MrBell, Malcolmxl5, Fadesga, Yobot, Erik9bot, Look2See1, Extfor, WikiU2013, Fireguy389 and Anonymous: 8

59.4.2 Images

- **File:(1)Central_building_Broadway_Sydney-1.jpg** *Source:* https://upload.wikimedia.org/wikipedia/commons/a/aa/%281% 29Central_building_Broadway_Sydney-1.jpg *License:* CC BY 3.0 *Contributors:* Own work *Original artist:* Sardaka (<a href='//commons.wikimedia.org/wiki/User_talk:Sardaka' title='User talk:Sardaka'>talk</a>) 08:28, 8 July 2014 (UTC)

- **File:Urban_subsystem.jpg** *Source:* https://upload.wikimedia.org/wikipedia/commons/d/d3/Urban_subsystem.jpg *License:* CC BY-SA 4.0 *Contributors:* Own work *Original artist:* Massimo.tadi

- **File:Veranotrigo.jpg** *Source:* https://upload.wikimedia.org/wikipedia/commons/4/4f/Veranotrigo.jpg *License:* CC BY 2.0 *Contributors:* http://www.flickr.com/photos/soilscience/2513807337/ *Original artist:* Soil-Science.info

- **File:VineyardDrip.JPG** *Source:* https://upload.wikimedia.org/wikipedia/commons/6/6b/VineyardDrip.JPG *License:* Public domain *Contributors:* ? *Original artist:* ?

- **File:Water-conservation-stamp-1960.jpg** *Source:* https://upload.wikimedia.org/wikipedia/commons/2/23/Water-conservation-stamp-1960.jpg *License:* Public domain *Contributors:* Transferred from en.wikipedia to Commons. *Original artist:* The original uploader was H2O at English Wikipedia

- **File:Western_Bluebird_leaving_nest_box.jpg** *Source:* https://upload.wikimedia.org/wikipedia/commons/4/42/Western_Bluebird_leaving_nest_box.jpg *License:* CC BY 2.0 *Contributors:* Western Bluebird using Nest Box. Brood unknown. *Original artist:* Kevin Cole from Pacific Coast, USA (en:User:Kevinlcole)

- **File:Wiki_letter_w.svg** *Source:* https://upload.wikimedia.org/wikipedia/en/6/6c/Wiki_letter_w.svg *License:* Cc-by-sa-3.0 *Contributors:* ? *Original artist:* ?

- **File:Wikibooks-logo.svg** *Source:* https://upload.wikimedia.org/wikipedia/commons/f/fa/Wikibooks-logo.svg *License:* CC BY-SA 3.0 *Contributors:* Own work *Original artist:* User:Bastique, User:Ramac et al.

- **File:Wikiquote-logo.svg** *Source:* https://upload.wikimedia.org/wikipedia/commons/f/fa/Wikiquote-logo.svg *License:* Public domain *Contributors:* ? *Original artist:* ?

- **File:Wikiversity-logo.svg** *Source:* https://upload.wikimedia.org/wikipedia/commons/9/91/Wikiversity-logo.svg *License:* CC BY-SA 3.0 *Contributors:* Snorky (optimized and cleaned up by verdy_p) *Original artist:* Snorky (optimized and cleaned up by verdy_p)

- **File:Wiktionary-logo-en.svg** *Source:* https://upload.wikimedia.org/wikipedia/commons/f/f8/Wiktionary-logo-en.svg *License:* Public domain *Contributors:* Vector version of Image:Wiktionary-logo-en.png. *Original artist:* Vectorized by Fvasconcellos (talk · contribs), based on original logo tossed together by Brion Vibber

- **File:Wiktionary-logo.svg** *Source:* https://upload.wikimedia.org/wikipedia/commons/e/ec/Wiktionary-logo.svg *License:* CC BY-SA 3.0 *Contributors:* ? *Original artist:* ?

- **File:Wind-turbine-icon.svg** *Source:* https://upload.wikimedia.org/wikipedia/commons/a/ad/Wind-turbine-icon.svg *License:* CC BY-SA 3.0 *Contributors:* Own work *Original artist:* Lukipuk

- **File:Windmills_D1-D4_-_Thornton_Bank.jpg** *Source:* https://upload.wikimedia.org/wikipedia/commons/f/f9/Windmills_D1-D4_-_Thornton_Bank.jpg *License:* CC BY-SA 4.0 *Contributors:* Own work *Original artist:* Hans Hillewaert

- **File:Yard_Sale_Northern_CA_2005.JPG** *Source:* https://upload.wikimedia.org/wikipedia/commons/e/e8/Yard_Sale_Northern_CA_2005.JPG *License:* Public domain *Contributors:* Transferred from en.wikipedia to Commons. *Original artist:* The original uploader was Jimmyjazz at English Wikipedia

59.4.3 Content license

www.ingramcontent.com/pod-product-compliance
Lightning Source LLC
Chambersburg PA
CBHW080759180526
45168CB00006B/2267